Public Health, Disease and Development in Africa

The closure of the Millennium Development Goals (MDGs) in 2015 prompted the need for a book of this kind. An interdisciplinary group of global health scholars contribute to the understanding of the emerging and fast-growing problem of the dual burden of communicable and non-communicable diseases (NCDs) in Africa.

This book is timely, as the international community has moved from the MDGs to adopt the Sustainable Development Goals (SDGs) as the blueprint for a new human development agenda. Contributions and case studies are situated in the revised Epidemiologic and Nutrition Transition Model to capture the current situation, referencing communicable and NCDs on the African continent. The case studies encapsulated aim to help minimize negative health outcomes and improve population health, well-being, and equity in the future.

This book will be significant in policy circles to assist international organizations, governments, and United Nations agencies. It aims to chart the future for health in Africa in light of recently adopted SDGs This book is also a useful complementary reader for global public health related courses.

Ezekiel Kalipeni is a population and health geographer with research interests in demographic, health, environmental, and resource issues in sub-Saharan Africa. He has carried out extensive research on the population dynamics of Malawi, and Africa in general, including concentrating on the issues of fertility, mortality, migration, and health care. He is currently researching the drivers and consequences of the HIV/AIDS pandemic in Africa.

Juliet Iwelunmor is an associate professor at St. Louis University, Missouri. She explores social, cultural, behavioural, and policy factors that influence the health of individuals, families, and communities across the lifespan. She is particularly interested in global health issues, having previously worked as the Culture Sector coordinator for the UNESCO Intersectoral Platform on HIV and AIDS.

Diana S. Grigsby-Toussaint is an associate professor in the College of Applied Health Sciences at the University of Illinois at Urbana–Champaign. As a social epidemiologist, she attempts to elucidate how and why socio-environmental factors influence health in local, national, and international contexts. Her research explores the three pillars of health (diet, physical activity, and sleep) on non-communicable disease risk among vulnerable populations by utilizing an interdisciplinary approach spanning epidemiology, geography, and nutrition. Dr. Grigsby-Toussaint's research has been supported by the *Robert Wood Johnson Foundation, USDA*, and *NSF*, and her work has been featured in the *Chicago Tribune* and *The Huffington Post.*

Imelda K. Moise is an applied health geographer and monitoring & evaluation (M&E) specialist, expert in multi-method approaches, community-based participatory research, culturally responsive research, and Geographic Information System (GIS) analysis. Her research focuses on the interface of the social determinants of health, health disparities, and inequities particularly among adolescents and maternal and child health populations, infectious diseases (HIV), and linking research to practice or policy. She also supported USAID-funded health programs in low- and middle-income countries (LMIC) and spent five years in Illinois coordinating federally funded research projects and program evaluation for state agency initiatives and ongoing programs, and six years as a Peace Corps technical trainer in Zambia.

Geographies of Health

Series editors

Allison Williams, Associate Professor, School of Geography and Earth Sciences, McMaster University, Canada

Susan J. Elliott, Professor, Department of Geography and Environmental Management and School of Public Health and Health Systems, University of Waterloo, Canada

There is growing interest in the geographies of health and a continued interest in what has more traditionally been labeled medical geography. The traditional focus of 'medical geography' on areas such as disease ecology, health service provision and disease mapping (all of which continue to reflect a mainly quantitative approach to inquiry) has evolved to a focus on a broader, theoretically informed epistemology of health geographies in an expanded international reach. As a result, we now find this subdiscipline characterized by a strongly theoretically informed research agenda, embracing a range of methods (quantitative; qualitative and the integration of the two) of inquiry concerned with questions of: risk; representation and meaning; inequality and power; culture and difference, among others. Health mapping and modeling has simultaneously been strengthened by the technical advances made in multilevel modeling, advanced spatial analytic methods and GIS, while further engaging in questions related to health inequalities, population health and environmental degradation.

This series publishes superior quality research monographs and edited collections representing contemporary applications in the field; this encompasses original research as well as advances in methods, techniques, and theories. The *Geographies of Health* series will capture the interest of a broad body of scholars, within the social sciences, the health sciences and beyond.

Children's Health and Wellbeing in Urban Environments
Edited by Christina R. Ergler, Robin Kearns and Karen Witten

Non-Representational Theory & Health
The Health in Life in Space-Time Revealing
Gavin J. Andrews

Geographies of Plague Pandemics
The Spatial-Temporal Behavior of Plague to the Modern Day
Mark R. Welford

Public Health, Disease and Development in Africa
Edited by Ezekiel Kalipeni, Juliet Iwelunmor, Diana S. Grigsby-Toussaint, and Imelda K. Moise

For a full list of titles in this series, please
visit https://www.routledge.com/Geographies-of-Health-Series/book-series/GHS

Public Health, Disease and Development in Africa

Edited by Ezekiel Kalipeni,
Juliet Iwelunmor,
Diana S. Grigsby-Toussaint
and Imelda K. Moise

Routledge
Taylor & Francis Group
LONDON AND NEW YORK

First published 2018
by Routledge
2 Park Square, Milton Park, Abingdon, Oxon OX14 4RN

and by Routledge
711 Third Avenue, New York, NY 10017

Routledge is an imprint of the Taylor & Francis Group, an informa business

British Library Cataloguing-in-Publication Data
A catalogue record for this book is available from the British Library

Library of Congress Cataloging-in-Publication Data
A catalog record has been requested for this book

ISBN: 978-1-138-63125-0 (hbk)
ISBN: 978-1-315-20900-5 (ebk)

Typeset in Times New Roman
by Deanta Global Publishing Services, Chennai, India

Contents

Figures

Tables

Contributors

Felix K. Assah (PhD) is an epidemiologist at the Cameroon Ministry of Public Health. His research interests are in epidemiology of cardiovascular diseases, diabetes, obesity, hypertension, and their risk factors. His focus is on changing patterns of lifestyle risk factors in populations undergoing transition (urbanization and westernization) and their relations to patterns of disease.

Emmanuella N. Atanga (MSN) is a research assistant at the HoPiT Research Group in Yaounde, Cameroon. She has acquired extensive experience in community health from her work with the Cameroon Physical Activity Study which involved assessing diabetes and other non-communicable disease risk among over 1,000 rural and urban Cameroonian residents.

Ogugua C. Aworh (PhD), Professor at the University of Ibadan, is a fellow of the International Academy of Food Science and Technology and recipient of the Lifetime Achievement Award of the Nigerian Institute of Food Science and Technology. He has published on post-harvest physiology and handling of horticultural commodities, traditional food processing, food biotechnology, and food safety.

Elijah Bisung (PhD) is Assistant Professor in the School of Kinesiology and Health Studies at Queen's University, Ontario, Canada. His research activities focus on environment and health, particularly at the interface of water and health, environmental risk perceptions and psychosocial wellbeing, and collective action for environmental health.

Sarah R. Blackstone (PhD) is an assistant professor at James Madison University in the Department of Health Sciences. She is interested in the influence of social determinants of health on maternal and child health, sexual and reproductive health, and chronic diseases in underserved populations.

Jenna Bryfonski (BA) holds a Bachelor of Arts degree in Cellular Neuroscience from Colgate University, New York, where she earned the Dean's Award for Academic Excellence. She attends Ontario Veterinary College at the University of Guelph in Toronto, Ontario, Canada.

Mary J. Christoph (PhD, MPH) is a postdoctoral fellow in the Department of Pediatrics and the Division of Epidemiology and Community Health at the University of Minnesota. Her research focuses on the impact of nutrition policies, including nutrition labels on dietary intake and eating behaviors in adolescents and young adults.

Jenna Dixon (PhD) is Postdoctoral Fellow at the University of Waterloo, Canada. Her research centers on questions of health inequalities, development and health, gender and health, and knowledge translation. Together, the streams of her work intersect with a critical focus on issues of global social justice.

Beatrice Ekesa (PhD) is a scientist at Bioversity International with specialization in human nutrition. She has over ten years' experience assessing and optimizing the contribution of agri-food systems to food and nutrition security in East and Central Africa. Beatrice has a PhD in Foods and Nutrition and an MSc in Community Nutrition and Development.

Susan J. Elliott (PhD) is Professor of Geography and Public Health at the University of Waterloo, Canada. She has over 200 peer-reviewed publications related to environment and (public) health focused on Canada and the developing world. She is leading the development of GLOWING – a Global Index of Wellbeing.

Diana S. Grigsby-Toussaint (PhD, MPH) is an epidemiologist whose research focuses on how and why socio-environmental factors influence health and wellbeing in global contexts. She is currently an associate professor in the Department of Kinesiology and Community Health and the Division of Nutritional Sciences at the University of Illinois at Urbana–Champaign.

Anne Hoefler (BA) holds a Bachelor of Arts degree in Geography from Colgate University, New York, where she earned the Dean's Award for Academic Excellence. She is Professional Staff Member at the U.S. Senate Committee on Energy and Natural Resources.

Juliet Iwelunmor (PhD) is an associate professor at St. Louis University, Missouri. She explores social, cultural, behavioural, and policy factors that influence the health of individuals, families, and communities across the lifespan. She is particularly interested in global health issues, having previously worked as the Culture Sector coordinator for the UNESCO Intersectoral Platform on HIV and AIDS.

Vincent Johnson has worked as a grants officer and science writer/editor for Bioversity International since 2008. He previously worked as a UK agronomist (>25 years); an aid-worker in Cambodia (four years); and a lecturer/teacher in rural development, biology, and plant and soil science (two years). He is a musician, naturalist, and writer.

Evan de Joya is currently an undergraduate student (junior) in the Departments of Geography and Biology (double major) at the University of Miami. He has been admitted to the Medical Scholars Program, an accelerated combined BS/MD program at Leonard M. Miller School of Medicine. His research interests include infectious diseases, geographic analysis, and adolescent health.

Ezekiel Kalipeni (PhD) is a population and health geographer with research interests in demographic, health, environmental, and resource issues in sub-Saharan Africa. He has carried out extensive research on the population dynamics of Malawi, and Africa in general including concentrating on the issues of fertility, mortality, migration, and health care. He is currently researching the drivers and consequences of the HIV/AIDS pandemic in Africa.

Gina Kennedy (PhD) is a senior scientist with Bioversity International and team leader of the Healthy Diets from Sustainable Food Systems Initiative. Her research interests include measurement of dietary diversity and diet quality, including assessing the contribution of agricultural biodiversity to nutrition and healthy diets.

John Kinyuru (PhD) is a food and nutrition science researcher and lecturer in the Department of Food Science and Technology at the Jomo Kenyatta University of Agriculture and Technology, Kenya. His research interests focus on utilization of indigenous knowledge, with specific focus on the exploitation of edible insects in human nutrition and health.

Marena Manley (PhD) received her PhD from the University of Plymouth, UK and is currently appointed as professor at Stellenbosch University. Her research interests involve the application of near-infrared (NIR) spectroscopy, NIR hyperspectral imaging, and X-ray micro-computed tomography to study whole-grain microstructure in association with cereal functionality.

Jean Claude Mbanya (PhD) is Professor of Medicine and Endocrinology at the University of Yaoundé I. His major interests include diabetes and other non-communicable disease epidemiology and prevention, integration of diabetes care in the health care system of developing countries, diabetes complications, obesity, and physical activity as a risk factor of non-communicable diseases.

Imelda K. Moise (PhD, MPH) is an applied health geographer and monitoring & evaluation (M&E) specialist, expert in multi-method approaches, community-based participatory research, culturally responsive research, and GIS analysis. Her research focuses on the interface of the social determinants of health, health disparities, and inequities, particularly among adolescents and maternal and child health populations, infectious diseases (HIV), and linking research to practice or policy. She also supported USAID-funded health programs in LMIC and spent five years in Illinois coordinating federally funded research projects and program evaluation for state agency initiatives and ongoing programs, and six years as a Peace Corps technical trainer in Zambia.

Samuel Mpiira is a PhD fellow in agricultural economics at Maseno University, Kenya and a researcher at the National Agricultural Research Organization

(NARO) Uganda since 2002. He has ten years' experience in on-farm research, participatory evaluation, and development of multiple stakeholder innovation platforms.

John H. Muyonga (PhD) is Dean of the School of Food Technology, Nutrition and Bio-Engineering at Makerere University, Uganda. He has published widely on aspects of protein chemistry, fish processing waste utilization, physicochemical characterization of underutilized foods, phytochemicals, and traditional food processing methods. He is passionate about research translation to facilitate wider application.

Deborah Nabuuma (MA) is a research fellow at Bioversity International with specialization in human nutrition. She has over five years' experience in community nutrition, dietary diversity, and capacity building within Eastern Africa. She is pursuing a PhD in nutritional sciences at Stellenbosch University, and holds an MSc in applied human nutrition from Makerere University.

Sophie Nansereko (MSc) has over nine years' experience in community food, health, and nutrition projects, including working with the Ministry of Health and NGOs in Uganda. Her professional interests involve developing and implementing innovative interventions aimed at improving the health, nutrition, and food security of vulnerable communities.

Domina Nkuba is a nutritionist working with Ministry of Agriculture Livestock and Fisheries, based in Dar es Salaam, Tanzania. She has an MSc degree in Foods and Nutrition from Sokoine University of Agriculture, Morogoro, Tanzania and is currently pursuing a PhD in foods and nutrition at the same university

James M. Ntambi (PhD) is Professor of Biochemistry, Steenbock Professor of Nutritional Sciences at the University of Wisconsin–Madison, and Adjunct Professor of Biological Chemistry, Johns Hopkins University School of Medicine, Baltimore, Maryland. Ntambi received his BSc and MSc degrees in Biochemistry and Chemistry from Makerere University Kampala, Uganda and his PhD in Biochemistry and Molecular Biology from the Johns Hopkins University School of Medicine.

Ucheoma Nwaozuru (MS) is currently a PhD student in community health at the University of Illinois at Urbana–Champaign. Her research interests include maternal, child, and adolescent health, specifically how social enterprise and mobile technology can be utilized to improve health outcomes among these populations.

Dorothy N. Nyangena (MSc) holds a BSc and MSc in Food Science and Technology from Jomo Kenyatta University of Agriculture and Technology, Kenya. She is interested in research on underutilized foods and is currently working on edible insects as food and feed projects in East Africa.

Joseph R. Oppong (PhD) is Professor of Geography and the Environment at the University of North Texas. A medical geographer focusing on Africa and

North America, his research interests include neighborhood characteristics and HIV/AIDS, geographic distribution of tuberculosis genotypes, and applications of GIS to understand spatial patterns of disease and health.

Afolabi Oyapero (BDS, MPH, FMCDS) is a consultant in dental public health and a lecturer in the Department of Preventive Dentistry, Lagos State University College of Medicine. He has a special interest in quality of life outcomes of medical conditions such as diabetes mellitus and their effects on oral health.

Warangkana Ruckthongsook is a PhD candidate in environmental science and research assistant in medical geography at the University of North Texas. Her research interests include disease mapping and neighborhood characteristics. She uses remote sensing, GIS, and spatial statistics to understand spatial patterns of disease, environment, and health.

Jude Saji (MPH) is a public health professional with a background in nursing and public health research. He works as research associate at the University of Montreal's School of Public Health. He is interested in health promotion to reduce the burden of non-communicable diseases and their risk factors in rural and urban Africa.

Linda L. Semu (PhD) is Associate Professor of Sociology at McDaniel College, Westminster, Maryland. Her work focuses on the intersection of gender and: globalization, immigration, and race; social, economic, political, and cultural change; family, marriage, and motherhood; land rights and food security; HIV/AIDS; urbanization; and research methods.

Alexandra Shapiro (BA) holds a Bachelor of Arts degree in Environmental Geography from Colgate University, New York, where she earned the Dean's Award for Academic Excellence. She is in the Postbaccalaureate Premedical Program at Columbia University in the City of New York.

Charles Staver (PhD) has been a scientist at Bioversity International since 2004, working on the ecological intensification of smallholder banana production systems. Previously he worked for CATIE on integrated pest management in coffee, bananas, and food grains. He has over 30 years of research experience in Central America, Latin America, and Africa.

Eric Y. Tenkorang (PhD) is Associate Professor of Sociology at Memorial University of Newfoundland. A social demographer, his broad research interests focus on the health of marginalized populations in low-income settings, in particular, sub-Saharan Africa (South Africa, Ghana, Kenya, Nigeria, and Malawi).

Leo C. Zulu (PhD) is an associate professor at Michigan State University. His research interests focus on nature-society interactions and health geography focusing on Africa, particularly southern Africa. They include environmental governance and development, community-based natural resources management, social forestry, solid biomass energy (including charcoal), extractive resources management, land use and environmental change, and climate change governance, vulnerability, and adaptation targeting rural communities. He also examines spatial temporal patterns in infectious diseases focusing on HIV, and some work on disparities in access to health services.

Acknowledgements

The conference in Urbana–Champaign in May 2015 that originally brought together many of the contributors to this book would not have been possible without grants from the National Science Foundation (NSF, Award 1461724) and many other departments at the University of Illinois at Urbana–Champaign which included the Illinois Strategic International Partnership; International Programs and Studies; the School of Earth, Society and Environment; the Department of Geography and Geographic Information Science; the Department of Kinesiology and Community Health; the Department of Human and Community Development; Women and Gender in Global Perspective; the College of Liberal Arts and Sciences; the Center for Advanced Study and the College of Business, just to mention a few. We want to note categorically that any opinions, findings, and conclusions or recommendations expressed in this material are those of the authors and do not necessarily reflect the views of the organizations that provided financial support. We want to take this opportunity to thank those individuals who attended the meeting and gave their time and expert thoughts on the subject which subsequently formed the focus of this book.

The editors would like to thank all of the contributors for their time and expertise, and for their patience in seeing this project through to fruition. Several anonymous reviewers provided invaluable comments on the issues and questions raised in this book. To these individuals we give our hearty thanks and hope that the final product comes somewhere close to their expectations. We would also like to gratefully acknowledge the help received from many individuals which inevitably led to the successful completion of this book. In particular, we would like to mention the assistance received from Lori Baker, Stephanie Cresap, and Matthew Cohn in the School of Earth, Society and Environment as well as Sarah R. Blackstone, Linda West, Dayanna Reeves, Ucheoma Nwaozuru, and Jong Cheol Shin in the Department of Kinesiology and Community Health, and Juan Andrade from the Department of Food Science and Human Nutrition who were very instrumental in organizing, supporting, and running the meeting that took place in May 2015 at the University of Illinois.

Finally, we wish to express great thanks to Melissa Heil of the Department of Geography and Geographic Information Science for initial edits on the contributions contained in this book and for making sure that the contributors responded in a timely fashion to external reviewers' comments on their submissions. But most of all we thank Ruth Anderson of Routledge Publishers for expediting the publication of this book and for her willingness to work with us. Without her, this book would not have happened. The dedicated and experienced staff at Routledge Publishers facilitated the timely production of this book. We are very grateful to all of them for ensuring that an exceptionally high standard of professionalism was involved in the production of the book.

Part I

Introduction

1 Introduction

Africa's epidemiologic transition of dual burden of communicable and non-communicable diseases

Ezekiel Kalipeni, Juliet Iwelunmor, Diana S. Grigsby-Toussaint, and Imelda K. Moise

Introduction

The chapters contained in this book are an offshoot of a symposium convened at the University of Illinois at Urbana–Champaign, May 20–22, 2015. The symposium, titled "Health in Africa and the Post-2015 Millennium Development Agenda", was sponsored by a grant from the National Science Foundation (NSF, Award 1461724) and the University of Illinois at Urbana-Champaign. An interdisciplinary group of scholars was brought together at this symposium to interrogate health conditions in Africa after 15 years of the implementation of the Millennium Development Goals (MDGs) that were adopted in 2000 by the United Nations. The result of the symposium is a collection of chapters that examine the current status of communicable and non-communicable diseases in Africa.

The coming to closure of the MDGs in 2015 prompted the need for a book of this kind. At this point in time the international community was in the process of formulating a new human development agenda to improve the lives of the world's growing population of over seven billion people. Questions on what the landscape of the Post-2015 Millennium Development Agenda would look like were being debated. The symposium was particularly concerned with socio-environmental and geo-political factors affecting health, wellbeing, and disease in Africa. The dual burden of diseases (i.e. the presence of both communicable or infectious diseases and non-communicable diseases [NCDs]) has already become a reality in many low- and middle-income countries (LMIC), including many countries in Africa (Miranda *et al.* 2008). For example, while highly developed countries have an 85% burden of NCDs versus communicable diseases, many countries in Africa experience an equal burden of communicable (44%) and NCDs (44%) (Aikins *et al.* 2010, Boutaye and Boutaye 2005). This burden makes already fragile, resource-constrained, fragmented, and siloed health systems on the continent of Africa even weaker. The scale of the challenge posed demands an extraordinary response which prompted us to bring together a leading interdisciplinary group of scholars and researchers to interrogate these issues, and examine the changing patterns of disease on the continent. As Aikins *et al.* (2010) points out, this dual burden of disease requires a careful rethinking about health care in Africa.

To echo Aikins *et al.* (2010), there is an urgent need for African health policymakers and governments to prioritize the development and implementation of chronic disease policies as this burden continues to grow alongside the infectious disease burden.

The scholarly significance of the dual burden of diseases in Africa cannot be overemphasized. First, the scale of the challenge posed by the double burden has devastating effects for a continent already facing catastrophe in human health resources due to a shortage of health workers (Kalipeni *et al.* 2012, Moise *et al.* 2017). There are also demographic, economic, social, and epidemiological impacts posed by the double burden of diseases. Demographically, life expectancy is on the rise with increases in population size and aging in Africa (Murray and Lopez 2013). According to available evidence, life expectancy at age 60 is becoming a reality for an increasingly large number of Africans (Aboderin 2012). While most of sub-Saharan Africa is still in earlier stages of the demographic transition and will remain younger than other world regions, the population share of older people (aged 60 years and over) will, nonetheless, see a sharp four-fold rise from 5% today to 19% by the end of the century and in the same time span, the absolute size of the older population will grow a massive 15-fold from 43 million to 644 million (Aboderin 2012). This is a sharper increase than for any other world region or age group. Yet population aging in Africa is given little attention in policy debates, which raises a serious health equity concern as it is among the key drivers contributing to increases in the burden of NCDs on the continent (Murray and Lopez 2013). While infectious diseases still account for 69% of deaths on the continent, many African health systems which are under-funded and under-resourced will have to struggle to cope with the dual burden of infectious diseases and NCDs in the coming years (Aikins *et al.* 2010).

It is in this regard that the contributions contained in this book are situated in the revised Epidemiologic and Nutrition Transition Model (Popkin 2002) in order to capture the current situation with reference to communicable and NCDs on the African continent. The epidemiologic transition model has been of great appeal to demographers, sociologists, population geographers, epidemiologists, public health care researchers, and others. This model, first described by Abdel Omran (1971), focuses on the shifting web of health and disease patterns on population groups and their links with several demographic, social, economic, ecologic, and biological changes as is currently happening in Africa. Essentially, in this model there are shifts in health and disease patterns as mortality moves from high to low rates. In this model it is argued that as a society experiences a decline in death rates due to socioeconomic changes, the major causes of death shift from communicable or infectious to degenerative or chronic diseases. With reference to the nutrition transition, large shifts have occurred in dietary and physical activity/inactivity patterns reflected in nutritional outcomes such as average stature and body composition. Modern societies have diets high in saturated fat, sugar, and refined foods and low in fiber, the so-called "Western diet" which is associated with high levels of chronic/degenerative disease (Popkin 2002).

Barry Popkin (2002) has revised the demographic, epidemiologic, and nutrition transitions to show that these three models are really one and the same. In his conceptualization,Popkin (2002)argues that the nutritional shifts are related to the demographic and epidemiological transitions.Both the epidemiologic transition and nutrition transition are shown as being intertwined with the demographic transition. As we move into the Post-2015 Millennium Development Agenda and the newly adopted Sustainable Development Goals (SDGs), it is imperative that we ascertain the position of Africa with reference to these transitions (demographic, epidemiologic, and nutrition). As noted above, Africa appears to be a special case where both degenerative and communicable diseases continue to ravage its peoples. The authors of the chapters contained in this book realize the importance of these frameworks to examine the current health status and conditions in Africa, i.e. the communicable/NCD complex. Borrowing on the concepts contained in the revised epidemiological and nutrition transition framework, this book interrogates and offers case studies of the dual burden of communicable and NCDs in Africa, obstacles posed by rising risk-factors in a continent with fragmented and siloed health systems and lack of financial resources. Although this will be a common theme and thread running through all the chapters, the contributions in this book represent different perspectives and approach the issues from different angles through its assemblage of interdisciplinary and transdisciplinary teams of experts.

Guided by the Popkin (2002) revised Epidemiologic and Nutritional Transition framework, the main body of this book is divided into the following parts: Part I: Introduction (two chapters); Part II: Emerging and re-emerging infectious diseases (five chapters); Part III: Non-communicable/degenerative disease complex (five chapters); and Part IV: Food security, nutrition and health (two chapters). Chapters in Parts II, III, IV are organized based on the Barry Popkin (2002) revised Epidemiological Transition, also referred to as the "Stages of Health, Nutritional and Demographic Change Model".

The second chapter (Chapter 2) in the Introduction section offers an index on how to measure a country's advancement in development (Elliot *et al.* in this book). Using Kenya as a case study, Elliott *et al.* apply the Canadian Index of Wellbeing (CIW) as a useful model to measure wellbeing given the just ended MDGs. These authors argue that such an index can be useful in both assessing change over time vis-à-vis the wellbeing of a population, as well as change introduced through (policy) interventions. The chapter offers an overview of existing measures and assesses their robustness and application to an LMIC context. The chapter concludes with next steps and policy implications for developing a robust index that captures wellbeing.

The second part of the book (Part II) offers a set of chapters that examine emerging and re-emerging infectious diseases, particularly Ebola, HIV/AIDS, Malaria, and Buruli ulcer (Chapters 3–7). Shapiro *et al.* in Chapter 3 employ the disease ecology framework to understand the factors that led to the Ebola outbreak starting in February of 2014 in West Africa. They outline a political background to provide the historical context of the three most severely affected countries: Guinea, Liberia, and Sierra Leone. With an overview of the historical context, considering Ebola through a disease ecology lens highlights various aspects of human activity that explain potential avenues by which Ebola entered the human population. Additionally, these authors argue that historical context can lend insight into which activities may have fueled this Ebola outbreak to become a globally threatening epidemic. The chapter identifies political instability, the mining industry, deforestation and climate change, and the cultural context as the key human activities that led to the rapid proliferation of the Ebola epidemic in the most affected countries. The chapter concludes by offering a set of proposed policy options which are informed by the identified human activities revealed through the use and analysis of the disease ecology framework.

In Chapter 4, Moise *et al.* examine the trajectory of the HIV/AIDS epidemic in Africa over the past three decades. They note that a greater understanding of the magnitude, trends, and diversity of the HIV/AIDS pandemic show some hope for the African continent. New estimates from the Population HIV Impact Assessment (PHIA) show a significant decrease in the rates of new infections, "stable numbers of people living with HIV, and more than half of all those living with HIV showing viral suppression through use of antiretroviral medication". The chapter ends on a positive note that if current trends continue, Africa is on the right track towards attaining 90:90:90, an ambitious global commitment to reduce the HIV/AIDS pandemic by the end of 2020. Also tackling the implications of the HIV/AIDS epidemic on women's lives, Semu in Chapter 7 questions whether an international development consensus is sufficient to overcome sociocultural and systemic processes that drive maternal mortality, HIV/AIDS, and gender inequality. She uses the case of Malawi's maternal mortality, morbidity, and HIV/AIDS as indicative of gender-based violence and disempowerment, poverty, and systemic process where ill-health and lack of sexual and reproductive health services lead to high risk for HIV infection, early-onset, frequent, and unwanted pregnancies and childbearing. Semu's chapter concludes that while the recently adopted SDGs reinforce the commitment to achieve MDGs, an integrated approach that is cognizant of structural and sociocultural processes is critical for the realization of transformative development and gender equality espoused by the SDGs.

Blackstone *et al.* in Chapter 5 examine progress towards combatting malaria using Nigeria as a case study. Their findings show that there is still a lot that needs to be done with respect to diagnosis of malaria between parents and physicians. The chapter examines the concordance between physician diagnosis of malaria and mothers' identified symptoms with mothers' perceptions of children's illness. They find that only 29% of children diagnosed with malaria by a physician were correctly diagnosed by their mothers. The policy implication of this finding is obvious; there is great need for effective local surveillance and education programs in order to improve the accuracy of at-home and clinical diagnoses and prompt referral to health care providers. Oppong and Ruckthongsook in Chapter 6 use satellite images and spatial analyses to determine environmental risk factors of the distribution of Buruli ulcer in Ghana. The results of their analysis show that Buruli ulcer rates are positively correlated to closed-forest and inversely correlated to grass land, soil, and urban areas, an indication that forest is the most important environmental risk factor with reference to this disease.

In the third part of the book (Part III), a set of five chapters examine the rise of non-communicable/degenerative disease complex in Africa and their challenge to development activities. In Chapter 8, Tenkorang examines the relationships between nutrition and physical activity and the risks of living with hypertension in Ghana and South Africa. His findings demonstrate that nutrition and physical activity are significantly associated with risk of living with hypertension. In Chapter 9, Tenkorang offers a case study of South Africa which examines whether racial distinctions exist among people living with NCDs and the extent to which such differences are explained by physical activity, nutrition, and socioeconomic inequality. The results of his analysis indicate significant differences in the burden of NCDs for the various racial groups. Interventions targeted at preventing chronic diseases in South Africa should pay attention to racial and ethnic needs.

In Chapter 10, Saji *et al.* note that excess weight, a problem that once was perceived as exclusively affecting the affluent, is fast becoming a major issue of concern for all population segments in Cameroon and many other parts of the developing world. This chapter offers an insightful review of the pattern of weight status and related risks over time among Cameroonians, possible reasons for the current state of affairs, and likely future trends. It particularly explores the role of uncontrolled urbanization in driving the current trend and potential intervention opportunities for public health workers. On the other hand, Chapter 11 by Oyapero offers a case study that examines maternal perception about early childhood caries in Nigeria. Their findings show that the majority of mothers, over 75%, were ignorant about infecting their babies with cariogenic bacteria due to uninformed bottle-feeding practices, which raises the need for health promotion activities that target mothers at antenatal clinics and in pediatric outpatients.

The last chapter in Part III, Chapter 12 by Christoph *et al.*, offers a case study of Uganda with reference to health data collection, particularly data on NCDs. While in recent decades, there has been a focus on collecting data on infectious diseases through surveys such as the Demographic and Health Survey (DHS), the collection of data on NCDs is almost non-existent. The authors in this chapter argue the increasing prevalence of NCDs and call for the establishment of a more comprehensive and timely surveillance system. The chapter encourages policymakers and practitioners in Uganda to take a multi-sector approach in order to strengthen health information systems used to track NCD risk. As NCDs and risk factors including overweight and obesity become more prevalent in Uganda and other countries in sub-Saharan Africa, it will be increasingly important to implement health information systems that adequately measure risk factors to account for the shifting burden of disease.

The final part of the book (Part IV) contains two chapters which address issues of food security, nutrition, and health. Muyonga *et al.* examine the nutritional and nutraceutical properties of traditional African foods. They note that some traditional African foods have exceptional nutritional value and supply health-promoting bioactive compounds. Increased consumption of such foods may, therefore, contribute to the alleviation of over- and under-nutrition. The chapter provides an excellent review of the nutritional value and health benefits associated with vegetables, fruits, cereals, edible insects, small fish species, mushrooms, legumes, sesame, and tuber and root crops traditionally consumed in Africa. The last chapter in this section and in this book, Chapter 14 by Ekesa *et al.*, explores two participatory approaches providing nutritious foods for infants and young children in Uganda, namely, "Multi-Sector Stakeholder Engagement for Learning, Feedback, and Scaling" and "Household Experimentation and Learning", each addressing infants and young children dietary shortfalls. The first approach engaged actors from several sectors at community and district levels. The second approach combined monitoring food intake and child nutritional status, farmer test plots for more diverse nutritious foods, and participatory development of nutrient-rich dishes. This increased food production and improved access to and consumption of nutritious diverse diet.

The chapters in this volume contribute to our understanding of the current pattern of disease in Africa at a stage when the world is moving from MDGs to SDGs. The book offers well-conceived case studies that are empirically, methodologically, and theoretically grounded. The book should be significant in policy circles to assist international organizations, governments, and United Nations agencies chart the future for health in Africa in light of the recently adopted SDGs. Indeed, the case studies contained in this book are informative to policy-makers and practitioners as they continue to tackle the emerging and fast-growing problem of the dual burden of communicable and NCDs to minimize the negative health outcomes and improve community level, population health, wellbeing, and equity. It should also prove useful as a complementary reader in global public health-related courses.

References

Aboderin, I. 2012. Global poverty, inequalities and ageing in sub-Saharan Africa: A focus for policy and scholarship. *Journal of Population Ageing*, 5(2), 87–90.

Aikins, A.d.-G., Unwin, N., Agyemang, C., Allotey, P., Campbell, C., and Arhinful, D. 2010.Tackling Africa's chronic disease burden: from the local to the global. *Globalization and Health*, 6(5), 1–7. Available online at http://www.globalizationandhealth.com/content/pdf/1744-8603-6-5.pdf [Accessed January 16, 2018].

Boutaye, A., and Boutaye, S. 2005. The burden of non-communicable diseases in developing countries.*International Journal for Equity in Health*, 4(2). Available online at http://www.equityhealthj.com/content/4/1/2 [Accessed January 15, 2018].

Kalipeni, E., Semu, L. L., and Mbilizi, M. A. 2012. The brain drain of health care professionals from sub-Saharan Africa: A geographic perspective. *Progress in Development Studies*, 12(2–3), 153–171.

Miranda, J. J., Kinra, S., Casas, J. P., Davey Smith, G., and Ebrahim, S. 2008. Non-communicable diseases in low- and middle-income countries: Context, determinants and health policy. *Tropical Medicine and International Health*, 13(10), 1225–1234.

Moise, I. K., Verity, J. F., and Kangmennaang, J. 2017. Identifying youth-friendly service practices associated with adolescents' use of reproductive healthcare services in post-conflict Burundi: a cross-sectional study. *International Journal of Health Geographics*, 16:2. Available online at https://doi.org/10.1186/s12942-016-0075-3 [Accessed January 20, 2018].

Murray, C. J., and Lopez, A. D. 2013. Measuring the global burden of disease. *New England Journal of Medicine*, 369(5), 448–457.

Omran, A. 1971. The epidemiologic transition: A theory of the epidemiology of population change. *The Milbank Memorial Fund Quarterly*, 49, 509–538.

Popkin, B. M. 2002. An overview on the nutrition transition and its health impacts. *Public Health Nutrition*, 5(1A), 93–103.

2 Taking it global

Toward an index of wellbeing for low- to middle-income countries

Susan J. Elliott, Jenna Dixon, and Elijah Bisung

Understanding how the world works

We live in a world faced with unprecedented change (Deaton 2013, Gallup World Poll 2017,World Economic Forum 2015). According to Klaus Schwab, CEO of the World Economic Forum (2015: iii),

> In the coming decade...our lives will be even more intensely shaped by transformative forces than are under way already. The effects of climate change are accelerating, and the uncertainty about the global geopolitical context and the effects it will have on international collaboration will remain. At the same time, societies are increasingly under pressure from economic, political, and social development, including rising income inequality.

Indeed, in the most recent annual assessment of global risks, the World Economic Forum uses a *Global Risks Perception Survey*, administered to over 900 stakeholders, to produce metrics to measure both likelihood and impact of a range of global risks (Figure 2.1).

These risks are organised by category (economic, geopolitical, environmental, societal, and technological) and situated on the graph according to the intersecting constructs of perceived likelihood and perceived impact. According to the data, the top ten risks over the next decade are as shown in Table 2.1.

In the face of such intense and rapid change, it is difficult to fathom how we might monitor related impacts on the wellbeing of population(s) affected. In the past, the world has typically relied upon measures of economic health or wellbeing such as Gross Domestic Product (GDP). As the world ends its commitment to the Millennium Development Goals (MDGs) and embarks on a commitment to the Sustainable Development Goals (see www.sustainabledevelopment.un.org/) questions about where we as a global society should continue our investments in wellbeing and efforts to measure those outcomes are now up for debate.

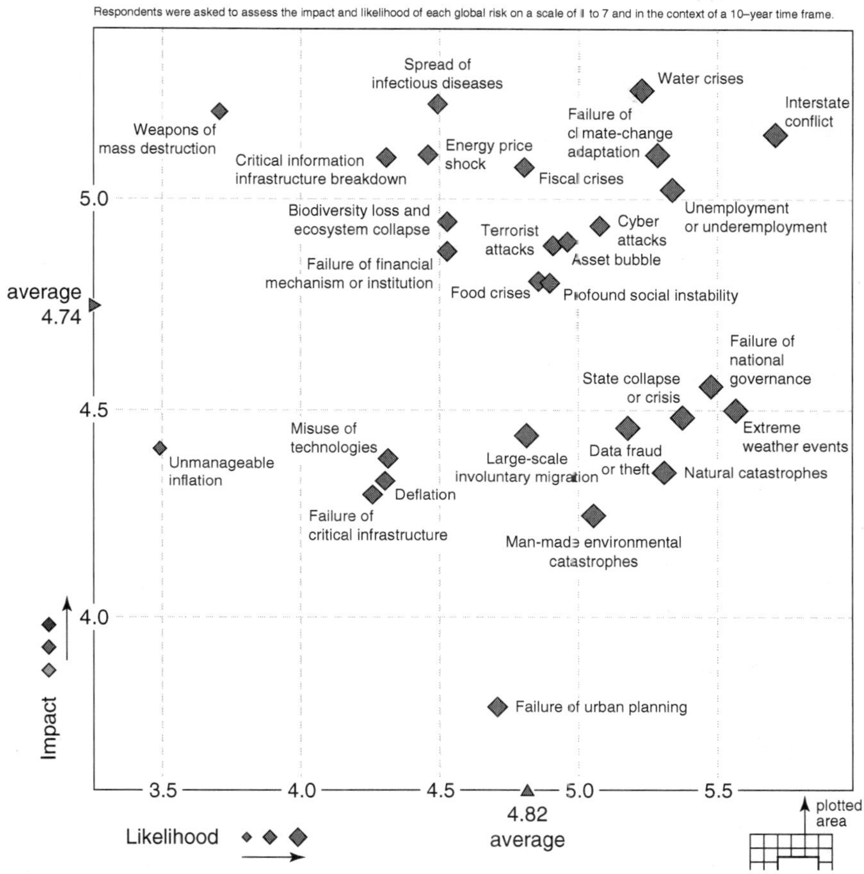

Figure 2.1 The global risks landscape (World Economic Forum, 2015).

These questions are especially poignant for those populations most vulnerable to change: low- to middle-income countries (LMICs). To date, this has proven a significant challenge (Tiliouine *et al.* 2006) particularly as these LMIC populations are also faced with the dual burden of the disease complex through their demographic/epidemiologic/nutrition transition (Popkin 2002). That is, as economies develop, "diseases of affluence" (i.e., cardiovascular disease, stroke, hypertension, overweight/obesity) begin to mark populations. Populations in LMICs must learn to deal with these heavy disease burdens while at the same time continuing to struggle with epidemics of infectious diseases (e.g., malaria, a range of neglected tropical diseases [see here for a list: http://www.who.int/neglected_diseases/diseases/en/], and the continuing challenges of HIV/AIDS). Challenges abound.

Table 2.1 The likelihood and impact of the top ten global risks

	Likelihood	*Impact*
1	Interstate conflict	Water crises
2	Extreme weather events	Spread of infectious diseases
3	Failure of national governance	Weapons of mass destruction
4	State collapse or crisis	Interstate conflict
5	Under and unemployment	Failure of climate change adaptation
6	Natural catastrophe	Energy price shock
7	Failure of climate change adaptation	Critical information infrastructure breakdown
8	Water crises	Fiscal crises
9	Data fraud or theft	Under and unemployment
10	Cyber attacks	Biodiversity loss and ecosystem collapse

This chapter discusses the limitations of using GDP-type tools in measuring the wellbeing of populations, reviews the recent development of alternative measures, and suggests a metric going forward – the Global Index of Wellbeing (GLOWING), based upon the successful Canadian Index of Wellbeing (see https://uwaterloo.ca/canadian-index-wellbeing/). GLOWING can be used to measure the impacts of rapid change – economic, climatic, demographic, health, cultural, social – felt most strongly by LMICs. The paper then describes the research agenda for the development, piloting, and scaling up of GLOWING.

Measuring how the world works: GDP as a measure of population wellbeing?

GDP or Gross Domestic Product represents the market value of all goods produced within a country over a period of time based on the simple assumption that the higher the GDP, the better off the population of the country that produced said goods. The concept was developed by Simon Kuznets for a US Congress Report in 1934 (Kuznets 1934) and subsequent to the Bretton Woods conference in 1944 became the measure of a country's economic health (that is, for all UN member states) (Costanza *et al.* 2009). Interestingly, however, even Kuznets was aware of the limitations of GDP:

> Economic welfare cannot be adequately measured unless the personal distribution of income is known. And no income measurement undertakes to estimate the reverse side of income, that is, the intensity and unpleasantness of effort going into the earning of income. The welfare of a nation can, therefore, scarcely be inferred from a measurement of national income.
>
> (Kuznets 1934: 6–7)

Approximately 30 years later, Robert F. Kennedy shared similar sentiments:

> the gross national product does not allow for the health of our children, the quality of their education, or the joy of their play. It does not include the beauty of our poetry or the strength of our marriages, the intelligence of our public debate or the integrity of our public officials…It measures neither our wit nor our courage, neither our wisdom nor our learning, neither our compassion nor our devotion to our country. It measures everything in short, except that which makes life worthwhile.
>
> (1968, from Kennedy 2012)

To illustrate, if we measure the progress of countries solely by growth in wealth and consumer spending, we might conclude that many countries are indeed doing well – but there is more to wellbeing than wealth (Smits and Steendijk 2015). If we broaden our measurement of progress to include social and economic inequalities, environmental, cultural, and governance issues that perhaps matter more to citizens, then our results might be quite different. For example, the Canadian Index of Wellbeing illustrates that Canada experienced impressive growth in GDP between 1994 and 2010; concomitantly, however, Canadians' confidence in their federal parliament as well as voter turnout in federal elections were at their lowest (Canadian Index of Wellbeing 2012). Treating GDP as a measure of economic wellbeing can, for a number of reasons, provide misleading indications about how well-off people are. First, non-market values and benefits, such as volunteer work, are often excluded from GDP calculations though they form an important part of how well society fares. Second, it fails to account for depletion of natural resources or the state of the natural environment. Third, it is difficult to capture inequalities in the distribution of wealth, resources, and opportunities in society. Finally, GDP does not distinguish between economic productions that are beneficial to societal wellbeing and those that are harmful:

> Too much and for too long, we seemed to have surrendered personal excellence and community values in the mere accumulation of material things…Gross National Product counts air pollution and cigarette advertising, and ambulances to clear our highways of carnage. It counts special locks for our doors and the jails for the people who break them. It counts the destruction of the redwood and the loss of our natural wonder in chaotic sprawl.
>
> (Kennedy 2012)

The notion of questioning the validity of GDP for measuring global progress is not new (see the insightful commentary by Costanza and colleagues (2014] in *Nature* as the concept applies to the development of the Sustainable Development Goals). The key issue is: what to do about it? In response, many countries are beginning to explore alternative mechanisms to measure the wellbeing of national populations. Indeed, the Istanbul World Forum (27–30 June 2007) concluded with international organisations (including the European Commission, Organisation for

Economic Co-operation and Development [OECD], Organisation of the Islamic Conference, United Nations, United Nations Development Programme [UNDP], World Bank, and many more) all affirming in a declaration their commitment to measuring and fostering the progress of societies in all dimensions, with the ultimate goal of improving policy making, democracy, and citizen wellbeing.[1] This does not entail throwing the baby out with the bath water; rather, it moves us toward a greater depth of understanding and breadth of knowledge for policy makers to make evidence-informed decisions, centred on what matters most to the citizens of the nation and the sustainability of our common future (Canadian Index of Wellbeing 2012, Stiglitz *et al.* 2009).

Toward a definition of wellbeing

Although the phrase "wellbeing" is widely used both in the literature and in common speech, there is no commonly agreed upon definition for the term (Matthews 2012). It has often been used interchangeably with quality of life, happiness, and life satisfaction (Hall *et al.* 2010,Matthews 2012), and these definitions are underpinned by diverse philosophical and competing traditions of inquiry ranging from the behavioural, social, economic, and health sciences (Wiseman and Brasher 2008, Hall *et al.* 2010). Further, some have eschewed formal definitions for a list of key characteristics. For example, the OECD (2011) argues that defining wellbeing must focus on four key pillars:

1 The wellbeing of people in each country, rather than on the macroeconomic conditions of economies.
2 The wellbeing of different groups of the population, in addition to average conditions.
3 Wellbeing achievements, measured by outcome indicators, as opposed to wellbeing drivers measured by input or output indicators.
4 Objective and subjective aspects of people's wellbeing as both living conditions and their appreciation by individuals are important.

As described below, a number of global initiatives have adopted wellbeing as an outcome worth measuring and have variously informed its development through existing frameworks or through a consultative approach in which the components, dimensions, or domains of wellbeing are developed through citizen consultation, dialogues, or political processes (Hall *et al.* 2010, Kroll 2011). Regardless of the route, we see a great deal of similarity in the crosscutting themes used in these measurement tools: issues of material living conditions, quality of life, socioeconomic performance, and sustainability of natural systems are some examples. From our perspective, we feel – going forward – that simplicity is and should be a key characteristic of the definition and measurement of wellbeing; if it is not simple, people will not understand it nor will it be measurable. As such, our proposed Global Index of Wellbeing (GLOWING) adopts Angus Deaton's (2013: 24) notion of wellbeing:

> I use the term *wellbeing* to refer to all of the things that are good for a person, that make for a good life. Wellbeing includes material wellbeing, such as income and wealth; physical and psychological wellbeing, represented by health and happiness; and education and the ability to participate in civil society through democracy and the rule of law.

We add to this the *context* within which wellbeing happens; for example, the characteristics of the physical and built environments and the level of vitality attached to one's community. Indeed, others have shown repeatedly how important context is to our understanding of the wellbeing of populations (Matthews 2012). Further, we intend to employ a fully consultative mechanism for informing the domains and indicators for measurement used within GLOWING (more on this below).

Progress in measuring wellbeing

Prior to embarking upon the development of yet another "index of wellbeing", it behoved the authors to undertake a systematic environmental scan of existing measures. Twenty-three "beyond GDP" measures of wellbeing (see Appendix A) were selected based on three primary inclusion criteria. First, measures had to be focused on societal level wellbeing (that is to say, made some attempt to discuss how society as a whole is doing, not solely focused on individuals[2]). Second, measures had to be conducted at the national or international scale. Third, measures had to have some empirical grounding. The resulting measures identified are categorised in Table 2.2.

Since the intent of this exercise is to evaluate measures that look "beyond GDP", we have not included measures that would fit exclusively into the "GDP" category;

Table 2.2 A classification of existing wellbeing measures (adapted from Vemuri and Costanza, 2006)

Classification	*Meaning*
GDP	Gross Domestic Product (or Gross National Product)
GDP+	GDP *plus* other basic indices (e.g., education, health)
GDP++	GDP+ *plus* broader economic welfare indicators that combine wealth distribution adjustments, and natural, social, and human capital adjustments
Objective Wellbeing	Derived from a broad range of domains and indices that rely on objective measures of wellbeing typically sourced from secondary data sources
Subjective Wellbeing	Derived from domains and indices that require an individual to reflect on and evaluate their overall wellbeing, happiness or life satisfaction; these indices are typically based on the collection of primary data

*Note: Two measures (State of the USA Key National Indicator system and Gallup-Healthways Well-Being Index, UK) were deemed to be either incomplete or too early in their development to be assessed and therefore are not included in this review.

rather, it remains as a point of reference. In some cases, the measure did not fit neatly within a single category and was therefore coded as belonging to primary and secondary (a/b) categories. With this, we end up with seven separate clusters (Appendix A). In addition to providing key aspects of each measurement tool (scale, type, year of first publication, domains, indicators, data sources, countries applied), we assess the usefulness of these existing tools in developing an index of wellbeing that can be used in LMICs to measure the impacts of rapid change (economic, demographic, environmental, health, social, cultural) on population wellbeing.

As a quick summary, at the international level, the OECD and the UNDP have been the major stakeholder organisations leading efforts toward measuring global wellbeing and social progress. Since 2007, the OECD has been leading international wellbeing initiatives and supporting processes toward measuring wellbeing in their member countries. Prior to the OECD's wellbeing initiatives, in 1990 the UNDP launched the Human Development Report together with the Human Development Index with the single goal of putting people at the centre of economic growth and development in terms of debate, policy, advocacy, and decision-making. The Human Development Index along with other indices such as the Multidimensional Poverty Index, Gender Inequality Index, and Inequality Adjusted Human Development Index continue to influence many regional and country-focused UN reports and initiatives on development and progress.

At the regional level, the European Union has been the only body with a clear initiative aimed at measuring progress "beyond GDP" in its member countries. In 2007, the European Commission, together with other stakeholders including the European Parliament, Club of Rome, and OECD, held a high-level conference on "Beyond GDP" to shed light on the most appropriate indices for measuring progress, and how best to integrate such indices into public policy and decision-making (Kroll 2011). Further resulting from this conference, the Commission in 2009 issued a roadmap with key actions to improve indicators of progress in ways that meet citizens' concerns as well as address global challenges of the twenty-first century, such as climate change, resource management and depletion, human health, and quality of life.

Several countries have since launched similar progress and wellbeing-related initiatives. There are common approaches in terms of how countries develop their thematic areas and frameworks for measuring – as well as reporting – the wellbeing and progress of citizens. The most notable examples include the Canadian Index of Wellbeing, Measures of Australia's Progress, New Zealand Social Report, Taiwan National Well-being Index, and the UK Measures of National Well-being. Further, Bhutan has often been cited for leading the way in terms of measuring the wellbeing and happiness of its citizens. Bhutan's Gross National Happiness Index was introduced in 1972 as an alternative method of measuring progress and has since been a central philosophy in the country's development process.

Limitations of "beyond GDP" measures

In principle, all "beyond GDP" approaches report some information about the wellbeing of populations albeit with some limitations. First, the GDP+ and

GDP++ measures only make adjustments to GDP to reflect other common aspirations of society and social outcomes (e.g., health, education, natural resource depletion, etc.). Though these measures mark some progress from GDP as an index of wellbeing, their foundations are rooted in wealth and capital as key indicators of wellbeing. Two GDP ++ measures that move a step away by excluding GDP as an indicator of wellbeing (Gender Inequality Index and Multidimensional Poverty Index) are also limited to gender inequalities and individual deprivations that affect human development. Notwithstanding the limitations of these measures, reporting them in addition to other wellbeing measures gives a more comprehensive story about how populations are faring, especially given the ease with which they can be incorporated into national statistical systems.

A second set of wellbeing measures are those based on subjective wellbeing or a combination of subjective and objective measures of wellbeing. These measures focus on *subjective* wellbeing as experienced by individuals (or a combination of subjective experiences and objective output indicators of wellbeing) and how that experience informs society as a whole. Examples of purely subjective measures include the Australian Unity Wellbeing Index and the World Happiness Report. These have been criticised for producing differing outcomes, depending upon the measurement instrument chosen (Cramm and Nieboer 2012). Aside from narrowly focusing on happiness and subjective feelings to measure wellbeing, these measures rely on primary data or sources such as Gallup World Poll that are not consistently collected in LMICs. Their application and relevance for timely policy making, therefore, remain limited in resource poor settings, especially at subnational levels. Further, objective/subjective and subjective/objective measures offer some opportunities to address the limitations of purely subjective measures, and are strongly supported by some mainstream wellbeing researchers (see, for example, La Placa *et al.* 2013) but remain conceptually difficult for policy application in LMICs for two reasons. The first relates simply to the cost and complexity of collecting primary data in LMICs; thus GLOWING will be built – as is the Canadian Index of Wellbeing – on the sole use of objective indicators. Second, there is empirical evidence that subjective wellbeing has a direct impact across a broad range of objective wellbeing and life outcomes such as health, productivity, and education (De Neve *et al.* 2013). The value of subjective measures of wellbeing for policy making thus lies primarily with empirically determining objective outcomes that emanate from happiness/life satisfaction/flourishing in different cultural contexts. Subjective indicators of wellbeing may be important for evaluation of policy measures (Kroll 2011) but remain difficult to apply in LMICs due to data constraints and complex social and economic challenges that face populations in these countries.

A third group of measures that are useful for the LMIC context are objective measures of wellbeing. These include Australian National Development Index, Measure of Australia's Progress, New Zealand Social Report, OECD Better Life Index, the Sustainable Society Index, Well-Being of Nations, and Social Progress Index. However, measures such as Sustainable Society Index and the Well-Being of Nations are centred on sustainable development and the wellbeing of ecosystems and the environment, which limits their focus as compared to the other objective wellbeing measures. Though these "environmentally focused" measures remain helpful for assessing the state of the environment and progress toward sustainable development in countries, moving beyond GDP requires thinking broadly about what matters to individuals and the opportunities available for people to express their full capabilities. Further, the Sustainable Society Index includes GDP per capita as an indicator in the final index, which draws it closer to other GDP ++ measures.

Given the limitations discussed above, we consider the broad frameworks used to develop the following country-specific measures – Canadian Index of Wellbeing, Australian National Development Index (currently under development), Measure of Australia's Progress, and New Zealand Social Report – as appropriate for LMICs for two primary reasons. First, they largely use domains and indicators that people value as important for their daily life *beyond GDP*, domains and indicators developed based on broad consultations among major stakeholders and citizen groups. The final indices thus become relevant for local policy making and highlight important gaps in the wellbeing of citizens and other identifiable populations (such as indigenous peoples) in these countries. For example, unlike the Social Progress Index that uses the same domains and indicators for all countries, the Measure of Australia's Progress and New Zealand Social Report are developed to suit the aspirations and what matters to Australian and New Zealand citizens respectively. Thus, an important step in developing wellbeing measures for any LMIC is to use a set of indicators to measure what people value as important for their daily life, and the environments in which they live, grow, and work. This consideration is currently missing in the Social Progress Index and Wellbeing of Nations as researchers and citizens in LMICs have little input into the domains and indicators used to measure their wellbeing. Second, the country-specific measures rely on secondary data that are readily available without any significant extra burden or sampling issues that arise in the case of primary data. With regards to LMICs, our proposed global index – GLOWING – could and should rely on secondary data from national statistics offices, relevant ministries, and other publicly available databanks from UN agencies, the World Bank, Multi Indicator Cluster Surveys, and regional agencies such the African Development Bank and African Union.

Using the Canadian Index of Wellbeing Framework for GLOWING

The framework of the Canadian Index of Wellbeing (Appendix B), recognised as a global leader at the forefront of wellbeing initiatives (Canadian Index of Wellbeing 2011), remains attractive and particularly useful among the other national objective wellbeing measures for application to LMICs for a number of reasons. First, though all the national objective wellbeing measures use multiple domains to provide a holistic picture of wellbeing, there are significant differences in how these domains and their indicators are presented. While the Canadian Index of Wellbeing aggregates domains and their indicators into a composite index, others, such as Measures of Australia Progress and New Zealand Social Report, are presented in the form of a "dashboard" that reflects diverse domains/indicators and their performances as a visual spread. There are those who would strongly support the dashboard model of presenting wellbeing measures (Foregeard *et al.* 2011, Matthews 2012); however, as others have noted (Stiglitz *et al.* 2009, Kroll 2011), a common advantage of a composite index is the ability to provide information on wellbeing with a single value, which can easily be communicated, interpreted, and compared across space and time. The wellbeing of a society can thus easily be compared over time and across space.

Whether wellbeing is presented as a dashboard or index, the decision of what to include is paramount. That is, incorporating indices that reflect the society and strike a balance between information and parsimony. That this will exclude some aspects of a society's aspiration, or include unimportant ones, remains a challenge. Moreover, the domains and indicators of wellbeing are not static but continue to change as more becomes known about society's aspirations (and how to measure them) and as more sources of data become available (Canadian Index of Wellbeing 2011). Furthermore, compiling social, economic, and environmental indicators into a single measure involves making decisions on appropriate weights. These decisions are value laden and difficult to justify. The Canadian Index of Wellbeing argues that the absence of such a clear justification is enough reason to treat all indicators equally at the present time (Canadian Index of Wellbeing 2012). As progress is made on measuring wellbeing on both national and international fronts, sufficient reason for different weights may become apparent as new knowledge and greater understanding of the relationships among indicators, domains, and thematic areas is developed (Canadian Index of Wellbeing 2012). Using a composite index within GLOWING will facilitate comparison between wellbeing vis-à-vis GDP performance in any chosen LMIC – or group of countries in a region (e.g., East African Region) – over a period of time as well as give a comprehensive picture of performance on each indicator and domain over time, across space, and in the face of global environmental change, broadly defined. While we agree that making a choice between an index and dashboard is subordinate to the establishment of a broad statistical system that captures as many relevant dimensions of wellbeing as possible in any country (Stiglitz *et al.* 2009), the usefulness of a single headline composite index vis-à-vis GDP as a communication instrument to policy makers cannot be ignored (Kroll 2011).

Another strength of the Canadian Index of Wellbeing is that its framework, domains, and indicators were created through combined efforts of national and international organisations, experts, community groups, and citizens after several years of extensive consultation and research. Many of the country level initiatives have a great deal of public involvement and stakeholder participation. There are significant similarities in the ways political leaders, experts, community groups, and civil society organisations combined efforts, with extensive public consultations, dialogues, and national roundtables, to achieve some form of general consensus on wellbeing domains/indicators, their measurement, and monitoring. However, there are some notable differences regarding *who* leads the process of measuring and reporting. For example, the Canadian Index of Wellbeing and Australia National Development Index are both led by collaborations between universities and NGOs, while national statistics offices and agencies lead other initiatives such as Measures of Australia Progress, New Zealand Social Progress Report, and UK Index of Well-being. Arguably, there are issues of legitimacy and citizen acceptance associated with government institutions (i.e., national statistics offices) leading the process of measuring wellbeing in some jurisdictions (Kroll 2011). Further, the extent and depth of citizen participation and engagement and the ability of indicators to adequately reflect citizen aspirations may also influence legitimacy and acceptance of reporting. As in the case of many LMICs where governance issues remain an ongoing challenge, it is important to adapt the Canadian Index of Wellbeing approach where local researchers and experts use publicly available data to report the wellbeing of citizens. As noted by Noll (2011), many of the principles and approaches suggested by social indicators researchers have been incorporated into statistical information systems of international agencies and national statistical offices which can provide aggregate level information to other bodies and research institutions interested in measuring and monitoring wellbeing and social progress.

Finally, two important global initiatives and their core dimensions remain central and form the overarching perspective for developing country level domains. These are 1) OECD's Better Life Initiative core domains: quality of life, material living conditions, and sustainability and 2) the Stiglitz-Sen-Fitoussi Commission core domains: revised economic indicators, quality of life, and sustainability. Though specific country level domains and indicators differ in order to reflect different cultural aspirations, different identities, and differences in data availability, the framework behind the Canadian Index of Wellbeing and its domains/indicators broadly reflect these recommendations and guidelines. As mentioned earlier, the OECD (2011) proposed two important focus areas for constructing domains that serve as a guide for other initiatives. The first is that economic indicators should focus on the wellbeing of people rather than on the macroeconomic conditions in a country. For example, efforts should focus on measuring and reporting progress in incomes, jobs, and living standards (as in the case of the Canadian Index of Wellbeing) instead of standard indicators of macroeconomic performance such as GDP, productivity, innovation, extraction, etc. Second, wellbeing should be measured by outcome indicators, as opposed to input indicators that only measure drivers of wellbeing. For example, instead of using spending or investment in education, the Canadian Index of Wellbeing uses indicators such as childcare spaces, student to educator ratio in public schools, basic knowledge and skills, high school and university completion, etc., to measure progress in education (Canadian Index of Wellbeing 2012).

Conclusion – The way forward

This chapter has made the case for the need for a composite measure of wellbeing useful in LMICs as they face unprecedented global environmental change (broadly defined) over the next decade and beyond. In so doing, we have provided a (brief) overview of contemporary progress in wellbeing initiatives with potential use in LMICs, along with their associated strengths and limitations, highlighting the strengths of the Canadian Index of Wellbeing as a model for adaptation to measure wellbeing in LMICs. Though questions around progress in wellbeing vis-à-vis GDP growth have assumed global importance, it is obvious that most of the country initiatives are based in high income countries. The time to transfer them from their comfort zone to LMICs – where wellbeing also matters and populations are most vulnerable to the impacts of global environmental change – is long overdue. This becomes especially important as LMICs enter the post-MDG world.

Going forward, the logical next step is to ground truth these ideas in a proof of concept. As such, we have begun a pilot of GLOWING in East Africa with a transdisciplinary team of researchers that includes geographers, economists, epidemiologists, ethicists, and psychologists. The East Africa Community (see http://www.au.int/en/recs/eac) is poised to develop economically and is experiencing rapid environmental, social, cultural, and health changes as a result. Kenya's GDP has virtually doubled between 1990 and 2012. The Lake Victoria Basin Commission is doing well to document the impacts of these changes (see http://www.lvbcom.org) and is keen to partner in the development of measures of the impacts of growth and change. Working with partners on the ground in Kenya and other parts of East Africa, we have discerned through reconnaissance that it is feasible to develop socially, culturally, and geographically relevant indicators across the existing domains (Appendix B), but this must be done in consultation with local partners, using a mixed-methods approach (Matthews 2012). In this way, we will begin to be able to "measure what matters", mark change, and guide evidence-informed policy making. Once the proof of concept has been assessed, we will undertake comparative analyses in other parts of Africa (west and south) and then head to the Caribbean to assess the reliability and validity of the tool in a substantially different context. In so doing, a critical task will be to identify key indicators that highlight the ways in which life and wellbeing are getting better or worse with regards to poverty, access to basic services, the epidemiologic transition, and environmental sustainability. Further, an important recommendation of the Stiglitz-Sen-Fitoussi Commission is the need for indicators to highlight the inequalities in individual experiences. This is important, as progress depends both on the *average* conditions in society as well as *inequalities* in people's conditions (Stiglitz *et al.* 2009). Further, while it is important to highlight inequalities across domains, certain inequalities may be mutually reinforcing (e.g., income and gender), and as such their combined effect must be assessed.

A major prerequisite for measuring both objective and subjective wellbeing is to build the capacity of official statistical agencies to meet the demand for data. These sets of data, including data in areas related to economic performance and environmental sustainability, are needed in high-, middle-, and low-income countries in order to build comprehensive and comparable measures of wellbeing for policy making. Looking into the future, measuring a complex and multifaceted concept such as wellbeing is not an end in itself but a means for informed policy making. Thus, the challenge is not only how to create and share knowledge about how communities, groups, and countries are flourishing, thriving, and using their capabilities to achieve their full human potential but how such knowledge is used to create healthy, just, and sustainable communities and nations (Wiseman and Brasher 2008, Hone *et al.* 2014, Krishnakumar and Nogales 2015). As the world struggles to pin down the Sustainable Development Goals and their measures, we learn from the recent Addis Ababa Action Agenda that it is all about building capacity in LMICs – through the incentivising of science, investment in education, and knowledge sharing – in order to make good decisions to support strong and healthy global populations (The Lancet 2015). As Matthews (2012: 99) points out, "following Stiglitz's advice, scientists can help governments 'do the right thing' by assisting them in measuring the right thing."

Notes

1 A copy of this declaration can be found at http://www.oecd.org/newsroom/38883774.pdf

2 Though not the concern of our review, there is a large body of literature that focuses on individual wellbeing and the inputs that determine individual wellbeing. For example, the highly cited work of Helliwell and Putnam (2004), which finds an individual's income, family-level social capital, marital status, religious beliefs, etc., all influence how that individual perceives their wellbeing. Thus, individuals within a society are expected to have varying levels of wellbeing and we are left with a poor understanding of how this reflects on society as a whole.

References

Australia Bureau of Statistics, 2012. *Aspirations of our nation: A conversation with Australians about progress*. Belconnen, Australia, catalogue No. 1370.0.00.002.

Canadian Index of Wellbeing, 2011. *How are we Canadians really doing? Highlights: Canadian Index of Wellbeing 1.0.*Waterloo, ON: Canadian Index of Wellbeing and University of Waterloo.

Canadian Index of Wellbeing, 2012. *How are we Canadians really doing? The 2012 CIW report.* Waterloo, ON: Canadian Index of Wellbeing and University of Waterloo.

Center for Sustainable Economy, 2012. *Measuring genuine progress: Towards global consensus on a headline indicator for the new economy* [Online]. Available from http://genuineprogress.net/wp-content/uploads/2013/01/Measuring-Genuine-Progress-Final.pdf [Accessed 30 April 2014].

Costanza, R., Hart, M., Posner, S., and Talberth, J., 2009. *Beyond GDP: The need for new measures of progress*. Boston: Boston University, The Pardee Papers No. 4.

Costanza, R. *et al.*, 2014. COMMENT: Time to leave GDP behind. *Nature*, 505, 283–285.

Cramm, J.M. and Nieboer, A.P., 2012. Differences in the association of subjective wellbeing measures with health, socioeconomic status, and social conditions among residents of an Eastern Cape township. *International Journal of Wellbeing*, 2(1), 54–67.

Deaton, A., 2013. *The great escape: Health, wealth and the origins of inequality*. Princeton, NJ: Princeton University Press.

De Neve, J. E., Diener, E., Tay, L., and Xuereb, C., 2013. The objective benefits of subjective well-being. In Helliwell, J. F., Layard, R., and Sachs, J., eds. *World Happiness Report*. New York: Earth Institute.

Forgeard, M.J.C., Jayawickreme, E., Kern, M.L, Seligman, M.E.P., 2011. Doing the right thing: measuring wellbeing for public policy. *International Journal of Wellbeing*, 1(1), 79–106.

Gallup WorldPoll, 2017. [Online]. Available from http://www.gallup.com/services/170945/world-poll.aspx [Accessed 15 June 2017].

Giovannini, E., Hall, J., Morrone, A., and Ranuzzi, G., 2010. *A framework to measure the progress of societies*. Paris: OECD, Statistics Directorate Working Paper No 34.

Helliwell, J.F. and Putnam, R.D., 2004. The social context of wellbeing. *Philosophical Transactions-Royal Society Of London Series B Biological Sciences*, 1435–1446.

Hone, L.C., Jarden, A., Schofield, G.M., and Duncan, S., 2014. Measuring flourishing: The impact of operational definitions on the prevalence of high levels of wellbeing. *International Journal of Wellbeing*, 4(1), 62–90.

Kennedy, R. F., 2012. *Remarks of Robert F.Kennedy at the University of Kansas on March 18, 1968* [Online]. Available at: http://www.jfklibrary.org/Research/Research-Aids/Ready-Reference/RFK-Speeches/Remarks-of-Robert-F-Kennedy-at-the-University-of-Kansas-March-18-1968.aspx [Accessed 15 June 2017].

Krishnakumar, J. and Nogales, R., 2015. Public policies for wellbeing with justice: A theoretical discussion based on capabilities and opportunities. *International Journal of Wellbeing*, 5(3), 44–62.

Kroll, C., 2011. *Measuring progress and well-being: Achievements and challenges of a new global movement*. Berlin: Friedrich Ebert Foundation.

Kuznets, S., 1934. *National income, 1929–1932: Letter from the acting secretary of commerce transmitting in response to senate resolution No. 220*. National Bureau of Economic Research.

La Placa, V., McNaught, A., and Knight, A., 2013. Discourse on wellbeing in research and practice. *International Journal of Wellbeing*, 3(1), 115–125.

Lancet, The, 2015. Editorial: Financing global health: the poverty of nations. *The Lancet*, 386 (9991), 311.

Matthews, G., 2012. Happiness, culture and context. *International Journal of Wellbeing*, 2(4), 299–312.

Noll, H.H., 2011. The Stiglitz-Sen-Fitoussi report: Old wine in new skins? Views from a social indicators perspective. *Social Indicators Research*, 102(1), 111–116.

Organisation for Economic Co-operation and Development, 2011. *Compendium of OECD well-being indicators*. Paris: OECD.

Popkin, B.M., 2002. An overview on the nutrition transition and its health implications: the Bellagio meeting. *Public Health Nutrition*, 5(1A), 93–103.

Smits, J. and Steendijk, R., 2015. The international wealth index (IWI). *Social Indicators Research*, 122, 65–85.

Social Policy, 2001. *The social report 2001*.Wellington, New Zealand: Ministry of Social Development.

Stiglitz, J., Sen, A., and Fitoussi, J.P., 2009. Report of the commission on the measurement of economic performance and social progress. Paris: OECD.

Tiliouine, H., Cummins, R.A., and Davern, M., 2006. Measuring wellbeing in developing countries: The case of Algeria. *Social Indicators Research*, 75, 1–30.

United Nations Development Programme, 2013. 2013 Human development report: The rise of the south: progress in a diverse world. New York: United Nations Development Programme.

Vemuri, A.W. and Costanza, R., 2006. The role of human, social, built, and natural capital in explaining life satisfaction at the country level: Toward a national well-being index (NWI). *Ecological Economics*, 58(1), 119–133.

Wiseman, J. and Brasher, K., 2008. Community well-being in an unwell world: trends, challenges, and possibilities. *Journal of Public Health Policy*, 29, 353–366.

World Economic Forum, 2015. *Global risks 2015*. Geneva: World Economic Forum.

Websites

Adjusted Net Savings (ANS): http://web.worldbank.org/WBSITE/EXTERNAL/TOPICS/ENVIRONMENT/EXTEEI/0,,contentMDK:20502388~menuPK:1187778~pagePK:210058~piPK:210062~theSitePK:408050,00.html

Australian National Development Index (ANDI) http://www.andi.org.au/

Australian Unity Well-Being Index http://www.australianunity.com.au/about-us/Well-being

Canadian Index of Well-being (CIW) https://uwaterloo.ca/canadian-index-well-being/

Gallup-Healthways Well-Being Index (UK) http://well-being.healthways.com/overview Uk.asp

Gallup-Healthways Well-Being Index (USA) http://www.healthways.com/solution/default.aspx?id=1125

Gender Inequality Index (GII) http://hdr.undp.org/en/statistics/gii

Gross National Happiness (GNH) http://www.grossnationalhappiness.com/articles/

Happy Planet Index (HPI) http://www.happyplanetindex.org/about/

Human Development Index http://hdr.undp.org/en/statistics/hdi

Inequality Adjusted HDI (IHDI) http://hdr.undp.org/en/statistics/ihdi

Measure of Australia's Progress http://www.abs.gov.au/ausstats/abs@.nsf/mf/1370.0

Multidimensional Poverty Index (MPI) http://hdr.undp.org/en/statistics/mpi

New Zealand Social Report http://socialreport.msd.govt.nz/documents/the-social-report-2010.pdf

OECD Better Life Index (BLI) http://www.oecdbetterlifeindex.org/

Social Progress Index http://www.socialprogressimperative.org/data/spi

State of the USA Key National Indicator System (KNIS) http://www.gao.gov/assets/320/317346.pdf

Sustainable Society Index (SSI) http://www.ssfindex.com/ssi/framework/

Taiwan National Well-being Index http://eng.stat.gov.tw/ct.asp?xItem=33831&ctNode=3274.

The Genuine Progress Indicator (GPI) http://genuineprogress.net/genuine-progress-indicator/

UK Index of Well-Being http://www.statistics.gov.uk/hub/people-places/communities/societal-well-being/index.html

Wellbeing of Nations http://sedac.ciesin.columbia.edu/data/set/cesic-wellbeing-of-nations

World Happiness Report http://unsdsn.org/resources/publications/world-happiness-report-2013/

World Values Survey http://www.worldvaluessurvey.org/wvs.jsp

Appendix A

Table 2.3 Wellbeing measures

Measures Of Wellbeing	*Description*	*Type*	*Domains And Indicators*	*Application*
GDP+ Measures of wellbeing				
1. Human Development Index (HDI) Scale: International Year: 1990 Data: Secondary	The HDI was created by UNDP as a summary measure of average achievement in key dimensions of human development: a long and healthy life, being knowledgeable, and have a decent standard of living. The HDI is the geometric mean of normalized indices for each of the three dimensions.	Index	Health • Life expectancy at birth Knowledge • Mean years of schooling • Expected years of schooling Standard of living • GNI per Capita (PPP $)	187 countries were included in the 2013 index
2. Inclusive Wealth Index (IWI) Scale: International Year: 2012 Data: Secondary	The IWI is joint initiative of the United Nations University International Human Dimensions Programme (UNU-IHDP) and the United Nations Environment Programme (UNEP) in collaboration with the United Nations Educational, Scientific, and Cultural Organization (UNESCO). It is based on the assumption that other key inputs are important components of the productive base of the economy, such as natural capital, human capital and social capital.	Index	Domains: • Manufactured • Capital Human Capital • Natural Capital	20 countries with different levels of income

(*continued*)

Table 2.3 Continued

Measures Of Wellbeing	*Description*	*Type*	*Domains And Indicators*	*Application*
GDP+ Measures of wellbeing				
1. Adjusted Net Savings (ANS) Scale: International Year: 2011 Data: Secondary	ANS measures the true difference between production and consumption, taking into account investments in human capital, depreciation of fixed capital, depletion of natural resources, and damages caused by pollution.	Index	Adjusted Net Savings = gross savings – consumption of fixed capital + education expenditures – energy depletion, mineral depletion, net forest depletion, and particulate emissions and carbon dioxide damage	150 countries
2. The Genuine Progress Indicator (GPI) Scale: Sub-national, National, International Year: 1995 Data: Secondary	The GPI uses the same personal consumption data as GDP but makes deductions to account for income inequality and costs of crime, environmental degradation, and loss of leisure and additions to account for the services from consumer durables and public infrastructure, as well as the benefits of volunteering and housework.	Index	The GPI has three domains and 26 indicators Domains: • Economy • Social • Environment	Australia, Austria, Canada, Chile, Germany, Italy, the Netherlands, Scotland, Sweden, and the United Kingdom
3. Inequality Adjusted Human Development Index (IHDI) Scale: International Year: 2010 Data Sources: Secondary	The IHDI is the HDI adjusted for inequalities in the distribution of achievements in each of the three dimensions of the HDI (health, education, and income). The IHDI will be equal to the HDI value when there is no inequality, but falls below the HDI value as inequality rises.	Index	The same HDI domains (with adjustment for inequality) are used in calculating the IHDI	145 countries

4. Gender Inequality Index (GII) Scale: International Year: 2011 Data Source: Secondary	The GII index shows loss in potential human development due to inequality between female and male achievements in three dimensions – reproductive health, empowerment, and the labour market.	Index	Health • Maternal mortality ratio • Adolescent fertility rate Empowerment • Female and male population with at least secondary education • Female and male shares of parliamentary seats Labour market • Female and male labour force participation rates	The 2012 human development report contains GII for 186 countries
5. Multidimensional Poverty Index (MPI) Scale: International Year: 2010 Data Source: Secondary	The MPI identifies multiple deprivations at the individual level in education, health, and standard of living that affect human development.	Index	Health • Nutrition • Child mortality Education • Years of schooling • Children enrolled Standard of living • Cooking fuel • Toilet • Water • Electricity • Floor • Assets	91 Countries

(*continued*)

Table 2.3 Continued

Measures Of Wellbeing	*Description*	*Type*	*Domains And Indicators*	*Application*
GDP+ Measures of wellbeing				
1. Australian National Development Index (ANDI) Scale: National Year: In progress Data Source: Primary	ANDI is guided by a conceptual framework that shifts the focus solely from the economy to include critical domains of people's lives that lead to enhanced wellbeing. The ANDI composite index is a single number that moves up or down, giving a quick snapshot of whether the overall quality of life of Australians is getting better or worse.	Index	ANDI is based on 12 domains with 12 headline indicators each measuring specific factors directly contributing to wellbeing. Domains: • Children and young people's wellbeing • Community and regional life • Culture, recreation, and leisure • Governance and democracy • Economic life and prosperity • Education, knowledge, and creativity • Environment and sustainability • Justice, fairness, and human rights • Health • Indigenous wellbeing • Work and work-life balance • Subjective wellbeing and life satisfaction	Australia

2. ***Measure of Australia's Progress (MAP)*** Scale: Sub-national, National Year: 2002 Data Source: Secondary	The MAP is published by the Australian Statistical Service and provides evidence about whether life in Australia is getting better.	Dashboard	MAP based on four domains and 26 themes and headline indicators. Domains: • Society • Economy • Environment • Governance	Australia
3. ***New Zealand Social Report (NZSR)*** Scale: Sub-National National Year: 2001 Data Source: Secondary data and surveys	The report shows how people are faring in New Zealand over time and how social outcomes vary for different groups (women, minorities, and indigenous people) in the population.	Dashboard	The 2010 NZSR was based on 10 domains and 43 indicators. Domains: health, education, standard of living, safety, leisure and recreation, cultural identity, paid work, life satisfaction, social connectedness, civil and political rights	New Zealand
4. ***OECD Better Life Index (BLI)*** Scale: International Year: 2011 Data Source: Secondary and Gallup Polls	An initiative of OECD that measure people's material conditions and quality of life in the member countries.	Index	The 2014 OECD's BLI has 10 domains and 24 indicators Domains: housing, income, jobs, community, education, environment, governance, health, life satisfaction, safety, and work-life balance.	34 countries of the OECD plus other "key partners" such as Brazil and Russia

(*continued*)

Table 2.3 Continued

Measures Of Wellbeing	*Description*	*Type*	*Domains And Indicators*	*Application*
GDP+ Measures of wellbeing				
5. Sustainable Society Index (SSI) Scale: International Year: 2006 Data Sources: Secondary	The SSI was developed by the Sustainable Society Foundation, a non-profit organisation based in the Netherlands with the objective of stimulating and assisting societies in their development toward sustainability.	Index	The 2012 SSI comprises of three levels: 3 wellbeing dimensions (economic wellbeing, environmental wellbeing, and human wellbeing); 7 categories; and 21 indicators	151 countries
6. Wellbeing of Nations (WBN) Scale: International Year Authored: 2001 Data: Secondary	The Wellbeing of Nations portion of the Compendium of Environmental Sustainability Indicator Collections contains a subset of 123 variables assembled from the Wellbeing of Nations, which assesses human and ecosystem wellbeing. The data are distributed by the Columbia University Center for International Earth Science Information Network (CIESIN).	Dashboard	The WBN is comprised of two equally weighted indices (Human Well-being Index and Ecosystem Well-being Index) with 9 domains each and 58 indicators in total.	180 countries grouped into 14 regions
7. Social Progress Index (SPI) Scale: International Year: 2012 Data Sources: Secondary data	The SPI was developed by the Social Progress Imperative, a nongovernmental organisation in the USA. It offers a framework for measuring the multiple dimensions of social progress and benchmarking success over a period of time.	Index	The SPI is based on three social progress dimensions. Each dimension, in turn has the four components and each component is made of between three and six indicators.	132 countries are included in the 2014 SPI

			Dimensions: • Basic needs • Foundations of wellbeing • Opportunity	
8. Canadian Index of Wellbeing (CIW) Scale: National Year: 2011 Data: Secondary	The first CIW published in 2012 by the University of Waterloo.	Index	The CIW has 8 domains and 64 indicators Domains: Community vitality, democratic engagement, education, environment, healthy populations, leisure and culture, living standards, and time use.	Canada
Objective/subjective measures of wellbeing				
1. UK Index of Well-being Scale: National Year: 2012 Data Source: Surveys and secondary data	The UK Index of Well-being is headed by the Office of National Statistic which has released two "Life in the UK" reports that highlights snapshot of the UK's wellbeing.	Index	The UK Index of Well-being is based on 10 domains and 41 indicators. Domains: Personal well-being, relationships, health, what we do, where we live, personal finance, economy, education and skills, governance, and natural environment.	United Kingdom

(*continued*)

Table 2.3 Continued

Measures Of Wellbeing	*Description*	*Type*	*Domains And Indicators*	*Application*
GDP+ Measures of wellbeing				
2. Gross National Happiness (GNH) Scale: National Year: 1972 Data Sources: Survey	The GNH considers happiness as multidimensional – not measured only by subjective wellbeing and "not focused narrowly on happiness that begins and ends with oneself and is concerned for and with oneself". GNH has often been explained by its four pillars: good governance, sustainable socio-economic development, cultural preservation, and environmental conservation.	Index	GNH has 9 domains and 33 indicators Domains: Psychological wellbeing, standard of living, good governance, health, education, community vitality, cultural diversity and resilience, time use, and ecological diversity and resilience.	Bhutan
3. Taiwan National Well-being Index Scale : National Year: 2012 Data Sources: Primary and Secondary	Taiwan National Well-being Index is based on the OECD's Better Life Index (BLI) but *also* runs a parallel set of indicators that are domestic in focus.	Index	Based on OECD's BLI domains but has 38 local indicators	Taiwan
4. Gallup-Healthways Well-Being Index (USA) Scale: National (USA) Year: 2008 Data Source: Primary	The analysis is based on data from the Gallup-Healthways Well-Being Index, a definitive measure and empirical database of real-time changes in wellbeing throughout the world.	Index	The Gallup-Healthways Well-Being Index is calculated from six indices: Life Evaluation Index, Emotional Health Index, Physical Health Index, The Healthy Behavior	USA

			Index, Work Environment Index, and Basic Access Index. Each of these indices is based on several indicators.	
Subjective wellbeing /objective wellbeing				
1. Happy Planet Index (HPI) Scale: International Year: 2006 Data Source: Secondary data and Gallup polls	The HPI was created by Nic Marks, Founder of the Centre for Wellbeing at NEF (the New Economics Foundation) and uses data from the HDI and the Ecological Footprint by the World Wildlife Fund together with other primary data.	Index	The HPI is constructed from three indices: Experienced well-being, life expectancy, and ecological footprint.	The 2012 HPI report ranks 151 countries
2. National Well-Being Index Scale: International Year: 2006 Data Source: World Values Survey and other secondary sources	Based on Vemuri and Constanza, 2006, which "aims to combine data on national levels of mean Subjective Well-being (SWB) with data on objective measures of built, human, social, and natural capital in order to better explain the determinants of national SWB"	Index	Domains: Social capital, natural capital, and subjective wellbeing.	56 countries included in the regression model; 172 countries included in the life satisfaction values
Subjective wellbeing				
1. Australian Unity Well-Being Index Scale: Sub-national, National Year: 2001 Data Source: Surveys	The Australian Unity Well-being Index regularly measures the "subjective well-being" of the Australian population by asking people to rate their satisfaction with aspects of their lives.	Index	Personal Wellbeing Index is created based on responses to questions about: • Personal health • Personal relationships	Australia

(continued)

Table 2.3 Continued

Measures Of Wellbeing	*Description*	*Type*	*Domains And Indicators*	*Application*
GDP+ Measures of wellbeing				
			• Safety • Standard of living • Achievement in life • Feeling part of the community • Future security	
2. World Happiness Report Scale: International Year: 2012 Data Source: Gallup World survey	WHR is published by the Sustainable Development Solutions Network (SDSN).	Indices	The WHR has three elements: positive affect, negative affect, and happiness (yesterday) with several indicators.	The 2013 report covers 150 countries

Indicators for most wellbeing measures are not included in order to tighten the matrix. Please refer to relevant websites in Appendix A for comprehensive lists of indicators for each wellbeing measure. Measures that utilize an index approach present results as a single numerical value, while measures that utilize a dashboard approach present a spread of results simultaneously (i.e., does not integrate domains).

Appendix B

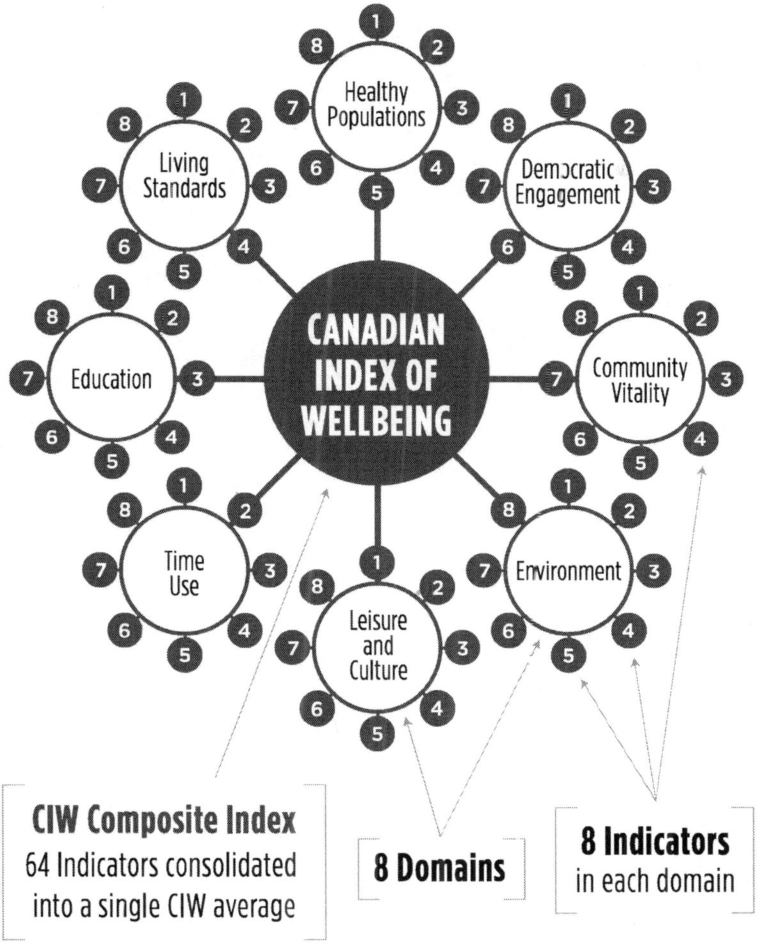

Figure 2.2 Canadian Index of Wellbeing framework (2012:13).

Part II

Emerging and re-emerging infectious diseases

3 Examining the West African Ebola outbreak through the application of the disease ecology framework

Alexandra Shapiro, Anne Hoefler, Jenna Bryfonski, and Ezekiel Kalipeni

Introduction

Since the discovery of the Ebola virus near the Ebola River in Zaire (Democratic Republic of the Congo) in 1976, the global community has been fascinated and horrified by the deadliness of the virus. For almost 40 years, specialists around the world have investigated the function and transmission of the Ebola virus, rushing to the location of each sporadic outbreak to glean more information. The Ebola virus has been costly in both medical resources and human lives, with fatality rates climbing as high as 90% at peaks of past outbreaks (World Health Organization 2004). Despite efforts to combat the virus, outbreaks of Ebola in Africa have emerged and spread rapidly, raising concerns regarding implications of this infectious disease in our increasingly interconnected world. The virus has potential to cause many fatalities near the epicentre of the outbreak and across political boundaries. Despite advances in medicine, Ebola has yet to be eradicated, and each new outbreak strikes harder than the previous.

The Ebola virus outbreak in Western Africa, which began in December 2013 and ended in June 2016, was among the first times that Ebola spread across international boundaries. Adding to the lethality of the 2013–2016 outbreak, access to real-time information and social media kept Ebola in the media spotlight, thereby contributing to widespread hysteria. A defining aspect of the 2013–2016 outbreak is that it highlighted the need for a shift in how to view and respond to infectious disease control in isolated regions. An infectious disease in one location is a threat to the immediate vicinity and to the global community. Interconnected lifestyles enabled by modern technologies generally prevail in both developed and developing nations, making a threat to one location an emergent threat worldwide. Therefore, it is crucial to examine possible ways to mitigate emerging infectious disease with the hopes of minimising casualties and suffering inflicted by disease outbreaks.

This chapter aims to examine the history of Ebola, the causes of the virus' high mortality rate, and patterns of human transmission and behaviour that facilitated the spread of the disease during the 2013–2016 Ebola outbreak, all within the context of Barry Popkin's demographic, epidemiological and nutrition transition model (2002). This chapter discusses this outbreak of Ebola under the epidemiological transition piece of the model, though each component of the transition is interconnected. Exploring these topics illustrates the potential for Ebola to be a disease burden that hinders further development within the affected region. Studying Ebola in the context of the outbreak of 2013–2016, along with its historical and scientific classifications, can serve as a model for how the principles of the human ecology of a disease can contribute to the understanding of a virus. Better understanding of a disease is an integral component in developing tactics to prevent further infectious disease spread.

Integrating Meade and Emch's (2010) disease ecology framework into Barry Popkin's demographic, epidemiologic, and nutrition transition model enables a better understanding of the interplay of populations, habitats, and behaviours within the spatial and temporal scale. By focusing on the epidemiologic transitional aspect of Popkin's model, it becomes clear that the disease ecology framework can be applied to understand the spread of viruses in developing countries. Outlining the salience of the relationship between humans and their environment ultimately is a major step towards addressing future health challenges in Western Africa and around the world.

Biological context of Ebola/Marburg viruses

Viral function and symptoms of disease

The Ebola virus disease, also known as Ebola Haemorrhagic Fever, is a disease caused by a specific strain of *Filoviridae* virus classified under the genera *ebolavirus*, of which five known strains exist: Zaire, Bundibugyo, Sudan, Reston, and Taï Forest. The viral families and genera are named after the locations in which they first occurred (Easter 2002). In order to fully understand the Ebola virus in the discussed framework, it is important to understand how the virus acts within the host.

Filoviridae derive their name from their long, filamentous structure, initially discovered in 1967 in a primate laboratory in Marburg, Germany (Funk and Kumar 2014). Like all *Filoviridae*, the Ebola virus enters cells in the body and uses cellular machinery to make virus-specific proteins. These proteins replicate and burst from the cell to infect other cells, killing the initial host cell in the process (Kiley *et al.* 1982). Specific to the Ebola virus is its success in infecting a large variety of cells in the body. This is done by means of a virus-specific glycoprotein, which binds the virus particle to the cell (Bhattacharyya *et al.* 2010). Viral machinery then encodes viral genes, which the cell itself turns into proteins, and burst from the cell to infect other cells in the body. Most importantly, this virus can successfully replicate in white blood cells, which participate in the immune response of the body (Funk and Kumar 2014). The death of white blood cells increases the body's susceptibility to the infection, worsening the condition (Funk and Kumar 2014). The Ebola virus also actively kills liver cells, resulting in irregular and failed mechanisms for normal blood clotting (Easter 2002). It also has the ability to target the adrenal gland cells, leading to sharp decreases in blood pressure (Easter 2002). Additionally, the Ebola virus triggers a release of protein signals into the blood that induce swelling, while the endothelial cells lining all the blood vessels of the body are attacked and dismantled by the virus (Wickelgren 1998, Easter 2002). This initial attack on the immune system and associated inflammatory responses result in the beginning stages of flu and malaria-like symptoms, which include, but are not limited to fever, muscle aches, fatigue, headache, and sore throat. Symptoms then progress to diarrhoea, vomiting, and chest and abdominal pain as well as swelling, red eyes, skin rash, and confusion (World Health Organization 2014a, Funk and Kumar 2014). The infection of the vascular endothelium results in the most recognisable haemorrhagic symptoms of Ebola (Funk and Kumar 2014). Capillary leaking due to systemic inflammatory responses deplete vascular volume and increase coagulation processes. This reduces the body's overall coagulation factors and increases the likelihood of bleeding out, usually from the gastrointestinal tract. These extreme haemorrhagic events lead to the descriptions of Ebola literally "liquefying" its victims (Funk and Kumar 2014). Death, if it occurs, is commonly due to fluid loss and/or organ failure and can happen anytime between 6 and 16 days after the onset of symptoms. Recovery can also occur during this window if treatment begins as soon as the first sign of a fever is detected and if internal bleeding remains minimal. Patients in a 1995 outbreak who made it to day 14 had a 75% likelihood of surviving the illness (Funk and Kumar 2014). The speed of the infection, severity of symptoms, and instances of death occurring within one week of onset classifies the Ebola virus as a Biosafety Level 4 Hot Agent, or a "high-risk, life-threatening infectious disease for which there is no vaccine or therapy" (Centers for Disease Control and Prevention 2011).

Ebola as an epizootic disease

Important in the transmission of the Ebola virus is its classification as an epizootic disease, as shown in Figure 3.1 (Centers for Disease Control and Prevention 2014c). An epizootic disease is one that is naturally occurring, or endemic, in an animal population of a specific area and jumps to human hosts to cause a disease in what is called a spillover event (Centers for Disease Control and Prevention 2014b). Humans are thus not the natural hosts or carriers of Ebola, which is why outbreaks in the human population are sporadic and deadly. A natural reservoir of the disease remains unknown, though it is speculated that African fruit and insectivorous bats, as well as rodents, are some of the many possible reservoirs of Ebola (Funk and Kumar 2014, Chowell and Nishiura 2014). Other animals, such as primates and forest mammals, tend to succumb to the infection and thus are also not likely to be the natural hosts. Figure 3.1 illustrates the hypothesis that Ebola circulates among bat populations, spreading to animals preceding human outbreaks, which occur when humans intrude among infected intermediate host animal populations, either by hunting or working in activities like deforestation or mining (Centers for Disease Control and Prevention 2014b, Chowell and Nishiura 2014). An Ebola epidemic in the human population can be caused by a single

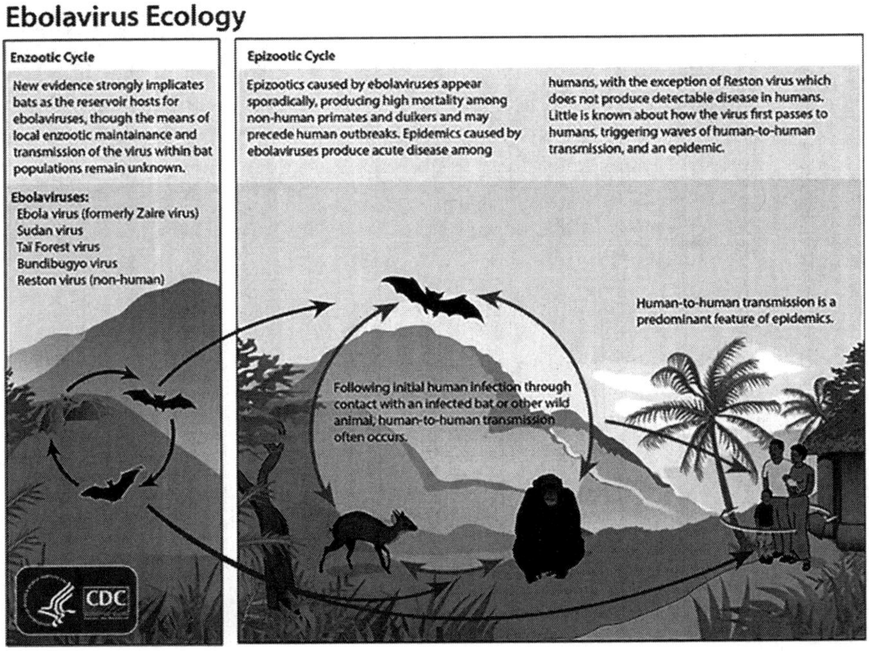

Figure 3.1 Ecology and transmission pathways of the Ebola virus (Centers for Disease Control and Prevention 2015).

spillover event from an infected intermediate host animal to a single human, spreading among the human population through human-to-human transmission (Centers for Disease Control and Prevention 2014b, 2014c). This can be shown by the examination of chains of transmission, which are often traced to a single individual at the onset of an outbreak. This indicates that the major spreading of such a disease is caused almost entirely by human contact and behaviour after the initial infection of the first patient (Chowell and Nishiura 2014).

Methods of transmission

The transmission cycle of the Ebola virus is especially pertinent to how and why the virus spreads during an outbreak. Most important of these factors is the relatively varying incubation period, which can last anywhere between 2 and 21 days after exposure before symptoms begin (World Health Organization 2014a). Variation in incubation may be due to the lack of reliable information regarding symptom onset in outbreak regions, especially due to the wide range of infectious diseases that begin with similar influenza-like symptoms. Therefore, it can be difficult to confirm an Ebola diagnosis at the first sign of symptoms (Funk and Kumar 2014). Although the patient is not contagious during the incubation period, the individual can spread the virus as soon as the first symptoms of the infection present (World Health Organization 2014a,Centers for Disease Control and Prevention 2014b). Studies have shown that Ebola can only pass from person-to-person by direct contact with bodily fluids from an infected individual. For example, the virus can access a healthy person via broken skin or mucous membranes by means of surfaces or objects that have been contaminated by an infected individual (World Health Organization 2014a, Funk and Kumar 2014). The survival of the disease outside of the host depends on conditions such as temperature, amount of light, and humidity (World Health Organization 2014a). Alarmingly, Ebola can survive for up to 50 days within a deceased host (World Health Organization 2014a). The resilience of this disease makes contact tracing of infected persons exceedingly important to control efforts.

Healthcare workers or family members who treat infected individuals are most at risk of contracting the disease. Healthcare workers that fail to wear personal protective equipment such as HAZMAT suits, gloves, and eye protection, or wear such equipment improperly, are also at high risk when handling infected materials (Centers for Disease Control and Prevention 2014b). Deceased bodies remain infectious, and many outbreaks in Africa have spread due to the direct contact with the deceased during ceremonial burial rituals (World Health Organization 2014a). Transmission increases in areas where proper medical isolation techniques do not exist, such as hospitals without running water or quarantine procedures. These risk factors form one of the more important aspects of why Ebola outbreaks typically arise in poor regions that may lack adequate healthcare (World Health Organization 2014a,Chowell and Nishiura 2014).

Historical context of past Ebola outbreaks

Death toll by country

Between the years 1976 and 2012, a total of 2,408 human cases of Ebola were reported, resulting in 1,681 deaths (Kuhn *et al.* 2010). The epidemiologic history of Ebola is complex due to the various strains and locations of occurrences (Kuhn *et al.* 2010). The first major outbreak of Ebola occurred in 1976 in central sub-Saharan Africa, in the town of Nzara, Sudan. The first victims of the disease were employees of a large cotton factory, whose intense haemorrhagic symptoms captivated the attention of the World Health Organization (World Health Organization 1976, Funk and Kumar 2014). Other patients and healthcare workers in the Nzara hospital soon fell ill. A total of 151 people died, with a case fatality of 53% (World Health Organization 1976). Post-mortem investigations revealed that the disease was viral in nature, with symptoms and features resembling that of the well-known German Marburg virus, but results were not exclusively diagnostic (World Health Organization 1976, Funk and Kumar, 2014).

Shortly after the conclusion of the Sudan outbreak, a second and more aggressive strain of Ebola broke out in the Mongala District of the northern Democratic Republic of the Congo (DRC), previously named Zaire. The first known patient was a school headmaster that had travelled along the Ebola River of the Central African Republic and received an injection to help combat malaria. He later fell ill along with other patients who had all received similar malarial injections. The World Health Organization (WHO) later discovered that the disease was spread by contaminated needles in a local hospital, along with direct contact with infected persons (World Health Organization 1978). The similarities with the viral outbreak in Sudan led WHO officials to conclude that the Zaire outbreak was due to the same strain as the one in Sudan, and thus the disease got its name from the nearby Ebola River (World Health Organization 1978, Funk and Kumar 2014). Zaire ebolavirus is the most deadly sub-strain of all Ebola viruses, with an average fatality rate around 68%. In a 2003 outbreak, the fatality rate of the Zaire ebolavirus strain reached 90% (World Health Organization 2004, Funk and Kumar 2014). This deadly strain is the one seen in the 2013–2016 outbreak, and referred to through this chapter simply as the Ebola virus in the context of the 2013–2016 outbreak.

From 1976 until 2012, the Zaire ebolavirus strain would go on to kill 809 more people in Africa and elsewhere, including those in South Africa, Gabon, the DRC, and Russia. Outbreaks occurred in groups of gold miners, forest workers, bushmeat hunters, and healthcare workers (Georges *et al.*1999, World Health Organization 2004, League of Nations 2007). Meanwhile, the Sudan ebolavirus strain killed 421 people total in Uganda and Sudan, spread primarily by close contact with the infected via family members during care or post-mortem funeral rituals (World Health Organization 1976, Baron *et al.* 1983, Okware *et al.* 2002, World Health Organization 2012). During this time period, three other viral strains of Ebola were also discovered, all with separate epidemiological histories and varying levels of severity: the Bundibugyo ebolavirus of western Uganda which killed 103 people in 2008 and 2012, and the Ivory Coast/Taï Forest ebolavirus and Reston ebolavirus, both of which broke out in primate laboratories and did not have human fatalities (Funk and Kumar 2014). A graphical summary of the lethality of the five strains of the ebolavirus is shown in Figure 3.2.

The 2013–2016 outbreak

The genesis of the 2013–2016 outbreak in Western Africa was traced to a two-year-old boy named Emile Ouamouno in the rural village of Meliandou, Guinea (Stylianou 2014). The outbreak began when he contracted the virus in December of 2013 after close contact with fruit bats; it spread and claimed the lives of his sister and pregnant mother (Stylianou 2014). Health officials traced the spread of the outbreak from Ouamouno and his family to healthcare workers in Meliandou and surrounding villages (Stylianou 2014). From there, Ebola was transferred to the families of healthcare workers and has spread across international borders to become a global health crisis.

The 2013–2016 outbreak was centred in the Western African countries of Guinea, Liberia, and Sierra Leone. The outbreak began in the forest region in Guinea (Beukes 2014), but was quickly transmitted throughout the country. Individuals travelling from Guinea to Liberia, assumed to be conducting business, brought the Ebola virus with them resulting in the first reported Liberian

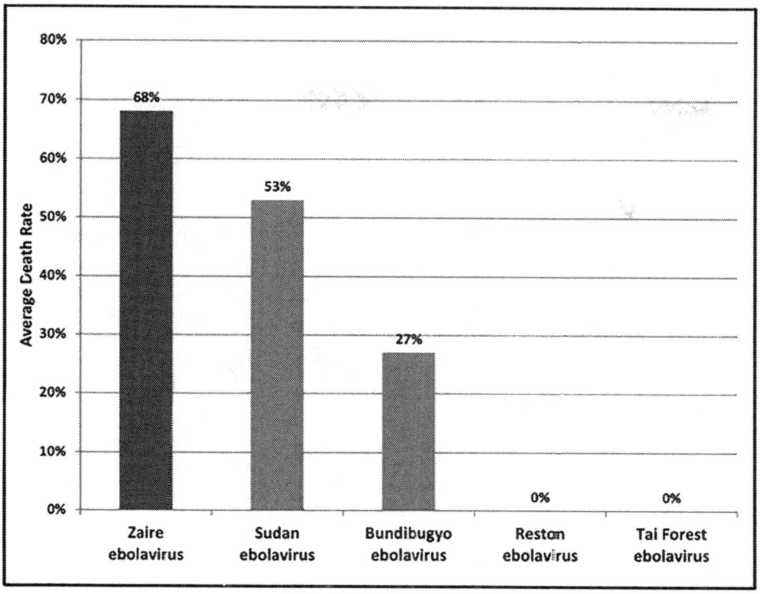

Figure 3.2 Average death rates from the five types of Ebola virus species (Centers for Disease Control and Prevention and WHO, June 2017).

Note: These data are not including the most recent outbreak. Death rates in this figure were calculated from records of all known outbreaks prior to 2013. The Zaire ebolavirus strain can be cited as having death rates as high as 90%, yet individual outbreaks may vary.

case in mid-March 2014. Experts traced the spread from Guinea to Sierra Leone to a funeral for a revered traditional healer from Sierra Leone in late May 2014 (Global Alert and Response 2014). The healer had been asked to travel to Guinea to heal individuals infected with Ebola (Global Alert and Response 2014). She fell ill and passed away a few days later. Hundreds of people attended her funeral, holding to tradition by touching and kissing her body; 365 deaths from Ebola are believed to be linked to that funeral (Global Alert and Response 2014).

It is difficult to quantify the exact number of people who were infected with Ebola, as well as the number of deaths, due to the remoteness of many rural African communities. Officials have tried to determine the number of infected individuals, yet it is believed that for every reported case, approximately 1.5 cases were unreported (Meltzer *et al.* 2014). On August 8, 2014, the epidemic was declared “a public health emergency of international concern” (WHO Ebola Response Team 2014, Funk and Kumar 2014). As of June 2016, the number of reported cases was 28,616, and the death toll reached 11,310 people (World Health Organization 2016) with nine total countries affected, including the United States (Grady and Fink 2014, World Health Organization 2015b). This makes the 2013–2016 outbreak almost seven times as deadly as all of the previous outbreaks combined.

The disease ecology framework

The disease ecology framework outlines human activity in three interconnected sectors: population, habitat, and behaviour which can be displayed in a simple triangle (Meade and Emch 2010). The interplay between humans and their environment ultimately leads either to the production of, or prevention of, a disease. The three vertices all exist in dynamic equilibrium when no disease is present. As soon as the balance is skewed, either by a change in the environment, gene pool in a population, or a new cultural behaviour, a disease can manifest (Meade and Emch 2010). Each of these three sectors can be expanded to observe further subdivisions that highlight more specific aspects encompassed in the framework. Meade and Emch (2010) state, "The customs, beliefs, and socioeconomic structures that characterise each global culture realm and local ethnic group create the environmental conditions and exposure patterns that result in geographic distribution of health and disease" (p. 65). Figure 3.3 depicts the disease ecology framework graphically to illustrate the interconnected nature of the three vertices: population, habitat, and behaviour.

Population

The subdivisions of population focus on genetics, age, and gender. Examples of diseases that are influenced by population factors include Alzheimer's disease and breast cancer. Alzheimer's disease is most commonly found in individuals over the age of 65 (Hoyert and Rosenberg 1997). Similarly, breast cancer is most prevalent among the female population.

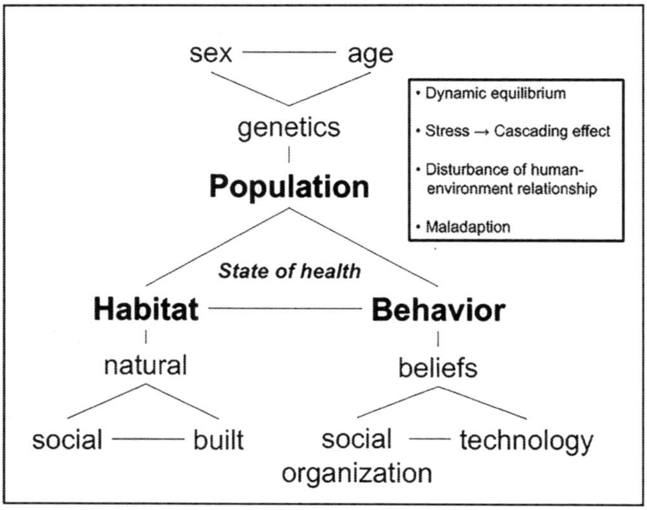

Figure 3.3 Disease ecology framework (Meade and Emch 2010).

Habitat

The subdivisions of the habitat component of the disease ecology framework include natural, social, and built. The natural component refers to the physical environmental surroundings of the habitat. The social component refers to societal institutions in the affected habitats that might contribute to or hinder the spread of the disease. The built component refers to infrastructural aspects of an affected habitat that might influence the spread of the disease.

These components of the disease ecology approach have been effective in the past to identify the roots of various disease outbreaks and to control the outbreaks. For example, malaria is an illness caused by a parasite that enters the bloodstream through mosquito bites (Centers for Disease Control and Prevention 2010). When considering malaria, the natural component of the environment suitable for the mosquitos that transmit the disease plays a significant role in determining how the disease spreads. Since mosquitos need stagnant water and certain temperatures to survive, tracking the disease is closely tied with understanding the natural environments in which the mosquitos might harbour malaria (Centers for Disease Control and Prevention 2010). An additional example of the disease ecology approach application to better understand disease outbreaks is in the context of schistosomiasis. The prevalence of the disease schistosomiasis in subtropical areas of Africa increased following the construction of new dams (Kassa 2005). Schistosomiasis, also known as bilharzia, is an acute parasitic disease caused by trematode worms (World Health Organization 2014b). These parasitic worms live a portion of their life cycle in certain species of freshwater snails (Kassa 2005). This disease was influenced by dam building because the dams create a suitable habitat for these snails, and allow ample resources for the schistosomiasis parasite to thrive (Kassa 2005). The constructed dam exemplifies the built component, as the infrastructure facilitated disease spread by creating an ideal environment for the disease agent.

Behaviour

The subdivision of behaviour evaluates beliefs, social organisations, and technologies. The beliefs component refers to religious and ritual practices that might influence disease spread. Social organisation includes migration patterns and mobility, urbanisation, and social norms. The technology component refers to technologies that might be related to disease spread. The behaviour subdivision wields the most importance in the context of the 2013–2016 Ebola outbreak.

One can analyse human behaviour when examining the spread of HIV/AIDS in Africa. HIV/AIDS is a viral disease transmitted through sharing of bodily fluids (Lurie 2004). Migration is one of the factors that influenced the spread of HIV/AIDS, as indicated by previous studies (Lurie 2004). One study found that in Uganda, people who have moved within the last five years are three times more likely to be infected with HIV than those who have lived in the same place for over ten years (Lurie 2004). In another study in South Africa, people who had recently changed their residence were three times more likely to be infected with HIV than those who had not (Lurie 2004). Clearly, migration patterns of the social organisation component can hold important information about the underlying causes contributing to the spread of disease, and these findings can be a useful tool to better understand the disease and ultimately control the spread.

Application of the disease ecology framework to the Ebola outbreak of 2013–2016

This section will examine the 2013–2016 outbreak from the disease ecology framework (Meade and Emch 2010) with a focus on Sierra Leone, one of the three countries that was hit the hardest by the Ebola epidemic in West Africa. Sierra Leone is a nation rich in mineral resources. Valuable minerals and metals such as diamonds, gold, bauxite, rutile, and iron ore exist in abundance. Natural resources can play a large role in triggering, prolonging, and financing civil wars (Bannon and Collier 2003). The reliance on natural resource exports to sustain the economy often leads to unusually high rates of poverty (Bannon and Collier 2003). Additionally, governments in resource-rich areas often fail to provide healthcare and education to their citizens (Bannon and Collier 2003).

These characteristics of resource-rich areas apply aptly to Sierra Leone. Although tremendous progress has been made to increase checks and balances of power, the system is relatively new and not well equipped to handle a serious humanitarian crisis such as an Ebola outbreak (Abdullah 1998). In order to effectively combat such an infectious disease, a strong central government or authority is needed to coordinate control initiatives. Sierra Leone is home to 16 different ethnic groups, each with its own culture (Columbus Travel Media 2014). The broad range of ethnicities and cultures in the region add to the challenge of creating a strong centralised government.

Population

When considering the deadly Ebola virus, population information is of minimal salience. The lethal virus has not been found to discriminate against any human host regardless of genetics, age, or gender (World Health Organization 2014a).

Habitat

Natural – The climate in Sierra Leone is tropical, and experiences a rainy and dry season each year (Central Intelligence Agency 2014). The majority of the population resides in rural villages in Sierra Leone, which rely on farming as a source of food due to plentiful rainfall and arable land (Columbus Travel Media 2014). Rural locations and small villages outside of densely populated urban areas have been identified as the epicentres of multiple outbreaks of Ebola (West and McDonnell 2014). Figure 3.4 shows a map of Ebola outbreaks in Africa since 1976, most of which broke out in remote rural areas.

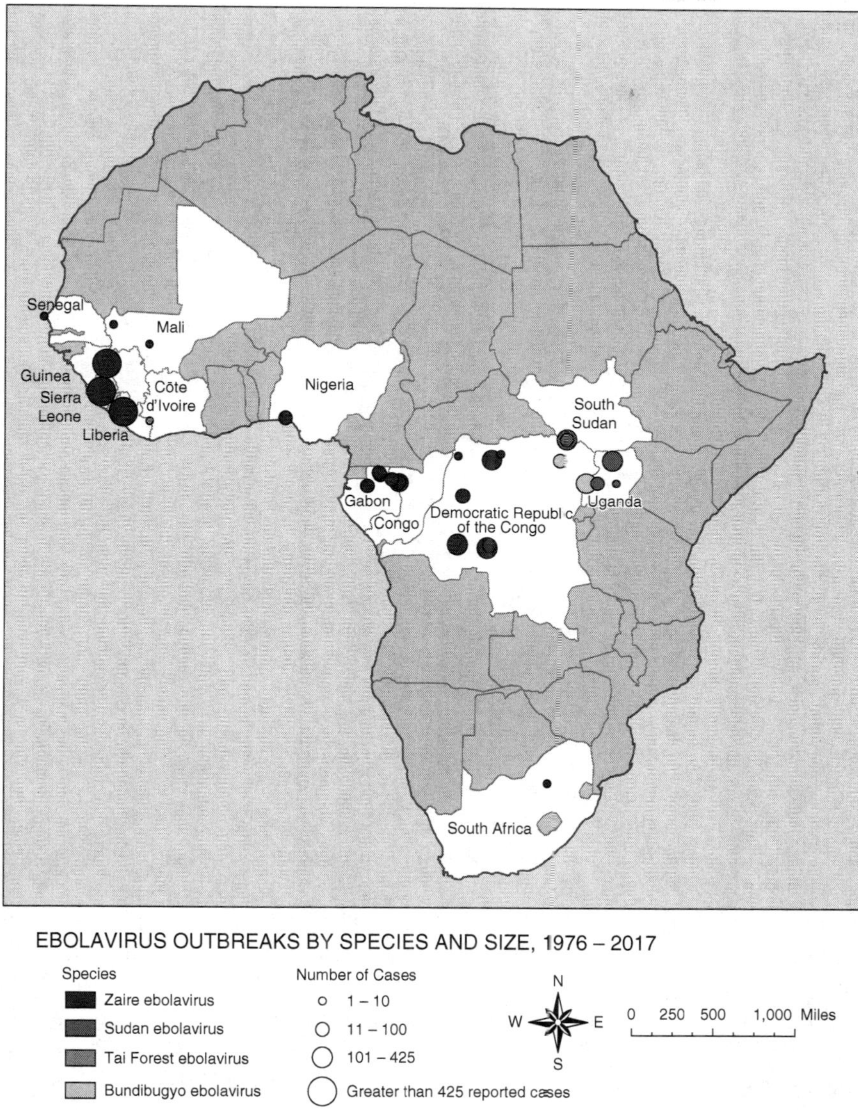

Figure 3.4 Map of Ebola outbreaks in Africa (WHO, Centers for Disease Control and Prevention, and HPA).

Social – Within Sierra Leone, there are 16 different ethnic groups, each with its own language and culture (Columbus Travel Media 2014). Sierra Leone is a religiously diverse country comprised of approximately 60% Muslims, 30% Christians, and 10% traditional and animists (Columbus Travel Media 2014). It is common for Muslims and Christians to live alongside each other and intermarry (Columbus Travel Media 2014). This practice increases contact among potentially infected people. The spatial separation of villages and cultures could explain why past outbreaks have often been confined to isolated rural villages.

Built – A large part of the population in Sierra Leone resides in rural areas (Larbi 2012). Individuals live in homes that vary in design based on the culture of the village (Larbi 2012). Building materials have shifted from mostly earth and wood to modern brick with corrugated tin roofs. Typically, families live in a single home and the village chief acts as a judge and healer (Larbi 2012). Since families usually live under one roof, the whole household is susceptible to Ebola if one family member contracts the virus (Baron *et al.* 1983). This housing format caused entire families to perish of Ebola due to close contact with the virus (Meade and Emch 2010).

In recent years, Sierra Leone's medical care system has undergone changes and reforms (Dodgeon 2010). In 2010, a new law declared healthcare free for pregnant and breastfeeding women and children under the age of five (Dodgeon 2010). This reform cost millions of dollars and was funded mostly by the United Nations and the United Kingdom (Dodgeon 2010). Because the healthcare infrastructure was so recently formed, it is not well equipped to deal with a crisis such as Ebola. Despite the recent reforms, the healthcare system is not well developed as a result of the recent emergence from tumultuous civil unrest and conflict (Larbi 2012). While the country is slowly gaining stability and growing the economy, rebuilding cities and villages after civil war consumes extensive amounts of time and resources. The lack of time to develop effective infrastructure creates an environment prone to infiltration by infectious disease. Healthcare infrastructure can only become a priority once a stable government has formed, and the swift attack of the Ebola virus has had crippling effects on this progress towards a stable region with effective infrastructural systems (Larbi 2012).

The aftermath of the war left farmers with little access to markets and market information (Larbi 2012). Farmers in remote communities still suffer from poor access to markets due to the collapse of infrastructure such as roads, processing facilities, and lack of standardised price information (Larbi 2012). The arable land allows for spatial isolation between villages to remain by growing sufficient food in each location (Larbi 2012). The lack of trade and large markets suggests another venue for Ebola spread. It is likely that past outbreaks have stayed localised because of rural isolation and minimal trade (Larbi 2012). This epidemic is much greater in scale than past outbreaks, suggesting that remote farming is not likely to be the underlying process that allowed disease to spread to greater scales.

Behaviour

Burial rituals – One practice that is problematic in the context of disease spread is funeral rituals (Global Alert and Response 2014, World Health Organization 2014, Dockins 2014). In many countries in Western Africa, funeral rituals involve close contact with the deceased person in a manner alarmingly conducive to rapid disease spread (Global Alert and Response 2014, World Health Organization 2014). Relatives are responsible for cleaning, bathing, and clothing the victim's body (Global Alert and Response 2014, World Health Organization 2014, Dockins 2014, Epatko 2014). Depending on the social status of the deceased person, a funeral can last for days as family, friends, and others pay their respects (World Health Organization 2014b, Dockins 2014). At the close of the funeral ceremony, a bowl is shared for ritual hand washing and a final touch or kiss is bestowed on the dead (Linton 2014,Dockins 2014). The body is often buried on family land in close proximity to the home and surface water reservoirs (Linton 2014). The goal of the ceremony is to make certain that the spirit of the deceased is happy and does not feel forgotten, and also to ensure a safe passage of the spirit to the afterlife (Linton 2014). If these funeral rituals are broken, it is commonly believed that harm and misfortune will come to the living family of the deceased person (Linton 2014).

A major concern with having funerals is that an infected person is most infectious 24 hours after death (Dockins 2014). Many resources have been devoted to safe disposal of bodies, but that often means that family members cannot participate in a traditional burial (Dockins 2014). While it is necessary to contain the spread of Ebola to prevent further outbreaks, prohibiting family members from participating in traditional burial practices adds to the trauma that they are already experiencing after the death of a loved one (Dockins 2014). Healthcare workers have been briefed on how to bury infected bodies while still being mindful of the life lost and the suffering that families are going through. However, healthcare providers struggled to communicate their mission to grieving family members (Epatko 2014). The low literacy rate within Sierra Leone as a result of lack of educational access and variety of different languages makes it difficult to communicate with the local populations and leads to distrust of external help, impeding the control efforts (Linton 2014). Scholars and aid workers have devised plans to educate communities in advance of outbreaks on how they will bury bodies, as well as aiding in the counselling and consideration of those grieving (Dockins 2014). These plans are intended to build trust and cooperation among the affected populations, both of which are essential to controlling disease spread (World Health Organization 2014a).

War – In periods of conflict, citizens can be displaced from their origin country to host countries seeking asylum and refugee status or within their home country as an internally displaced person. When people relocate due to war, they are often moved quickly and with very little provision or care. Refugees are forced into camps which have little to no resources to provide to the incoming groups, often exhausted and malnourished. The poor living conditions have limited sanitation capabilities and often little funding to provide food or bedding for the residents, leading to conditions which allow viruses to multiply and spread rapidly (Kalipeni and Oppong 1998). A lack of medical infrastructure and knowledge of various diseases and a limited ability to communicate with the international community inhibits the ability to eradicate and stop the spread of many viruses, especially Ebola (Linton 2014). There is significant concern that Ebola or another infectious disease will arise in a resettlement camp, making it virtually impossible to contain and control the virus (Kalipeni and Oppong 1998).

Mining – In many of the past Ebola outbreaks, the epicentres have been traced to locations in close proximity to mines since they provide suitable habitats for bats to live (West and McDonnell 2014). An outbreak of Marburg virus, in the same family as Ebola, in Durba, DRC has been linked to the Goroumbwa Mine in which 154 cases were reported, the majority of whom were young males working in the mine (Swanepoel *et al.* 2007). Another outbreak of Marburg virus was traced to a group of tourists who came into contact with bats at the Kitaka mine in Uganda, a popular tourist attraction (Amman *et al.* 2012). Both of the initial spillover events have been linked to bats in mines (Swanepoel *et al.* 2007,Amman *et al.* 2012). The 1994 outbreak of Ebola was traced to a small mining camp deep in the rain forest (West and McDonnell 2014). The epicentre of the current outbreak has been traced to the south-east of Guinea near reserves of iron ore (West and McDonnell 2014).

Mining provides a source of income for Sierra Leone, Guinea, and Liberia (West and McDonnell 2014). Foreign investment in the natural resources of this region spurred a 20% growth in GDP in Sierra Leone in 2013 (West and McDonnell 2014). As many are turning to employment in mining, there is greater potential for a large outbreak to occur in a mining camp. If a species spillover event took place at a mining camp where people often live in close quarters, the disease could spread through employees in the camp (West and McDonnell 2014). While this is an issue within the camp, the larger issue presents itself when miners return to their homes (West and McDonnell 2014). Since mining camps are often far removed from cities and villages, individuals migrate from home to the mines for work (West and McDonnell 2014). These miners can unknowingly bring home the deadly disease and spread it to their villages and families. Frequent movement, both spatially and temporally, enhances the probability that Ebola will contaminate individuals who do not work at the mine.

Deforestation – Economic and resource development pressures are driving people further into the dense forests in tropical and subtropical regions of Africa. Over half of Liberia's forests have been sold to industrial logging companies during the post-war government (West and McDonnell 2014). Logging to meet demand for firewood and slash and burn agriculture has the potential to contribute to the spread of Ebola in Sierra Leone (West and McDonnell 2014). First, greater contact with hosts is facilitated through the displacement of organisms, such as bats, from their native forested habitat into human populated areas, as well as the entering of humans into deeply forested areas (Wolfe *et al.* 2005, McCoy 2014). Second, humans are disturbing previously untouched host reservoirs and opening up new channels of transmission (Wolfe *et al.* 2005). Third, international populations of people are being moved into these locations with reservoirs of Ebola for work, who then travel back to their origin country and can transfer the virus to new locations (Wolfe *et al.* 2005). By altering the landscape and creating the need for individuals to venture further into the forest, individuals can have increased contact with infected animals (West and McDonnell 2014).

Hunting – Ebola originates naturally in animal reservoir populations, such as fruit bats, and in intermediate animal hosts in forested regions, including gorillas, chimpanzees, and other small mammals (Muyembe *et al.* 2012). In rural and poor areas, hunting bushmeat can be both a primary source of income and a dietary necessity. Hunters enter the forest to find food and come in contact with live, infected animals and/or infected carcasses. When hunters bring the carcass back to their home, they unknowingly transmit Ebola through consumption of bushmeat and by handling the infected carcasses. Once an individual contracts the virus and shows signs of symptoms, they have the ability to pass it to family members, others in the community, and aid workers.

The natural, rural location of many villages likely reinforces the practical social practice of eating bushmeat as it is an easily accessible food source (West and McDonnell 2014). The rural locations and consumption of bushmeat as a primary food source increase chances of contacting infected wildlife, and therefore increases the risk of contracting Ebola (Bradbury 2003, West and McDonnell 2014). Reports from the World Health Organization caution against eating bushmeat in an effort to control the outbreak (Bradbury 2003, World Health Organization 2014a).

Gender roles – Age and gender play a role in explaining the spread of Ebola with regards to social constructs. A study of an outbreak of Ebola in Sudan in 1979 suggested that family units are highly vulnerable to the spread of Ebola (Baron *et al.* 1983). This vulnerability can likely be explained by the resilience of the virus (Baron *et al.* 1983). In the context of the current outbreak, it is important to consider the implications of the growing number of children orphaned by Ebola (Ford and Elbagir 2014). The increasing number of orphaned children suggests that there are possibly more adults dying than children. This evidence supports the assertion that higher numbers of adults than children are perishing from the disease. In a smaller spatial context, if a family member is sick, it is likely that an adult in the family will be the primary caregiver rather than a child. It is also more likely that this caregiver is female, as females are generally responsible for maintaining the home and caring for children (Columbus Travel Media 2014). The caregiver is therefore in closer proximity to the bodily fluids of the infected person and has a higher chance of contracting the disease. Since it is likely that the person giving care is not a young child, younger children have a lower risk of contracting the disease due to decreased contact with bodily fluids of the infected person (Grady and Fink 2014).

On a larger spatial scale, adults are the ones working in mines, logging, and hunting animals for food (Columbus Travel Media 2014). These tasks are typically carried out by the adult population rather than younger children. Each of these jobs poses a high risk of contact with an infected animal or contact with another human that has contacted an infected animal. Additionally, when individuals have spoken about relatives dying from Ebola, nearly all of the accounts describe adults contracting the virus more often than or prior to children (British Broadcasting Corporation 2014). Though the Ebola virus can affect anyone, the human activities discussed above show that the adult population can be in situations where they are more likely to contract Ebola than children.

Sanitation – The technology subdivision of the disease ecology framework encapsulates factors such as running water and the prospect of developing a vaccination against Ebola. In addition to lack of developed healthcare infrastructure as previously discussed, rural villages often do not have running water (Water and

Sanitation Program 2011). This issue transitions into the realm of technological aspects of behaviour within the disease ecology framework. The lack of municipal water systems and plumbing in the majority of rural homes causes water to be a precious commodity (Water and Sanitation Program 2011). The sparing use of water is not conducive to practices such as frequent hand washing, which often can help minimise the spread of disease (World Health Organization 2014). However, Ebola has the capacity to survive outside of a living host for anywhere from a few hours to 50 days, and hand washing would have likely had minimal impact on slowing the spread of this disease (World Health Organization 2014a). In short, lack of infrastructural capabilities renders the affected region essentially helpless in the face of this highly infectious disease. Combatting Ebola requires extensive resources and efforts. The new post-war governments were not yet in a position to effectively control the spread of the disease given the lack of infrastructure. It was virtually impossible to contact, trace victims, treat them, and put measures in place to avoid further contamination and spread of Ebola.

Vaccinations – There is currently no vaccine for Ebola, but many trials are underway with hopeful signs in some cases (World Health Organization 2014a). Some of these trials have seen some success thus far, but none have completed the trial stages yet (World Health Organization 2014a). Research conducted on survivors of the virus indicate that the earlier the body begins to work towards creating the proper antibodies to fight the disease, the stronger the chance of survival. However, in most cases, the initial immune response takes too long, leaving the body little time to work on producing an effective antibody (Easter 2002). This information is useful in the process of creating a vaccine. Unfortunately, the creation of vaccines takes extensive research and testing over a long period of time. The demand for an Ebola vaccine is urgent and has the potential to control the disease faster and more effectively.

Analysis

Population

When considering the deadly Ebola virus, the population aspect of the disease ecology framework is of minimal salience in determining who is most vulnerable to contracting the disease. This lethal virus does not discriminate against any human host regardless of genetics, age, or gender (World Health Organization 2014a). Current knowledge indicates that no human host is immune to contracting the Ebola virus, which adds to the importance of understanding all facets of disease spread. While this finding may seem insignificant, it is important to understand that every person, no matter gender, race, or age, in Ebola zones is at risk to contract the disease and potentially lose their life.

Habitat

With regards to habitat, the infrastructure of a region is extremely important. Many buildings in rural Africa are not insulated from the outside. It is easy for Ebola to be transmitted into households; therefore, houses do not necessarily protect individuals from Ebola. Many of the buildings are in poor shape as a result of minimal funding from governments to improve infrastructure. The civil war in Sierra Leone depleted the government of most of its funds, and hospitals and schools were some of the sectors that were hit the worst with decreased funding. Since Ebola is highly contagious, patients and victims need strict quarantine measures to prevent the spread of the disease. However, the hospitals were not in proper shape to treat Ebola patients and thus contributed to the spread of Ebola.

One of the themes that came to light while studying burial rituals was that healthcare workers had a hard time communicating with individuals in rural Africa because of language barriers and low literacy rates. Schools have also been hit with budget cuts, and the rundown buildings are inhibiting teachers from educating and improving literacy rates. This may translate into decreased knowledge of proper sanitation practices and inhibits the ability for healthcare workers to educate the population on how to protect oneself from contracting Ebola. The examples above highlight how social issues stemming from a war can have cascading effects on the surrounding community and infrastructure.

Behaviour

In the context of this Ebola epidemic, human behaviour is exceedingly important. A number of themes were revealed that are also evident in past outbreaks. Across the different patterns of behaviour, human migration was a primary source of Ebola transmission. Travel to and from mining and deforestation sites, funerals, displacement from war, and hunting for food were all methods of Ebola movement. Many of these movements are a result of individuals having to travel for work or to provide food for their families.

One of the most significant sources of transmission, if not the most, is cultural practices with respect to funerals and burial rituals. Many of the spillover events between countries were traced to a traditional healer or a family member of a victim crossing regional and international boundaries to attend a funeral. The combination of travelling long distance to attend the funerals, congregating in large groups, and coming into contact with the victim's body create a perfect scenario for transmission. When officials started targeting funerals as a source of transmission, family members experienced severe trauma that discouraged them from seeking help in the future. The feelings of distrust of officials and agony over losing a loved one have all enhanced the ability for Ebola to spread and ravage communities. Multiple different factors from the social sphere combined together to create the perfect combination of variables to make funerals a primary source of transmission.

Discussion and conclusion

All vertices of the disease ecology triangle are interconnected. In the case of this Ebola outbreak, the behaviour and the habitat components are the most important with a primary focus on behaviour. The most salient aspects of the behavioural component include a lack of schools and decreased literacy rates inhibiting communication of information that could help control the outbreak such as proper hygiene and dangerous burial rituals. One approach to bridging this communication gap is to target local social organisations such as funerals to spread information that could potentially save lives. In past outbreaks, technology has often been utilised as a primary vehicle for disease control. While technology can be a helpful tool, taking a more comprehensive disease ecology approach is more aptly suited for identifying the various factors contributing to disease spread, and ultimately provides a useful platform to stop disease outbreaks.

Proof of the effectiveness of the disease ecology framework lies in the course of disease control. Prior to control efforts utilising a broader disease ecology framework, efforts were not entirely successful and disease continued to spread. The root of the issue of disease spread in this epidemic was the funeral rituals, as most points of outbreak can be traced back to specific funerals. When conventional approaches to disease control were failing, officials were able to re-evaluate the situation and realise that the victims of the outbreak were not limited to individuals affected by the disease itself, but also extended to include the emotional trauma felt by loved ones of the affected and deceased. The emotional pain felt by loved ones resulting from inability to complete funeral rituals simply perpetuated the scope of the disease. This outbreak marks a turning point in addressing infectious disease outbreaks in that officials effectively incorporated human emotions and values into the disease control efforts. Moving away from the traditional approach of disease control that is detached from human emotions and behaviours appears to be an effective tactic as seen by this epidemic. The importance of creating a comprehensive view of infectious disease outbreak control cannot be emphasised enough, and the disease ecology framework provides an ideal platform for this course of effective action for this outbreak and also for future outbreak control.

The epidemiologic transitional aspect of Popkin's model further contextualises the findings of analysis from the disease ecology perspective. Popkin's model is an excellent mechanism to demonstrate how the interplay of humans and environmental factors is applicable to both a focused region and a broader spatial scale. Infectious diseases can transcend cultural and geographic boundaries and have the potential to affect us all. Infectious disease outbreaks such as Ebola can ravage populations and hinder development of a nation for generations. Popkin's framework aptly illustrates the demand for application of the disease ecology approach focused on the epicentre of an outbreak. Thus, the demographic, epidemiologic, and nutrition transition model allows us to expand on the implications of the regionally focused disease ecology framework to highlight the importance of the relationship between humans and their environment.

References

Abdullah, I., 1998. Bush path to destruction: the origin and character of the Revolutionary United Front/Sierra Leone. *The Journal of Modern African Studies*, 36(2), 203–235.

Amman, B.R. *et al.*, 2012. Seasonal pulses of Marburg virus circulation in juvenile *Rousettus aegyptiacus* bats coincide with periods of increased risk of human infection. *PLOS Pathogens*, 8(10), 1–11, e1002877. doi:10.1371/journal.ppat.1002877.

Bannon, I. and Collier, P., 2003. *Natural resources and violent conflict: Options and actions*. The International Bank for Reconstruction and Development. Washington D.C.: The World Bank.

Baron, R.C., McCormick, J.B., and Zubeir, O.A., 1983. Ebola virus disease in southern Sudan: Hospital dissemination and intrafamilial spread *Bulletin of the World Health Organization*, 61(6), 997–1003.

Beukes, S., 2014. Ebola: finding patient zero [Online]. UNICEF. Available from http://blogs.unicef.org/2014/10/27/ebola-finding-patient-zero/ [Accessed 5 December 2014].

Bhattacharyya, S.*et al.*, 2010. Ebola virus uses Clathrin mediated endocytosis as an entry pathway. *Virology*, 401(1), 18–28.

Bradbury, J., 2003. Risky meat from the forests. *Frontiers in Ecology and the Environment*, 1(6), 284.

British Broadcasting Corporation, 2014. Ebola fear: 'Having a small fever makes you very afraid' [Online]. Available from http://www.bbc.com/news/29331061 [Accessed 28 June 2017].

Centers for Disease Control and Prevention, 1990. Update: Filovirus infection in animal handles. *MMWR*, 39(13), 221.

Centers for Disease Control and Prevention, 2010. Malaria [Online]. Available from http://www.cdc.gov/malaria/about/distribution.html [Accessed 28 June 2017].

Centers for Disease Control and Prevention, 2011. Laboratory biosafety level criteria [Online]. Available from http://www.cdc.gov/biosafety/publications/bmbl5/BMBL5_sect_IV.pdf [Accessed 28 November 2014].

Centers for Disease Control and Prevention, 2014a. Ebola (Ebola virus disease): Case counts [Online]. Available from http://www.cdc.gov/vhf/ebola/outbreaks/2014-west-africa/case-counts.html [Accessed 11 December 2014].

Centers for Disease Control and Prevention, 2014b. Ebola (Ebola virus disease): Transmission [Online]. Available from http://www.cdc.gov/vhf/ebola/transmission/index.html [Accessed 30 November 2014].

Centers for Disease Control and Prevention, 2014c. Life cycles of the Ebolavirus [Online]. Available from http://www.cdc.gov/vhf/ebola/resources/virus-ecology.html [Accessed 20 November 2014].

Centers for Disease Control and Prevention, 2014d. What is contact tracing? [Online]. Available from http://www.cdc.gov/vhf/ebola/pdf/contact-tracing.pdf [Accessed 3 December 2014].

Central Intelligence Agency, 2014. The world factbook: Sierra Leone [Online]. Available from https://www.cia.gov/library/publications/the-world-factbook/geos/sl.html [Accessed 28 June 2017].

Chowell, G. and Nishiura, H., October 2014. Transmission dynamics and control of Ebola virus disease (EVD): a review. *BMC Medical*, 12(1), 196.

Columbus Travel Media, 2014. Sierra Leone history, language and culture [Online]. *World Travel Guide*. Available from http://www.worldtravelguide.net/sierra-leone/history-language-culture [Accessed 28 June 2017].

Dockins, P., 2014. WHO: Traditional burials hamper Ebola fight [Online]. *Voice of America*. http://www.voanews.com/content/traditional-burial-practices-hamper-efforts-to-fight-ebola/1970353.html [Accessed 5 December 2014].

Dodgeon, S., 2010. Sierra Leone's free healthcare initiative: responding to emerging challenges [Online]. *Health, Poverty, Action*. Available from http://www.healthpovertyaction.org/wp-content/uploads/downloads/2012/07/SierraLeoneFHIbriefingweb12.pdf [Accessed 28 June 2017].

Easter, A., 2002. Emergency: mass casualty: Ebola. *The American Journal of Nursing*, 102(12), 49–52.

Epatko, L., 2014. Bringing safer burial rituals to Ebola outbreak countries [Online]. *PBS Newshour*. Available from http://www.pbs.org/newshour/updates/bringing-safer-burial-rituals-ebola-countries [Accessed 5 December 2014].

Ford, D. and Elbagir, N., 2014. Inside the world's worst Ebola outbreak. *CNN*. Available from http://www.cnn.com/2014/10/19/world/africa/west-africa-ebola-outbreak/index.html [Accessed 28 June 2017].

Funk, D.J. and Kumar, A., 2014. Ebola virus disease: An update for anesthesiologists and intensivists. *Canadian Journal of Anaesthesia*, (2015)62, 80.

Georges, A.J. *et al.*, 1999. Ebola hemorrhagic fever outbreaks in Gabon, 1994–1997: Epidemiologic and health control issues. *The Journal of Infectious Diseases*, 179, S65–S75.

Global Alert and Response, 2014. Sierra Leone: A traditional healer and a funeral [Online]. *World Health Organization*. http://www.who.int/csr/disease/ebola/ebola-6-months/sierra-leone/en/ [Accessed 5 December 2014].

Grady, D. and Fink, S., 2014, August 9. Tracing Ebola's outbreak to an African 2-year-old [Online]. *The New York Times*. Available from http://www.nytimes.com/2014/08/10/world/africa/tracing-ebolas-breakout-to-an-african-2-year-old.html?_r=0 [Accessed 28 November 2014].

Hoyert, D.L. and Rosenberg, H.M., 1997. Alzheimer's disease as a cause of death in the United States. *Public Health Reports*, 112(6), 497–505.

Kalipeni, E. and Oppong, J., 1998. The refugee crisis in Africa and implications for health and disease: a political ecology approach. *Social Science Medicine*, 46(12), 1637–1653.

Kassa, L. *et al.*, 2005. *Schistosomiasis diploma program for the Ethiopian health center team* [Online]. Haramaya University. Available from http://www.cartercenter.org/resources/pdfs/health/ephti/library/modules/Diploma/SchistosomiasisDiploma.pdf [Accessed 28 June 2017].

Kiley, M.P. *et al.*, 1982. Filoviridae: A taxonomic home for Marburg and Ebola viruses?*Intervirology*, 18(1–2), 24–32.

Kuhn, J.H. *et al.*, 2010. Proposal for a revised taxonomy of the family *Filoviridae*: Classification, names of taxa and viruses, and virus abbreviations. *Archives of Virology*, 155(12), 2083-–2103.

Larbi, A., 2012. Country pasture/forage resources for Sierra Leone [Online]. Food and Agriculture Organization of the United Nations. Available from http://www.fao.org/ag/AGP/AGPC/doc/counprof/Sierraleone/Sierraleone.htm [Accessed 28 June 2017].

League of Nations, 2007. Outbreak news. Ebola virus haemorrhagic fever, Democratic Republic of the Congo--update. *Weekly Epidemiological Record / Health Section of the Secretariat of the League of Nations*, 82(40), 345–346.

Linton, M., 2014. Kissing the corpses in Ebola Country [Online]. *The Daily Beast*. Available from http://www.thedailybeast.com/articles/2014/08/13/kissing-the-corpses-in-ebola-country.html [Accessed 28 June 2017].

Lurie, M.N., 2004. *Migration, sexuality and the spread of HIV/AIDS in rural South Africa* [Online]. Southern African Migration Project. http://dspace.africaportal.org/jspui/bitstream/123456789/30708/1/Migration%20Policy%20Series%20No.%2031.pdf?1 [Accessed 28 June 2017].

McCormick, J.B., 2004. Ebola virus ecology. *The Journal of Infectious Diseases*, 190(11), 1893–1894.

McCoy, T., 2014, July 8. How deforestation shares the blame for the Ebola epidemic [Online]. *The Washington Post*. Available from http://www.washingtonpost.com/

news/morning-mix/wp/2014/07/08/how-deforestation-and-human-activity-could-be-to-blame-for-the-ebola-pandemic [Accessed 20 November 2014].

Meade, M.S. and Emch, M., 2010. The human ecology of disease. In *Medical geography: Third edition*. New York: The Guilford Press, 26–72.

Meltzer, M.I. *et al.*, 2014. Estimating the future number of cases in the Ebola epidemic – Liberia and Sierra Leone, 2014–2015 [Online]. Centers for Disease Control and Prevention. Available from http://www.cdc.gov/mmwr/preview/mmwrhtml/su6303a1.htm [Accessed 2 December 2014].

Muyembe-Tamfum, J.J. *et al.*, 2012. Ebola virus outbreaks in Africa: Past and present. *Onderstepoort Journal of Veterinary Research*, 79(2), 451. doi: 10.4102/ojvr.v79i2.451.

Okware, S. I. *et al.*, 2002. An outbreak of Ebola in Uganda. *Tropical Medicine & International Health*, 7(12), 1068–1075.

Popkin, B., 2002. An overview on the nutrition transition and its health implications: the Bellagio meeting. *Public Health Nutrition*, 5(1A), 93–103.

Stylianou, N., 2014. How is Ebola being treated on the ground? [Online]. *British Broadcasting Corporation*. Available from http://www.bbc.com/news/world-africa-29537156 [Accessed 28 June 2017].

Swanepoel, R. *et al.*, 2007. Studies of reservoir hosts of Marburg virus. *Emerging Infectious Diseases*. 13(12), 1847–1851.

Water and Sanitation Program–Africa Region, 2011. *Water supply and sanitation in Sierra Leone*. Nairobi: The World Bank.

West, J. and McDonnell, T., 2014. We are making Ebola outbreaks worse by cutting down forests [Online]. *Mother Jones*. Retrieved http://www.motherjones.com/environment/2014/07/we-are-making-ebola-worse [Accessed 28 June 2017].

WHO Ebola Response Team, 2014. Ebola virus disease in West Africa—the first 9 months of the epidemic and forward projections. *New England Journal of Medicine*, 371(16), 1481–95.

Wickelgren, I., 1998. A method in Ebola's madness. *Science*, New Series, 279(5353), 983–984.

Wolfe, N.D., Daszak, P., Kilpatrick, A.M., and Burke, D.S., 2005. Bushmeat hunting, deforestation, and prediction of zoonotic disease. *Emerging Infectious Diseases*, 11(12), 1822–1827.

World Health Organization, 1976. Ebola haemorrhagic fever in Sudan, 1976. *Bulletin of the World Health Organization*, 56(2), 246–270.

World Health Organization, 1978. Ebola haemorrhagic fever in Zaire, 1976. *Bulletin of the World Health Organization*, 56(2), 271–293.

World Health Organization, 2004. Ebola haemorrhagic fever in the Republic of the Congo – Update 6 [Online]. Available from http://www.who.int/csr/don/2004_01_06/en/ [Accessed 30 November 2014].

World Health Organization, 2012. End of Ebola outbreak in Uganda [Online]. Available from http://www.who.int/csr/don/2012_09_03/en/ [Accessed 28 November 2014].

World Health Organization, 2014a. Frequently asked questions on the Ebola virus disease [Online]. Available from http://www.who.int/csr/disease/ebola/faq-ebola/en/ [Accessed 20 November 2014].

World Health Organization, 2014b. Schistosomiasis fact sheet [Online]. Available from http://www.who.int/mediacentre/factsheets/fs115/en/ [Accessed 28 June 2017].

World Health Organization, 2016. Situation report. Available from http://apps.who.int/iris/bitstream/10665/208883/1/ebolasitrep_10Jun2016_eng.pdf?ua=1 [Accessed 22 February 2017].

4 Progress towards combatting HIV/AIDS in Africa

Imelda K. Moise, Evan de Joya, Leo C. Zulu, Ezekiel Kalipeni, and Diana S. Grigsby-Toussaint

Introduction

After three decades of HIV infection in Africa, signs of hope have emerged with a greater understanding of the magnitude, trends, and diversity of the HIV/AIDS pandemic (Bongmba 2007, McNeil 2017, Morris *et al.* 1995, Santos *et al.* 2014, UNAIDS 1998). New estimates from the Population HIV Impact Assessment (PHIA)[1] show a significant decrease in the rates of new infections, with "stable numbers of people living with HIV, and more than half of all those living with HIV showing viral suppression through use of antiretroviral medication" (Science News 2016). The Joint United Nations Programme on HIV/AIDS (UNAIDS) *Global AIDS Update Report 2016* shows that the number of individuals on treatment increased "from 24% [22–26%] in 2010 to 54% [50–58%] in 2015, reaching 10.3 million people" in the world's most affected sub-regions of eastern and southern Africa (ESA). During the same period, countries in the region reduced AIDS-related deaths by 36% (UNAIDS 2016a).

The trends are particularly encouraging for Sub-Saharan Africa (SSA), the most affected sub-region globally. The more than doubling in Antiretroviral Therapy (ART) coverage reported by UNAIDS (2016a) has resulted in significant reductions in health care costs and potential years of life lost (PYLL) in the SSA region. The 2016 UNAIDS report shows that the treatment coverage has reduced AIDS-related deaths in many SSA countries. By the end of 2015, South Africa alone had nearly 3.4 million people on ART, more than any other country in the world, followed by Kenya (approximately 900,000 people). Many SSA countries increased treatment coverage by more than 25% between 2010 and 2015, including Botswana, Eritrea, Kenya, Malawi, Mozambique, Rwanda, South Africa, Swaziland, Uganda, Tanzania, Zambia, and Zimbabwe. If current trends continue, Africa may be on track towards attaining the UNAIDS 90-90-90 treatment target, an ambitious global commitment to reduce the HIV/AIDS pandemic by 2020. This goal seeks to achieve a 90% rate of diagnosis among people living with HIV, place 90% of those diagnosed on ART, and maintain viral suppression for 90% of patients on ART. Meeting these targets is expected to provide 30 million people living with HIV/AIDS (PLWHA) access to antiretroviral (ARV) drugs by 2020, including 1.6 million children by 2018, and eliminate new infections in children by 2020 (UNAIDS 2016a).

Although SSA has made substantial progress in reversing the trend in HIV prevalence rates, major challenges remain, with marked differences by country, gender, and age. In 2016, global HIV incidence remained at 2.1 million people, raising the total number of people living with HIV to 36.7 million (UNAIDS 2016a). Stagnating declines and recent increases in new HIV infections among certain demographics in some SSA countries threaten to reverse the tremendous gains attained. In the sections that follow, we have created a chronology to highlight some of the key factors, events, and actors that have been pivotal in shaping the status of the HIV/AIDS pandemic in SSA and future prospects for ending it as a major public health problem.

The 1980s: HIV discovery and the early years

Mostly a 'Western' disease among gay men

Between July 1981 and April 1982, 19 cases of a rare lung infection and *Kaposi's sarcoma* were reported among young gay men living in Los Angeles and Orange Counties in California (Task Force on Kaposi's Sarcoma and Opportunistic Infections 1982). At the time, cases of *Kaposi's Sarcoma,* an unusually aggressive cancer, had only been seen in elderly men of Mediterranean descent. This cluster of cases was the first recorded manifestation of HIV in the United States. Positive identification of HIV-1 from a 40-year-old blood sample from a patient in a 1959 study in Leopoldville, Belgian Congo (now Kinshasa, the Democratic Republic of the Congo [DRC]) indicates a much earlier record (De Cock 2001).

In the early years in the United States, HIV was widely associated with men who have sex with men (MSM), and characterized as an affliction unique to this population. Early HIV terminology, such as gay-related immunodeficiency disease, and "gay cancer", reflects this association (Alcabes 2006). By mid-1982, the disease was detected among hemophiliacs and Haitian patients at Jackson Memorial Hospital in Miami, challenging the assumed confinement to MSM. Thus, in September 1982, it was assigned the name "Acquired Immune Deficiency Syndrome", or "AIDS", in a report by the U.S. Centers for Disease Control and Prevention (CDC) (Curran *et al.* 2011).

In Africa, the first signs of the impending pandemic arose in Kinshasa, Zaire (DRC) and rural Uganda. In 1983, following a report of an unidentified outbreak involving a two-fold increase in the cases of *Kaposi's Sarcoma* at Mama Yemo Hospital in Kinshasa, a team of European and American scientists was assembled to investigate (Iliffe 2006). Peter Piot, a Belgian microbiologist leading the team said of the experience, "I walked in and saw all these young men and women, emaciated, dying". Soon after, the mysterious outbreak was identified as AIDS (Malan 2001). From late 1982 through 1983 in Uganda, several men in Kasansero, a fishing village along Lake Victoria, and a number of patients in the Rakai and Masaka districts, died from a disease locally known as "slim", due to the severe weight loss experienced by sufferers. The disease was subsequently identified as AIDS (Serwadda *et al.* 1985).

HIV had hitherto spread undetected across Africa, likely for decades. Idi Amin, the President of Uganda at the time, reportedly told a journalist of an "incurable", "dangerous form of gonorrhea", in which "its victims waste away and lose all the hair", brought by invading Tanzanian soldiers during the 1978–1979 war (Kinsman 2010). HIV continued to spread rapidly, and by 1983 cases had been reported in the Central African Republic, Congo, Rwanda, Tanzania, and Zambia (Knight 2008, Iliffe 2006).

Theories about the source of HIV/AIDS

HIV is a general term describing two strains of the virus: HIV-1 and HIV-2. Both are thought to have originated from strains of Simian Immunodeficiency Viruses endemic in primates in SSA (D'arc *et al.* 2015). The two strains account for 95% of HIV infections today. HIV-1 is the dominant strain identified in the early 1980s while HIV-2 is less aggressive and limited geographically to a part of the West African coast (Marlink *et al.* 1994, Iliffe 2006). HIV-1 consists of four strains classified into groups M, N, O, and P. Group M accounts for most HIV-1 infections worldwide, while Group O constitutes less than 5% of infections in certain western and central African (WCA) countries. Groups N and P have been found in a small portion of infected people in Cameroon (Requejo 2006). HIV strains are further divided into genetically distinct subtypes, with at least nine known in Group M alone. Group M's Subtype C constitutes nearly half (48%) of all infections worldwide, and 95% of infections in southern and eastern Africa (SEA), while Subtype B is predominant in North America and Europe. Despite accounting for a mere 12% of infections globally, Subtype B is the most researched viral strain (Hemelaar 2012), reflecting the global axes of power and priorities. Phylogenetic analysis and the presence of relatively stable prevalence rates point to Central Africa as the early epicenter of the HIV-1 epidemic (Hemelaar 2012). HIV-2, a similar but antigenically different strain, was identified later (1986) in men with AIDS in Guinea Bissau and Senegal (Clavel 1986).

Despite significant expert agreement that HIV is a zoonotic disease that spreads from primates, theories on the mode of transmission vary. Jacques Pépin (De Cock 2012) posits inter-species transmission from chimpanzees during the early twentieth century in Léopoldville (Kinshasa in modern-day DRC). Others give the butchering of primates for consumption and the keeping of monkeys as pets as the transmission mechanism (Hemelaar 2012). The current HIV epidemic is likely the result of multiple cross-species transmissions in the WCA region. Subsequent transmissions to other parts of Africa have happened through urbanization, gender inequities, the sex trade, colonial disease-control policies, labor migration, cultural factors, poverty, extant disease burdens, and some recent global forces (Pépin 2011, Kalipeni and Zulu 2012). Denialism, inadequate resources, and bureaucratic inertia further slowed down responses.

Early HIV response in fits and starts

Denial and resistance from public health officials in much of SSA inhibited early efforts to prevent the spread of HIV. An American physician researching the disease was refused a visa to Zaire and threatened with expulsion from Zambia for reporting on what the president's press secretary called an "American disease" (Altman 1999). Among international development agencies, there was a widespread misconception of HIV as a prostitution-related issue concentrated in urban areas. Early efforts focused on containment, through the promotion of condoms among high-risk populations, rather than prevention of transmission (Pickett 2011). Both the denialism and narrow approach likely delayed the response and contributed to the explosion of HIV in SSA. A 1986 study found that 66% of sex workers in Nairobi, overwhelmingly of low-socioeconomic status, were infected with the disease (Kreiss *et al.* 1986), and similarly high rates were reported in this high-risk groups across many SSA cities (Morison *et al.* 2001).

Socially, HIV-positive people faced widespread stigma during the early years. Most of the world viewed it as a "gay disease", and many leaders in SSA refused to recognize it as a problem because of a belief that homosexuality was non-existent in their country. In some instances, religious organizations blocked efforts to distribute contraceptives (Bongmba 2007). Economic conditions also played a role in the spread of HIV throughout SSA. SSA carried nearly a quarter of the world's HIV disease burden, yet was among the lowest in the world in healthcare expenditure and physicians per capita, another major confounding challenge (World Bank 2018a).

By the mid-late 1980s, despite the growing realization of the potential dire impacts of HIV as a pandemic, responses in most SSA countries were slow and inadequate, allowing rapid spread and growth of the pandemic. Uganda, however, was one of the first countries in Africa to mount a national-scale response through the launch of the National AIDS Control Program in 1986. In January 1987, the main hospital serving the Rakai and Masaka districts in Uganda began screening its blood supply (Kinsman 2010). Prevalence rates in the screened donated blood were 15% and 21% for men and women, respectively – one potential new infection for every six blood transfusions (Carswell 1987).

The blood screening results in Uganda epitomized the rapid growth of the epidemic, with no serious control prospects through the end of the decade. By the end of the 1980s, the disease had spread mainly across eastern, central, and southern Africa, with the Uganda/Rwanda region having the highest prevalence based on rates among pregnant women attending designated antenatal clinics (Kalipeni and Zulu 2008).

The 1990s: A pandemic is born and rages on

The pandemic increases in intensity and geographic scope, peaking in prevalence and new infections

The 1990s were critical, marked by the rapid spread, growth, peak, and arrest of the epidemic in SSA. The beginning of the decade saw diverse HIV epidemics, differentiated in nature and extent (prevalence, incidence, death rates) across world regions, subregions of SSA, SSA countries, and sub-national regions. The global pandemic was in full swing, with SSA at the epicenter. Within SSA, the decade ushered in a shift from a split eastern (around Uganda) and southern (around Zimbabwe) African nuclei in 1990 to a decidedly southern African epicenter by 1999 (Kalipeni and Zulu 2008). SSA countries had the highest transmission rates globally (World Health Organization 2017).

National contrasts were similarly significant. HIV prevalence in Zimbabwe, for instance, increased from 15% in 1990 (the highest globally then) to nearly 26% by the end of the decade (Mugurungi *et al.* 2007). Prevalence in South Africa's 15–49 age group jumped 20-fold, from 1% in 1990 to nearly 20% ten years later, and by 1991, the number of new infections from heterosexual sex had overtaken that from homosexual sex, which remained the major mode of transmission (World Health Organization 2017). In contrast, the prevalence in the DRC, suspected origin (Kinshasa) of HIV, remained around 2% for the decade (World Bank 2018b). The relatively stable prevalence rates suggested a mature epidemic, with deaths in older populations balancing out high incidence rates in younger ones, providing further evidence of disease longevity and of the DRC as a possible origin of HIV/AIDS (Iliffe 2006).

HIV/AIDS is a very complex disease, and diverse factors "beyond epidemiology" drove its fast but social-spatially differentiated spread in Africa (Kalipeni *et al.* 2004). Drivers include sociocultural, political, economic, and environmental factors. The number and rate of sexual partner acquisition, prevalence of sexually transmitted diseases, poverty, inequitable gender and power relations, sexuality, culture, migration, education, and the activities of high-risk 'core groups' including IV drug users, sex workers, and truck drivers, can enhance HIV spread (Kalipeni *et al.* 2004, Kalipeni and Zulu 2008). Concurrence, or long-term overlapping sexual partnerships, is often blamed for the rapid HIV spread in SSA, but the issue is complex and the debate unsettled. One study reports Ugandan men as having higher frequencies and longer duration of concurrency compared to their American or Thai counterparts (Morris and Kretzschmar 1997, Morris *et al.* 1995), yet relatively significant levels of discordant couples, i.e., couples within which one is HIV positive and the other negative, partly counters the concurrency argument. A study found one in seven co-habiting couples in Kigali to be discordant (Allen *et al.* 1992).

The varying, contextualized effects of education also illustrate the complexity of explaining HIV spread in SSA. The worst hit countries had some of the lowest literacy levels, consistent with a statistically significant correlation between high rates of infection and low literacy rates, globally. However, in SSA, an inverse relationship was observed in urban areas where the disease remained concentrated. Here, high literacy rates correlated with high rates of infection (UNAIDS 1998).

Between 1994 and 1997, HIV spread began to slow down in countries which had taken early action against HIV/AIDS, such as Uganda and Malawi (UNAIDS 1998). Globally, the number of new infections peaked at nearly 3.5 million people by 1997 (Roser 2018; see Figure 4.1). HIV prevalence peaked at much higher rates in SSA than other regions. Prevalence rates in the United States and Thailand peaked below 1% and 2%, respectively, while the rate in Uganda peaked at over 13% (AVERT 2018). Numerous SSA countries saw a massive escalation in infections. Nevertheless, the slowing down and incipient reversal of HIV infections and prevalence in some SSA countries ended an otherwise dire decade with a ray of new

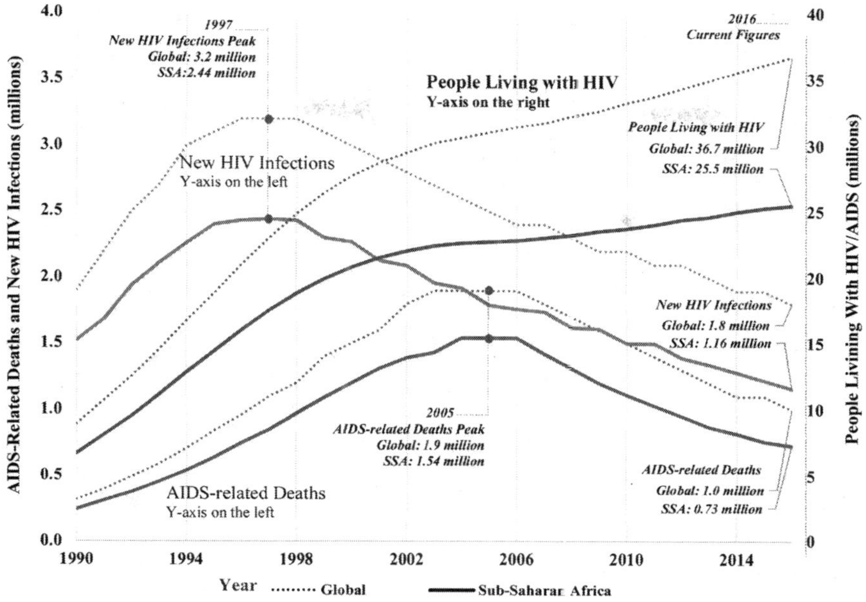

Figure 4.1 Temporal trends in people living with HIV/AIDS, new HIV infections and AIDS-related deaths in Sub-Saharan Africa, 1990–2014.

Source: UNAIDS (via www.aidsinfoonline.org)

hope. Uganda's 5% decline in HIV prevalence and a more than three-fold reduction in new infections during the 1990s (Government of Uganda 1992) illustrates dividends from the country's early aggressive response, including its ABC approach promoting **a**bstinence, **b**eing faithful to sexual partners and using **c**ondoms.

HIV deaths escalate

Global deaths from HIV rose linearly throughout the 1990s, climbing from 300 000 in 1990 to 1.4 million people in 1999 (UNAIDS 2012). Most (83%) of the worldwide AIDS deaths had by 1997 occurred in SSA. Increases in HIV-related deaths inflated overall mortality rates for young adults in the countries most affected. Between 1988 and 1994, the death rate for men aged 15–60 tripled in Zimbabwe and nearly doubled in Zambia (Timaeus 1998). A 1997 study in rural Tanzania found that AIDS accounted for 35% and 75% of all deaths in areas with 4% and 20% HIV prevalence rates among adults, respectively (Boerma *et al.* 1997).

Adverse socioeconomic impacts

The skyrocketing death rates had sweeping adverse effects on African societies (Ghosh and Kalipeni 2004). Surveys showed dramatic decreases in income among households having a family member with AIDS. Often, children in AIDS-affected households lived with relatives who could support them. This informal system helped to alleviate the burden of the rapidly swelling number of orphans due to increasing AIDS-related deaths. With fewer and fewer adults to serve as caregivers, the social system was overstretched (UNAIDS 1998). AIDS deaths in Africa during the 1980s and 1990s left millions of children orphaned (Grant and Yeatman 2012, Bicego *et al.* 2003). The orphaned children were at greater risk of malnutrition and illness than non-AIDS orphans were (Magadi 2011), and to this day, the AIDS orphans often face AIDS-related stigma and discrimination (McHenry *et al.* 2017). Despite intervention efforts by governments and NGOs to help these children, orphan-headed households became common in most affected countries like Zimbabwe, where 7% of children under 15 years old were orphaned by HIV/AIDS (Pisani *et al.* 2000). Adverse impacts permeated nearly all sectors of life, especially as AIDS deaths affected the most productive demographics. In the first ten months of 1998, 1,300 teachers in Zambia died, two-thirds of the total number of teachers trained each year (Mulkeen 2007). Agricultural output declined, decreasing food security among farming communities. In Zimbabwe, declines in 1998 of 61% and 47% in marketed maize and cotton, respectively, were attributed to the HIV/AIDS epidemic (Pisani *et al.* 2000). However, the latter 1990s brought some hope, particularly on prevention of mother-to-child transmission (PMTCT), the introduction of ARTs, and the operational start-up in 1996 of UNAIDS to mobilize and coordinate the global HIV/AIDS response within the UN system.

Prevention of mother-to-child transmission

In 1994, researchers announced that the anti-retroviral drug zidovudine (AZT) effectively prevented mother-to-child transmission (MTCT) of HIV, a major challenge then. By 1999, some 3.8 million children had died of AIDS before age 15, nearly 500 000 in 1999 alone (UNAIDS 2012). Another 1.3 million children, a majority of whom had contracted HIV from infected mothers, were living with HIV and awaited a similar fate. It was not until 1998 and 1999 that several trials in SSA and other developing countries showed that administering AZT to infected mothers during the last month in pregnancy reduced the likelihood of HIV transmission by a third to half, with or without breastfeeding. Additionally, administration of "nevirapine to the mother at the onset of labor and then to the infant after delivery" was found to produce similar results, at the reduced cost of $4 per HIV-infected woman, less than 10% of the monthly AZT cost (UNAIDS 2010).

ARVs introduced in SSA HIV programs

Despite the granting of approval by the U.S. Food and Drug Administration in March 1987 (U.S. Food and Drug Administration 2017), the prohibitive initial market cost of nearly $10 000, per person, annually (Ford *et al.* 2007), prevented the introduction of government distribution programs in SSA until the late 1990s (Topouzis 2004). Due to the high cost, SSA governments, international aid agencies, and researchers had advocated prioritizing funding for more cost-effective prevention programs rather than on ART (Ford *et al.* 2011). Early pressures on governments for lower-priced ARVs from patient groups, health providers, activists, and other stakeholders in many heavily affected countries, including South Africa, Kenya, and Uganda, were largely unsuccessful (von Schoen Angerer *et al.* 2001). Officials in South Africa rejected calls for the distribution of AZT in 1998, citing the high cost and the government policy of focusing on prevention rather than treatment (McNeil 2017). Among the international aid community, there was concern over adherence to ARTs in SSA, and the early dialogue was often marred with generalizations and stereotypes of Africans. In 2001, Andrew Natsios, the head of the United States Agency for International Development (USAID) was quoted saying that many Africans "don't know what Western time is. You have to take these (AIDS) drugs a certain number of hours each day, or they don't work. Many people in Africa have never seen a clock or a watch their entire lives" (Herbert 2001). With little access to ARVs in SSA, AIDS-related deaths continued to increase, becoming the top cause of death in Africa by 1999 (World Health Organization 1999). The Drug Access Initiative established in 1998 by UNAIDS and several pharmaceutical partners to pilot price-reduced ARV therapy in some developing countries affirmed the feasibility of publicly administered ART initiatives. It paved the way for differential pricing of ARTs, with lower pricing provided to developing countries.

The 2000s: Global partnerships, an HIV pandemic in transition, and new hope for a continent

The dawn of the new millennium brought new hope to Africa in several dimensions of the HIV/AIDS pandemic. There were significant reductions in the number of new HIV infections, HIV-related deaths, and growth in the number of people living with HIV. While the decade brought increased focus on treatment of HIV/AIDS in SSA, widespread provision of ARTs remained a major challenge. World Trade Organization (WTO) regulations helped to maintain the prohibitive ART prices by preventing the manufacture and import of generic drugs, delaying ART rollout in SSA. Using these regulations, 40 pharmaceutical companies even sued the South African Government in 1997 to stop the passing of a law that would allow the importation of cheap generic drugs including ARVs. In 2001, Cipla Inc, an Indian pharmaceutical company, began offering generic triple-combination ARV drugs for less than US$350 annually – less than US$1 per day (Binswanger 2001). Soon after, South Africa began importing generic ARV drugs, despite the pending lawsuits from pharmaceutical companies. The companies eventually withdrew the lawsuits following immense pressure from activists, opening the door for other SSA countries to import generic ARV drugs (Fisher and Rigamonti 2005). Although the private sector eventually joined the global effort to increase ART access, some activists blamed pharmaceutical companies for obstructing access to life-saving drugs in pursuit of profit.

For the duration of the decade, advances in the provision of HIV prevention services, care, and support expanded, increasing life expectancy and further reducing HIV-related deaths in most SSA countries.

A global contract against HIV/AIDS marks the turning point in the HIV fight

The new decade/millennium marked a turning point in global cooperation, political commitment, and resource mobilization in the fight against HIV/AIDS, and raised hopes for ending HIV/AIDS as a major public health problem. In early 2000, the UN Security Council discussed HIV/AIDS as a "major human security concern, and an obstacle to development", the first time a health issue was brought up for discussion (Knight 2008). Incorporation of HIV/AIDS as one of the eight Millennium Development Goals (Goal 6: combat HIV/AIDS, malaria, and other diseases) further elevated HIV on the global health and development agenda, and the level of political will. The World Health Organization (WHO) and UNAIDS were instrumental in mobilizing industrialized nations, development agencies, the private and NGO sector, affected governments, and resources into the fight against HIV/AIDS.

Mobilization and sustenance of global funding was the next major driver of the many gains achieved in HIV/AIDS research, monitoring, prevention, treatment, and control. The establishment of three global/regional funding programs or mechanisms catalyzed the successes won in the HIV fight (Windisch *et al.* 2011). In 2001, The World Bank launched an SSA-targeted Multi-Country HIV/AIDS Program (MAP), committing US$500 million in funding for the first three years. The Global Fund to Fight HIV/AIDS, Tuberculosis, and Malaria (GFATM), a public-private partnership organization founded in 2002, has raised approximately US$4 billion annually to support interventions and accelerate the elimination of AIDS, malaria, and tuberculosis as epidemics. In 2003, the U.S. government established the President's Emergency Plan for AIDS Relief (PEPFAR) with $15 billion in funding, focusing initially on delivering life-saving services in nations hardest hit by HIV/AIDS. Global funding for HIV/AIDS research and prevention grew from $1.6 billion in 2001 to $8.9 billion in 2006 (World Bank 2008). In 2008, the George Bush Administration tripled PEPFAR funding to $48 billion for the period 2008–2013, $39 billion of it designated for HIV and the Global Fund, and framed HIV/AIDS as a moral (religious) issue, not a national security threat (Flint 2011).

While global cooperation and funding alleviated many adverse dimensions of the pandemic, by the late 2000s, accelerated scale-up of life-prolonging ARTs was the most impactful strategy for SSA and the larger developing world. In 2003, the WHO and UNAIDS launched the "3 by 5" campaign with the goal of bringing ARVs to 3 million people by 2005 (Rassool 2004). The funding and efforts allowed the widespread provision of free ARVs to people who would otherwise not afford them. Thus, free ARVs were available to Ugandans by 2003 and to Ethiopians in 2004 (Görgens-Albino *et al.* 2007). At the start of PEPFAR in 2003, fewer than 50 000 people were on ARVs in SSA, but by 2010, the PEPFAR and the GFATM supported ART provision for nearly 4.7 million people globally, more than 3.2 million of them supported by PEPFAR alone. ART coverage reached 23% of PLWHA in the most affected ESA region in 2010, a decade after first ART availability at discounted prices (UNAIDS 2017b). By the end of the decade, ART coverage in SSA took off, increasing by 20% from 2009 to 2010 alone (UNAIDS 2011; also see Figure 4.1).

ARV therapies, the early HIV miracle

Increased access to ART became the early miracle in the fight against HIV/AIDS, serving as both a method of treatment and of prevention. By suppressing the viral load (and replication) in people with HIV infection, ART prevents further damage to and allows the immune system to rehabilitate, thereby reducing morbidity and AIDS-related deaths. Viral suppression also reduces the risk of HIV transmission to others (West *et al.* 2007). Early treatment with ART of an HIV-positive partner in a discordant couple can also reduce the relative risk of HIV transmission by up to 96% (Cohen *et al.* 2011). High levels of availability of ART helped to further reduce new HIV infections by breaking through the 'plateau' effect, an oft-observed stabilization and peaking of HIV-incidence rates following an initial reduction attained through behavior-based prevention measures (Mathers *et al.* 2008). PMTCT also utilizes ART for viral suppression in infected pregnant mothers to prevent HIV transmission before, during, and after birth.

Though ART scale-ups have improved prospects in SSA, challenges remain. ART delivery is largely contingent on wide-scale HIV counseling and testing, and an individual's knowledge of their status. Gaps in HIV-status awareness persist in certain demographics, including young people (ages 15–24) and critical at-risk groups, partly a reflection of continuing stigma and criminalization (UNAIDS 2017b). While poor adherence to the ART regimen hinders positive effects of the scale-up, and more progress is needed to reduce it, a major study of nearly 30 000 AIDS patients in North America and SSA found pooled adherence in SSA of 77%, more than 20% greater than in North America (55%) (Mills *et al.* 2006).

Turnaround in AIDS-related deaths and rebound in life expectancy

Accelerated ART scale-up also brought a changed perception of HIV/AIDS. It was no longer an automatic death sentence, but instead, a manageable chronic disease. Annual global AIDS-related deaths peaked at approximately 2.2 million people around 2004–2005 and began to decline, reaching 1.6 million by 2009 (UNAIDS 2011). The reduced deaths also reflect the rapid recovery in life expectancy among adults and children in SSA. UNAIDS statistics from ten ESA countries showed a stalling in life expectancy from the mid-1980s to mid-1990s, followed by a plunge and bottoming out in early 2000s (reversing steady gains by many SSA countries during 1960–1990), and a rapid rebounding by 2009 (UNAIDS 2017b). For Botswana, Lesotho, Swaziland, and Zimbabwe, life expectancy had gone from 60 years or more to around 45 years and below. A large cohort study in rural KwaZulu-Natal, South Africa, showed an 11-year gain in life expectancy from 49.2 years in 2003 to 60.5 by 2011, and found that the economic benefits of ART through reductions in PYLL were higher than treatment costs (Bor *et al.* 2013).

ARTs have also dramatically reduced deaths among children via PMTCT. From the mid-1990s to 2015, global PMTCT efforts averted 1.4 million infections among infants (UNAIDS 2015). AIDS-related deaths among children declined by 20% from 2005 to 2010 (UNAIDS 2011). In 2010 alone, PEPFAR supported more than 114,475 mothers from 31 countries on ARTs for PMTCT purposes (U.S. Department of State 2011). Rwanda reduced new HIV infections among children by 88% between 2009 and 2014, bringing down MTCT rates to less than 2% by 2014 (UNAIDS 2015). These gains are a reflection of the major increases in access to ART for PMTCT, from nearly zero in 2000 to nearly 60% by 2009, globally (UNAIDS 2017). Over the same period, new HIV infections among the 0–14 age group declined from over 450,000 to 300,000. Botswana, Lesotho, Namibia, South Africa, and Swaziland had reached universal ART access for PMTCT, versus a 48% coverage rate globally (UNAIDS 2011).

The decade saw remarkable reductions in new HIV cases in SSA. They dropped by more than 26%, from 2.6 million in 1997 to 1.9 million people in 2009 (UNAIDS 2011). In 22 of the 33 countries in which HIV incidence had fallen in SSA between 2001 and 2009, new infections declined by 25%. For Zimbabwe, HIV prevalence among the 15–49 age group declined sharply from 24% to below 16% during the decade (World Bank 2018b).

The 2000s registered other gains in prevention measures on the biomedical and behavioral change fronts. These included improved treatment and adherence to ART regimens; behavior-change programs (e.g., voluntary testing and counseling, delayed sex, minimizing sex partners, abstinence); voluntary early male circumcision (which can reduce HIV transmissions by up to 60%); and condom promotion and distribution (UNAIDS 2011). Awareness campaigns targeted social stigma minimization and the changing of harmful cultural practices, and focused on high-risk groups. The increase in the number of PLWHA also stabilized during the decade (Figure 4.1), increasing between 2009 and 2010 by just 300,000 in SSA (UNAIDS 2011). The relative stabilization was primarily attributable to ART scale-up.

Despite these gains, the disproportionate toll of HIV/AIDS on SSA has had many long-term implications. By 2007, over 22.5 million people in Africa were AIDS positive, more than 11.4 million under-18 children had lost one or more parents to AIDS, and 2.2 million children under the age of 15 were living with HIV (World Bank 2008). Companies in some countries had begun recruiting two workers for every open position to mitigate against deaths from HIV/AIDS. By 2010, though constituting only 12% of the world's population, SSA was home to 68% of all people living with HIV, 70% of new HIV infections, and nearly 75% of all AIDS-related deaths (UNAIDS 2011).

In sum, the turnaround during the 2000–2009 decade transformed HIV from a death sentence to a largely manageable, chronic disease mainly due to rapid ART scale-up in SSA (Figure 4.1). There was major progress in PMTCT, and reductions in new HIV cases and HIV deaths in most SSA countries, and the numbers of PLWHA stabilized. Major challenges remained, but the brunt of the adverse impacts of HIV/AIDS had begun to ease.

The 2010s: Beginning of the end of HIV/AIDS

The current decade (2010–2019) consolidates the tremendous gains made particularly during the previous decade in SSA (and globally) on HIV prevention and treatment. Major global initiatives encapsulate the acceleration, consolidation, and strategic transition from merely containing the epidemic to eliminating it as an epidemic. They include an emphasis on achieving the new global 90-90-90 targets (90% of PLWHA know their status, 90% of PLWHA are on ARTs, and 90% of people on ARTs have undetectable viral loads) by 2020.[2] The other is the ambitious UN Sustainable Development Goal to end AIDS by 2030 via a fast-track strategy and commitments from the global adoption of the 2016 United Nations Political Declaration on Ending AIDS. The decade further raises new hope of biomedical advancements in prevention (e.g., Pre-Exposure Prophylaxis or PrEP), and progress on HIV-vaccine trials and HIV cures.

Consolidating gains and accelerating HIV epidemic control

Consolidating gains is essential to accelerating the end of HIV/AIDS as a major public health problem. Latest UNAIDS (2017b) data show a dramatic 62% decline in AIDS-related deaths in the ESA region (versus 48% globally, 2005–2016) from a peak of 1.1 million in 2004 to 420 000 in 2016; 30% decline in west and central Africa (UNAIDS 2016a). Attainment of the steepest declines from 2010 to 2016, by 42%, provides evidence of such acceleration. The ESA region had the highest reduction in new HIV cases, 29% between 2010 and 2016 versus 16% globally, 9% in western and central Africa, and 4% in the Middle East and North Africa. However, current global progress falls short of the UN 2020 target of fewer than 500 000 HIV deaths.

Other promising trends include a 27% reduction in HIV-related deaths among women and girls in 2016, lower than among men globally despite females constituting 51% of all PLWHA (UNAIDS 2017b), suggesting a narrowing of the gender gap. However, much work remains among the youth (ages 15–24). New infections among young women (360 000) were 44% higher than among men (250 000) with similar rates of decline of 17% and 16% respectively during 2010 to 2016. ART treatment of female PLWHA for PMTCT increased from 47% to 76% during 2010 to 2016, making new HIV-infection declines the highest (47%) among children. These statistics show a mixed but promising 2016 progress report for the UNAIDS 90-90-90 targets: 76% – 79% – 83% for the ESA region and a lower 42% – 83% – 73% for western and central Africa. All the above illustrates the necessity of ART treatment within the global HIV strategy as well as the remarkable gains from rapid ART scale-up.

Malawi illustrates the scale-up potential of ART, which doubled the number of PLWHA receiving ARTs between 2010 and 2014 (UNAIDS 2015), and their prevention potency. Putting 10% of the PLWHA in Malawi's southern rural district of Chiradzulu on ART (27,000 people by June 2013) reduced new HIV infections to only four for every 1000 residents, despite a 17% HIV prevalence. Ending the epidemic nevertheless requires a renewed emphasis on *ending the virus* and on prevention (Senthilingam 2016a).

New techno-epidemiological advances to achieve epidemic control of HIV/AIDS

The 2010s have also seen encouraging signs of combining technological advances with other measures to rebalance attention towards prevention in a necessarily pluralistic approach to HIV/AIDS. Examples include using PrEP on uninfected people to protect themselves from infection, trials on vaginal microbicides, suppression of herpes simplex 2 infections, and progress on various vaccines and cures. Thus, the PEPFAR 2017–2020 strategy focuses on achieving epidemic control in 13 highly burdened areas (12 SSA countries and Haiti) through expanding ART provision (target ≥70% in each country by 2018) for viral load suppression among PLWHA, prevention and risk-avoidance among HIV-negative individuals, and voluntary medical male circumcision (VMMC) among HIV-negative young men (U.S. Department of State 2017).

To reach epidemic control, the total number of new infections has to become lower than total deaths among PLWHA. By 2017, five of PEPFAR supported countries were close to achieving HIV/AIDS epidemic control – Lesotho, Malawi, Swaziland, Uganda, Zambia, and Zimbabwe. Malawi, Zambia, and Zimbabwe had achieved an average 65% (51%–76%) reduction in new HIV infections since PEPFAR commenced in 2003. Further, 21 countries in southern Africa had adopted the WHO-recommended 'treat all' approach for PLWHA, regardless of the seriousness of the HIV infection or level of damage caused to the immune system, which is assessed by the density of white blood cells, CD4 count (UNAIDS 2017a). This is a shift in approach. Previously, ART treatment commenced only when HIV was symptomatic or the CD4 count had fallen below a certain range.

Recent progress on the development of HIV treatments and vaccines gives hope for ending HIV/AIDS. The debate has revolved around two main approaches: a "shock and kill" strategy that cures by killing the virus, or a "block and lock" strategy (also known as a "functional cure") that stops HIV spread by rendering the virus dormant (Deeks *et al.* 2016, Kessing *et al.* 2017). The "shock and kill" strategy has been the dominant research target in the past few years because the latter approach still requires patients to visit physicians regularly and does not remove the stigma of HIV. An example of a 'functional cure" includes a successful recent test of the Tat inhibitor didehydro-Cortistatin A (dCA) (Kessing *et al.* 2017). However, effective vaccines and cures are still years away.

Meanwhile, PrEP has been added to the prevention strategies for individuals at substantially high risk of HIV infection. PrEP uses ARVs administered orally (daily), or via a vaginal ring in the case of women, to protect uninfected people from contracting HIV. UNAIDS reported efficacies of 90% in reducing infections through oral PrEP (often a combination of tenofovir – TDF, 300 mg – and emtricitabine, FTC, 200 mg (UNAIDS 2016b). By March 2016, the drugs had been approved in France, Kenya, South Africa, and the United States (brand name Truvada), and UNAIDS set a global target of reaching 3 million people at high HIV-infection risk with PrEP by 2020. Several other countries have adopted it since. As for vaginal PrEP, even the modest 30% efficacy rate in early trials (in South Africa, Uganda, Malawi, and Zimbabwe) of a PrEP vaginal ring that slowly releases ARVs (dapivirine), are encouraging because the benefits go beyond prevention to empowering women to make reproductive-health decisions that protect them (Senthilingam 2016b). One major vaccine trial (HVTN 702) is underway in South Africa, with results expected in 2020.

Some challenges persist

Despite the tremendous progress achieved over the past two decades, major challenges persist. These include limited adherence to ART regimes; institutional capacity, including at community level; financial constraints; and a false sense of safety brought by life-saving ARTs (and potentially, PrEPs). Gaps between awareness and behavioral change, and gender and age disparities in the HIV/AIDS burden remain unclosed, including difficulty in reaching the youth, men, empowering women, and treating HIV-infected children. In 2016, SSA accounted for nearly two-thirds (64%) of the 5000 new HIV infections occurring daily, with 25.5 million PLWHA (19.4 million in eastern and southern Africa), nearly 1.2 million new infections, and 730 000 AIDS deaths (Figure 4.1) (UNAIDS 2017b).

Despite significant decreases in new infections in SSA, some countries registered increases during 2010–2016. These include South Sudan, Eritrea, Ethiopia, and Madagascar in eastern and southern Africa, and Benin, Togo, Central African Republic, Burkina Faso, Congo, Liberia, and Ghana in west and central Africa (UNAIDS 2017b). Further, the risk of HIV infection among young women aged 15 to 24 years remained disproportionately high. Such young women constituted only 10% of the population but accounted for 26% of new infections in eastern and southern Africa (22% in western and central Africa). Of the new HIV infections among these 15–24-year-olds in SSA (61 0000 in 2016), 59% were among young women, 44% higher than among young men (UNAIDS 2017b). This challenge is confounded by a considerable age gap in HIV-status awareness. Findings from a recent PEPFAR survey shows that fewer than 50% of young PLWHA in Malawi, Zambia, and Zimbabwe were aware of their HIV status, versus 78% in the 35–59 age group (UNAIDS 2017b).

Finally, resource-availability gaps in international commitments could prevent meeting the 2020 fast-track goals in middle- to low-income countries and slow down HIV-control efforts. UNAIDS (2017b) statistics show that national and international sources of funding increased from $10 billion in 2006 but peaked at a level below $20 billion in 2013 and has stagnated or declined slightly since 2016. Yet meeting the 2020 goals would require nearly $27 billion annually by 2020, even with the commendably increasing levels of national funding among the most affected countries (UNAIDS 2017b).

Conclusion and prospects for the 2020s – The post-pandemic era?

HIV/AIDS is one of the greatest global healthcare epidemics, historically. SSA has been at the epicenter of the HIV/AIDS pandemic, which remains a major inhibitor of socioeconomic development. Adverse effects include millions in HIV morbidity and deaths; a severe drain on healthcare systems; disruption of the social fabric through the millions of orphaned children; reduced quality of life and life expectancy; reduced human resources needed for individual, community and national socioeconomic development; and HIV-related discrimination from social stigma against PLWHA. Against such a dire status analysis, SSA has made impressive progress against HIV/AIDS over the past three decades. From its discovery in the early 1980s and throughout the decade, the spread of the disease initially associated with the gay community was characterized by confusion and even denial. During the 1990s, HIV/AIDS reared its ugly head in SSA and around the world as a devastating health crisis, a pandemic, largely transmitted via heterosexual sex. The pandemic spread rapidly, peaking in prevalence by the late 1990s after early denialism among some SSA governments turned into concerted, prevention-based responses. The 2000s brought much-needed hope to SSA and reduced HIV/AIDS from a death sentence to a manageable disease.

International recognition of HIV/AIDS as a global problem, commitment of resources reflected in the United Nations Millennium Development Goals, launches of the Global Fund, and PEPFAR, greatly contributed to both prevention and treatment with ARTs. The 2010s have seen continued consolidation of these gains, with a shift in approach from prevention-dominated to a treatment-focused approach driven by the scale-up of life-prolonging ARTs, augmented by several international political commitments such as the 90-90-90 targets by 2020 and the bold Sustainable Development Goal of ending HIV/AIDS by 2030. The pivot towards treatment in the 2000s has brought hope to the millions of people living with AIDS.

Despite the still high numbers of PLWHA and new infections; deficiencies in adherence to ART regimes, institutional capacity, targeting the youth, empowering women, and financial resources; and the risk of complacency due to life-saving ARTs and emerging preventions such as PrEPs; SSA appears set on a hopeful path to end HIV/AIDS if not by 2030, not long after. Prospects of finding effective vaccines or even a cure for HIV/AIDS will only help to accelerate the end of HIV/AIDS as a major health problem.

Notes

1 The PHIA Project is a unique, multi-country initiative funded by the U.S. President's Emergency Plan for AIDS Relief (PEPFAR).

2 Since the launch of a sister Fast-Track-Cities initiative in 2014 under the Paris Declaration – *Fast-Track Cities: Ending the AIDS Epidemic*, at least 200 cities/ municipalities the world over have committed to achieving the 90-90-90 targets by 2017.

References

Alcabes, P., 2006. The ordinariness of AIDS. *The American Scholar*. June 1. Available at: https://theamericanscholar.org/the-ordinariness-of-aids/#.WllVoeSWxUQ.

Allen, S. *et al.*, 1992. Effect of serotesting with counselling on condom use and seroconversion among HIV discordant couples in Africa. *BMJ*, 304(6842), 1605–1609.

Altman, L. K., 1999. The doctor's world: in Africa, a deadly silence about AIDS is lifting. Available at: https://www.nytimes.com/1999/07/13/health/the-doctor-s-world-in-africa-a-deadly-silence-about-aids-is-lifting.html.

AVERT, 2018. HIV/AIDS. Available at: https://www.avert.org/professionals/hiv-around-world/sub-saharan-africa/uganda.

Bicego, G., Rutstein, S., and Johnson, K., 2003. Dimensions of the emerging orphan crisis in sub-Saharan Africa. *Social Science Medicine*, 56(6), 1235–1247.

Binswanger, H. P., 2001. HIV/AIDS Treatment for Millions. *Science*, 292(5515), 221–223.

Boerma, J. T. *et al.*, 1997. Levels and causes of adult mortality in rural Tanzania with special reference to HIV/AIDS. *Health Transition Review*, 7, 63–74.

Bongmba, E. K., 2007. *Facing a Pandemic: The African Church and the Crisis of AIDS.* Waco, TX: Baylor University Press.

Bor, J., Herbst, A. J., Newell, M. L., and Bärnighausen, T., 2013. Increases in adult life expectancy in rural South Africa: valuing the scale-up of HIV treatment. *Science*, 339(6122), 961–965.

Carswell, J. W., 1987. HIV infection in healthy persons in Uganda. *Aids*, 1(4), 223–227.

Clavel, F., 1986. Isolation of a new human retrovirus from West African patients with AIDS. *Science*, 233, 343–346.

Cohen, M. S. *et al.*, 2011. Prevention of HIV-1 infection with early antiretroviral therapy. *New England Journal of Medicine*, 365(6), 493–505.

Curran, J. W., Jaffe, H. W., and CDC, 2011. AIDS: the early years and CDC's response. *MMWR Suppl*, 60(4), 64–69.

D'arc, M. *et al.*, 2015. Origin of the HIV-1 group O epidemic in western lowland gorillas. *Proceedings of the National Academy of Sciences*, 112(11), E1343–E1352.

De Cock, K. M., 2001. Epidemiology and the emergence of human immunodeficiency virus and acquired immune deficiency syndrome. *Philosophical Transactions: Biological Sciences*, 356, 795–798.

De Cock, K. M., 2012. The origins of AIDS. *Emerging Infectious Diseases*, 18(7), 1215–1215.

Deeks, S. G.*et al.*, 2016. International AIDS Society: Global scientific strategy towards an HIV cure 2016. *Nature medicine*, 22(8), 839–850.

Fisher, W. W. and Rigamonti, C. P., 2005. *The South Africa AIDS Controversy: A Case Study in Patent Law and Policy*. Cambridge, MA: Harvard Law School. Available at: https://cyber.harvard.edu/people/tfisher/South%20Africa.pdf.

Flint, A., 2011. *HIV/AIDS in Sub-Saharan Africa: Politics, Aid and Globalization.* New York: Palgrave Macmillan.

Ford, N., Calmy, A., and Mills, E. J., 2011. The first decade of antiretroviral therapy in Africa. *BMC Globalization and Health*, 7:33, doi:https://doi.org/10.1186/1744-8603-7-33. Available at: https://globalizationandhealth.biomedcentral.com/articles/10.1186/1744-8603-7-33.

Ford, N., Calmy, A., and von Schoen-Angerer, T., 2007. Treating HIV in the developing world: getting ahead of the drug development curve. *Drug Discovery Today*, 12(1–2), 1–3.

Ghosh, J. and Kalipeni, E., 2004. Rising tide of AIDS orphans in southern Africa. In E. Kalipeni, S. Craddock, J. Oppong, and J. Ghosh, eds. *HIV/AIDS in Africa: Beyond Epidemiology*. Malden, MA: Blackwell Publishers, 304–315.

Görgens-Albino, M., Mohammad, N., Blankhart, D., and Odutolu, O., 2007. *The Africa Multi-Country AIDS Program 2000–2006: Results of the World Bank's Response to a Development Crisis.*Washington, DC: World Bank. Available at: https://openknowledge.worldbank.org/handle/10986/6705.

Government of Uganda, 1992. *Uganda AIDS Commission Act, 1992*. Kampala, Uganda: Government of Uganda Press.

Grant, M. J. and Yeatman, S., 2012. The relationship between orphanhood and child fostering in sub-Saharan Africa, 1990s–2000s. *Population Studies*, 66(3), 279–295.

Hemelaar, J., 2012. The origin and diversity of the HIV-1 pandemic. *Trends in Molecular Medicine*, 18(3), 182–192.

Herbert, B., 2001. Refusing to save Africans: antiretrovirals are needed now. (anti-AIDS drugs) (Column). *The New York Times*, p. A19.

Iliffe, J., 2006. *The African AIDS epidemic: a history*. Athens: Ohio University Press.

Kalipeni, E., Craddock, S., Oppong, J. R., and Ghosh, J. 2004. *HIV and AIDS in Africa: Beyond Epidemiology*. Malden, MA: Blackwell Publishing.

Kalipeni, E. and Zulu, L., 2008. Using GIS to model and forecast HIV/AIDS rates in Africa, 1986–2010. *The Professional Geographer*, 60(1), 33–53.

Kalipeni, E. and Zulu, L., 2012. HIV/AIDS in Africa: a geographic analysis at multiple spatial scales. *GeoJournal*, 77(4), 505–524.

Kessing, C. F. *et al.*, 2017. InVivo suppression of HIV rebound by Didehydro-Cortistatin A, a "block-and-lock" strategy for HIV-1 treatment. *Cell Reports*, 21(3), 600–611.

Kinsman, J., 2010. *AIDS Policy in Uganda: Eviddence, Ideology, and the Making of an African Success Story*. New York: Palgrave Macmillan.

Knight, L., 2008. *The First Ten Years, 1996–2006*. Geneva: Joint United Nations Programme on HIV/AIDS (UNAIDS).

Kreiss, J. K. *et al.*, 1986. AIDS virus infection in Nairobi prostitutes: spread of the epidemic to East Africa. *New England Journal of Medicine*, 314(7), 414–418.

Magadi, M. A., 2011. Household and community HIV/AIDS status and child malnutrition in sub-Saharan Africa: Evidence from the demographic and health surveys. *Social Science and Medicine*, 73(3), 436–446.

Malan, R., 2001. AIDS in Africa. *Rolling Stone*, 882, 70.

MarlinkR. *et al.*, 1994. Reduced rate of disease development after HIV-2 infection as compared to HIV-1. *Science*, 265(5178),1587–1590.

Mathers, B. M. *et al.*, 2008. Global epidemiology of injecting drug use and HIV among people who inject drugs: a systematic review. *Lancet*, 372(9651), 1733–1745.

McHenry, M. S. *et al.*, 2017. HIV stigma: perspectives from Kenyan child caregivers and adolescents living with HIV. *Journal of the International Association of Providers of AIDS Care*, 16(3), 215–225.

McNeil, J., 2017. A history of official government HIV/AIDS policy in South Africa. HIV/AIDS in South Africa. www.sahistory.org.za: South African History Online. Available at: http://www.sahistory.org.za/topic/history-official-government-hivaids-policy-south-africa (Accessed: September 18, 2017).

Mills, E. J. *et al.*, 2006. Adherence to antiretroviral therapy in sub-Saharan Africa and North America: a meta-analysis. *JAMA*, 296(6), 679–90.

Morison, L. *et al.*, 2001. Commercial sex and the spread of HIV in four cities in sub-Saharan Africa. *AIDS*, 15Suppl 4, S61–69.

Morris, M., Epstein, H., and Wawer, M., 1995. Timing is everything: international variations in historical sexual partnership concurrency and HIV prevalence. *PLoS ONE*, 5(11), e14092.

Morris, M. and Kretzschmar, M., 1997. Concurrent partnerships and the spread of HIV. *AIDS*, 11(5), 641–648.

Mugurungi, O., Gregson, S., McNaghten, A., Dube, S., and Grassly, N., 2007. HIV in Zimbabwe 1985–2003: measurement, trends and impact. In Michel Caraël and Judith R.Glynn, eds. *HIV, Resurgent Infections and Population Change in Africa*. Dordrecht: Springer.

Mulkeen, A., 2007. *Recruiting, Retaining, and Retraining Secondary School Teachers and Principals in Sub-Saharan Africa.* Washington, DC: World Bank, Africa Regional Office.

Pépin, J., 2011. *The origins of AIDS*. Cambridge: Cambridge University Press.

Pickett, C., 2011. Encyclopedia of Africa. *Reference Reviews*, 25(1), 8.

Pisani, E., Schwartländer, B., Cherney, S., and Winter, A., 2000. *Report on the Global AIDS Epidemic*. Geneva: UNAIDS.

Rassool, G. H., 2004. Current issues and forthcoming events. 46(5), 567–570.

Requejo, H. I. Z., 2006. Worldwide molecular epidemiology of HIV. *Revista de Saúde Pública*, 40, 331–345.

Roser, M., 2018. Our world in data: HIV/AIDS. Available at: https://ourworldindata.org/hiv-aids.

Santos, M. M. D., Kruger, P., Mellors, S. E., Wolvaardt, G., and Ryst, E. V. D., 2014. An exploratory survey measuring stigma and discrimination experienced by people living with HIV/AIDS in South Africa: the people living with HIV stigma index. *BMC Public Health*, 14(1), 80.

ScienceNews, 2016. Significant progress against HIV epidemic in Africa. Available at: https://www.sciencedaily.com/releases/2016/12/161201120503.htm (Accessed: September 18, 2017).

Senthilingam, M., 2016a. HIV cure study provides insight into 2008 case. *CNN*. Available at: http://www.natap.org/2016/IAC/IAC_04.htm.

Senthilingam, M., 2016b. What will it take to end HIV? London: London School of Tropical Medicine. Available at: https://www.lshtm.ac.uk/research/research-action/features/what-will-it-take-end-hiv (Accessed: January 12, 2018).

Serwadda, D. *et al.*, 1985. Slim disease: a new disease in Uganda and its association with HTLV-III infection. *The Lancet*, 326(8460), 849–852.

Task Force on Kaposi's Sarcoma and Opportunistic Infections, 1982. A cluster of Kaposi's sarcoma and pneumocystis carinii pneumonia among homosexual male residents of Los Angeles and Orange Counties, California. *Morbidity and Mortality Weekly Report (MMWR-CDC)*, 31(23), 305–307.

Timaeus, I. M., 1998. Impact of the HIV epidemic on mortality in sub-Saharan Africa: evidence from national surveys and censuses. *AIDS*, 12Suppl 1, S15–S27.

Topouzis, D., 2004. Moving therapy to frontline of AIDS war. Available at: http://www.un.org/africarenewal/magazine/april-2004/moving-therapy-frontline-aids-war (Accessed: October 19, 2017).

UNAIDS, 1998. *Report on the Global HIV/AIDS Epidemic – June 1998.* Geneva: Joint United Nations Programme on HIV/AIDS. Available at: http://data.unaids.org/pub/report/1998/19981125_global_epidemic_report_en.pdf.

UNAIDS, 2010. *UNAIDS Report on the Global AIDS Epidemic 2010.* Geneva: Joint United Nations Programme on HIV/AIDS. Available at http://www.unaids.org/globalreport/documents/20101123_GlobalReport_full_en.pdf.

UNAIDS, 2011. *World AIDS Day Report 2011: How to Get to Zero Faster, Smarter, Better.* Geneva: Joint United Nations Programme on HIV/AIDS. Available at: http://www.unaids.org/sites/default/files/media_asset/JC2216_WorldAIDSday_report_2011_en_1.pdf.

UNAIDS, 2012. *2012 UNAIDS Report on the Global AIDS Epidemic.* Geneva: UNAIDS. Joint United Nations Programme on HIV/AIDS. Available at: http://www.unaids.org/en/resources/documents/2012/20121120_UNAIDS_Global_Report_2012.

UNAIDS, 2015. *On the Fast-Track to End AIDS by 2030: Focus on Location and Population.* Geneva: Joint United Nations Programme on HIV/AIDS. Available at: http://www.unaids.org/sites/default/files/media_asset/WAD2015_report_en_part01.pdf.

UNAIDS, 2016a. *Global AIDS Update, 2016.* Geneva: Joint United Nations Programme on HIV/AIDS. Available at: http://www.unaids.org/sites/default/files/media_asset/global-AIDS-update-2016_en.pdf.

UNAIDS, 2016b. *Oral Pre-Exposure Prophylaxis Questions and Answers.* Geneva: Joint United Nations Programme on HIV/AIDS. Available at: http://www.unaids.org/sites/default/files/media_asset/UNAIDS_JC2765_en.pdf.

UNAIDS, 2017a. *Update: South Africa Launches New Plan to Advance Progress Towards Ending AIDS.* Available at: http://www.unaids.org/en/resources/presscentre/featurestories/2017/april/20170403_south-africa-NSP.

UNAIDS, 2017b. *UNAIDS Data.* Available at: http://www.unaids.org/en/topic/data.

U.S. Department of State, 2017. *PEPFAR Strategy for Accelerating HIV/AIDS Epidemic Control (2017–2020).*Washington, DC: U.S. Department of State.

U.S. Food and Drug Administration, 2017. FDA-approved HIV medicines. Available at: https://aidsinfo.nih.gov/understanding-hiv-aids/fact-sheets/21/58/fda-approved-hiv-medicines (Accessed: January 12, 2018).

von Schoen Angerer, T., Wilson, D., Ford, N., and Kasper, T., 2001. Access and activism: the ethics of providing antiretroviral therapy in developing countries', *AIDS*, 15Suppl 5, S81–90.

West, G. R., Corneli, A. L., Best, K., Kurkjian, K. M., and Cates, W., 2007. Focusing HIV prevention on those most likely to transmit the virus. *AIDS Education and Prevention*, 19(4), 275–288.

Windisch, R., Waiswa, P., Neuhann, F., Scheibe, F., and de Savigny, D., 2011. Scaling up antiretroviral therapy in Uganda: using supply chain management to appraise health systems strengthening', *Globalization and Health*, 7:25, DOI:https://doi.org/10.1186/1744-8603-7-25. Available at: https://globalizationandhealth.biomedcentral.com/articles/10.1186/1744-8603-7-25.

World Bank, 2008. *The World Bank's Commitment to HIV/AIDS in Africa: Our Agenda for Action, 2007–2011.*Washington, DC: The World Bank.

World Bank, 2018a. Health expenditure, total (% of GDP). Washington, DC: The World Bank. Available at: https://data.worldbank.org/indicator/SH.XPD.TOTL.ZS (Accessed: January 11, 2018).

World Bank, 2018b. Prevalence of HIV, total (% of Population Ages 15–49).Washington, DC: The World Bank (Accessed: January 11, 2018).

World Health Organization, 1999. *The World Health Report 1999 – Making a Difference.* Geneva, Switzerland. Available at: http://www.who.int/whr/1999/en/.

World Health Organization, 2017. Global Health Observatory data repository: Number of people (all ages) living with HIV Estimates by WHO region. Geneva: World Health Organization. Available at: http://apps.who.int/gho/data/view.main.22100WHO.

5 Progress towards combatting malaria in Africa

Sarah R. Blackstone, Ucheoma Nwaozuru, and Juliet Iwelunmor

Introduction

Malaria in sub-Saharan Africa

In 2000, the United Nations developed eight Millennium Development Goals to improve health and wellbeing worldwide. Millennium Development Goal 6, "combat HIV/AIDS, malaria, and other diseases," is crucial for many countries worldwide, especially in sub-Saharan Africa which faces a unique dual burden of communicable and non-communicable diseases. Malaria, in particular, has plagued sub-Saharan Africa for decades, a region which carries a disproportionately high global burden of the disease. In 2015, sub-Saharan Africa bore 88% of all malaria cases and 90% of all malaria-related deaths. In areas with high malaria transmission, children under the age of five years old are the most vulnerable to infection, illness, and death, with 70% of malaria deaths occurring in this age group (World Health Organization 2016a). As a result of the substantial expansion of interventions globally, there has been a 58% decline in malaria deaths between 2000 and 2015 (World Health Organization 2016a). In sub-Saharan Africa, as well, the burden of malaria has declined as treatment and diagnosis initiatives have been scaled up. In this 15-year time period, over 6.2 million malaria deaths have been prevented, primarily among children in sub-Saharan Africa. Furthermore, the global incidence of malaria has fallen by 37% (United Nations 2015).

Malaria was first realised to have detrimental social and economic implications for regional development in the early 1900s when an emphasis on combating malaria was brought to light. In a review of studies of malaria over the last century, Snow, Trape, and Marsh (2001) grouped malaria research into three time periods: pre-1960s, 1960–1989, and the 1990s. Prior to 1960, malaria mortality in sub-Saharan Africa was approximately 9.5 per 1,000 children under five years of age. This mortality rate declined by 18% between 1960 and 1989, with mortality rates falling to 7.8 per 1,000 (Snow *et al.* 2001). While gains were made in combating malaria morbidity and mortality up through the 1980s, mortality began to increase again. By 1995, malaria mortality increased to 10.2 per 1,000. In contrast, all-cause mortality rates steadily declined during each of those time periods, leading to an increase in the proportion of death caused by malaria during those time periods: 18% pre-1960, 12% 1960–1989, and 30% in the 1990s.

Though malaria outcomes have improved since the launch of the Millennium Development Goals in 2000, the waxing and waning trends in malaria morbidity and mortality over the last several decades are concerning as we move to address the Sustainable Development Goals and implement solutions that can improve health in a sustainable way long-term. Sustainable Development Goal 3 aims to "ensure healthy lives and promote wellbeing for all at all ages" (United Nations 2015). Part of that goal is to end the epidemics of AIDS, tuberculosis, malaria, and neglected tropical diseases and combat hepatitis, water-borne diseases, and other communicable diseases… [and] end preventable deaths of newborns and children under five years of age, with all countries aiming to reduce neonatal mortality to at least as low as 12 per 1,000 live births and under-five mortality to at least as low as 25 per 1,000 live births [by 2030] (United Nations 2015).

Working within areas with high malaria transmission to understand the available resources, current practices, and feasible methods of disease control are needed as we move forward to ensure the success of Sustainable Development Goal 3. For malaria, in particular, an important indicator is care-seeking practices or how people recognise malaria symptoms, perceive these symptoms, and make decisions about seeking healthcare. Understanding care-seeking practices can provide insight for malaria education, care practices, and control strategies.

Care-seeking practices for childhood malaria

Currently, in sub-Saharan Africa, presumptive diagnosis by a physician based on clinical symptoms (e.g., febrile state) has been the protocol for malaria case management (Rafael *et al.* 2006). Though studies have shown that rapid diagnostic testing procedures are efficacious (Allen *et al.* 2011) and more accurate in the diagnosis of malaria and preventing over-diagnosis (Iwelunmor *et al.* 2013), many clinics have limited, if any, access to the materials needed for these types of tests. Sustaining interventions utilising these diagnostic procedures is difficult due to limited funding. Despite discordance between physician diagnosis and rapid diagnostic tests, presumptive diagnosis by a physician in a clinical setting is often the best option for many cases of malaria. Unfortunately, one of the barriers to effective malaria case management is the failure to bring a child to a clinic due to self-diagnosis at home. A better understanding of factors involved in at-home diagnoses can help improve malaria detection education and increase the number of children brought to see healthcare providers, thus improving diagnosis and treatment rates and potentially reducing the burden of disease.

There have been a number of studies on care-seeking for malaria in Africa using both quantitative and qualitative methodologies. A review of many of these studies from the past several years noted several common themes emerging from the various studies, including the importance of local community and/or folk perceptions, terminology, and explanations of illnesses with similar symptoms to malaria that distinguished fever and convulsions as distinct aetiologies that required immediate treatment (de Savigny *et al.* 2004). Many of these studies reported that care-seeking for simple fevers and otherwise uncomplicated cases of malaria were typically managed at home while more severe cases (i.e., convulsions) were more likely brought to health practitioners. It was common for caretakers to seek care multiple times and switch between types of providers. Uncomplicated cases were most typically brought to formal, modern medical centres. However, complicated cases involving convulsions were more often brought to traditional healthcare facilities or treated with traditional practices, with few cases being brought to modern care facilities. This hierarchy of events is likely to influence the timing of treatment, and ultimately outcomes. Thus, if complicated cases are first brought to traditional healers as opposed to biomedical health facilities, this delays the point of entry to antimalarial drugs and modern treatment to be prescribed. Because of the importance of rapidly accessing antimalarial drugs and the common use of modern health facilities, at least in uncomplicated cases, understanding factors involving in-home detection of malaria and the decision to consult with health professionals can assist us as we move forward to reduce the burden of malaria and improve childhood health and wellbeing.

Situating malaria care-seeking practices with the framework for epidemiologic transition

This chapter uses the revised demographic, epidemiologic, and nutrition transition model approach as a theoretical perspective to guide the discussion of childhood malaria and care-seeking practices. The premise of this framework recognises that scientists have begun to understand only recently that the demographic and epidemiological transitions occurring in developed countries are also occurring in developing, lower-income countries. Several major changes have emerged that contribute to the restructuring of nations' health including ageing populations, rapid urbanisation, the epidemiologic transition, and economic changes – all of which affect populations in varying and often disparate ways (Popkin 1994). The revised demographic, epidemiologic, and nutrition transition model offers a broader theoretical framework in which to situate the issue of childhood malaria transmission and care-seeking practices, as it conceptualises how populations' health changes over time, both prospectively and retrospectively, while focusing on how the interplay of environmental, socioeconomic, and other factors influence health. Particularly now, as the health field is at the end of the Millennium Development Goals and just beginning initiatives to promote the Sustainable Development Goals, understanding both retrospectively and prospectively how a multitude of individual, sociocultural, and structural factors influence malaria can help with understanding the shortcomings of the Millennium Development Goals in addressing malaria over the last several years, and how the Sustainable Development Goals can work to address these moving forward. The framework focuses on three large overarching topics: demographic transition, epidemiological transition, and nutrition transition. Specifically, the issues discussed in this chapter relate most to the epidemiological transition, with special attention to family planning for control of infectious disease. Thus, according to the framework, a high prevalence of infectious disease coupled with poor environmental conditions impacts families' abilities to properly control infectious diseases. These environmental conditions and epidemiological states of different regions can vary over time. The factors affecting environmental conditions and epidemiological states can vary as well (Popkin 2002). Paramount for controlling childhood malaria within families is caretakers' responses to and perceptions of childhood malaria that will lead them to seek treatment from physicians.

Previous studies have reported the importance of individual, sociocultural, and structural factors in predicting malaria treatment-seeking patterns in mothers, who comprise the majority of caretakers for young children (Glik *et al.* 1989). Individual factors are those which increase readiness to take action, including maternal age, maternal occupation, child's age and gender, and presence of items such as a radio in the household. Sociocultural factors refer to shared knowledge based on both traditional and modern healthcare systems that are linked to the recognition of symptoms of illness and prescribed courses of action. Sociocultural factors include perceptions of illness severity, familiarity with medications, ethnicity, and so on (Glik *et al.* 1989). Structural factors are those features of the healthcare system (i.e., distance to centre, affordability, and availability of medications) which enable or discourage use of recommended treatment (Glik *et al.* 1989). Over the past two decades since the study by Glik *et al.* (1989), a considerable amount of information has become available on the individual, sociocultural, and structural factors that influence child malaria incidence and treatment-seeking practices in endemic countries. For example, in Ethiopia, Peterson *et al.* (2009) found that age (specifically children in the age group 5–9) was statically associated with malaria incidence. A study in Ghana by Krefis *et al.* (2010) reported that sociodemographic factors such as ethnic group, parent's education and occupation, use of protective measures, and living standard of the family are suggested to be important factors contributing to malaria. Some religious denominations in Nigeria prohibit biomedical treatment for any illness; as such individuals that are affiliated with this religion do not seek medical help for any illness (Bedford and Sharkey 2014). Further, several misconceptions about malaria still exist, despite educational materials about the preventive measures that should be taken. Often, families' perceptions, beliefs, and attitudes are overlooked in efforts to control and treat malaria (Singh *et al.* 2014). Research suggests a need for targeted educational materials designed to inform communities about the importance of early diagnosis (Deressa *et al.* 2007, Singh *et al.* 2014). Though research has found that prompt diagnosis of malaria can assist with effective management of the disease while reducing mortality rates (Talisuna and Meya 2007), there are many obstacles preventing this in clinical settings. Among these barriers include self-diagnosis of children from family members, household incomes, perceptions and previous experience with health services (Amexo *et al.* 2004, Bedford and Sharkey 2014), access to effective treatment, and limited access to effective diagnostic tools and poor laboratory infrastructure leading to mis- and over-diagnosis (Rafael *et al.* 2006, Uzochukwu *et al.* 2009, Carneiro *et al.* 2010). Structural factors, such as the presence of a healthcare facility in the community, were predictive of treatment-seeking in Nigeria (Oresanya *et al.* 2008). However, it was found in Kenya that medicine sellers operating in drug shops and kiosks are generally closer to homes (87% of rural households live within 1km of a shop) than formal healthcare centres (Goodman *et al.* 2007). As a result, they form an important alternative supply of antimalarial drugs for patients despite concerns surrounding the appropriateness of the drugs they provide and that they may hinder treatment-seeking behaviours. These various factors coalesce to influence childhood malaria caretaking practices and ultimately the health and wellbeing of children with malaria.

Case study of Nigeria

Of the many countries in sub-Sahara Africa that are impacted by malaria, Nigeria has more reported cases of malaria and deaths due to malaria than any other country in the world (Centers for Disease Control and Prevention 2012). In Nigeria, an estimated 150 million people are at risk of infection with malaria parasites. The disease results in an estimated 20% of all hospital admissions, 30% of outpatients, and 10% of hospital deaths (Isiguzo *et al.* 2014). Malaria is also consistently recorded as one of the leading causes of mortality in children under five years old. Prompt and accurate diagnosis of childhood malaria has been emphasised throughout the literature as an important factor in reducing mortality in children, yet there are many barriers to prompt and accurate diagnosis of the disease in children. In continuing to meet the goals of Millennium Development Goal 6 and moving forward towards Sustainable Development Goal 3 of eradicating malaria by 2030, particularly among children, it is necessary to identify all factors including individual (maternal or physician characteristics), sociocultural (perceptions and beliefs), structural (availability or affordability of treatment regimens), or combined clinical factors (physician's diagnosis and the provision of malaria rapid diagnosis test) that are essential to malaria reduction among children in endemic countries. Moreover, a firm understanding of the impact of these factors on diagnosis and treatment practices would in all likelihood substantially contribute to improved case management of malaria among children.

In order to improve malaria diagnosis and treatment, it is necessary to develop a better understanding of factors that influence whether malaria treatment is sought. Particularly as care-seeking practices can influence the timing of treatment for malaria, the type of treatment (e.g., biomedical or traditional), and ultimately outcomes of treatment, understanding how, why, and when healthcare for malaria is sought out is necessary. Particularly, as children under five years old are most vulnerable to the negative consequences of malaria and malaria infection worldwide and in Nigeria, and as mothers are typically the primary caretaker of children in this age group, examining mothers' perceptions of malaria symptoms and when seeking treatment is appropriate can offer valuable insight to improve malaria outcomes in children. The purpose of the described case study was to investigate individual, sociocultural, and structural factors that are associated with treatment-seeking behaviours among mothers with febrile children in Nigeria. Additionally, we examined mothers' perceptions of symptoms of malaria in their children and whether they believed their child to have malaria in order to develop a more comprehensive understanding of case management for malaria in Nigeria. As the incorrect at-home diagnosis is one of the barriers to prompt and effective treatment of malaria in children, a better understanding of associated factors can assist with future education programmes regarding malaria symptoms and treatment-seeking.

Research framework

The theoretical framework for this case study is situated within the seminal work of the World Health Organization in describing social determinants of health, or "the conditions in which people are born, grow, work, live, and age, and the wider set of forces and systems shaping the conditions of daily life" (World Health Organization 2016b). These forces and systems include economic policies and systems, development agendas, social norms, social policies, and political systems (World Health Organization 2016b). These forces and systems are grouped into three categories of determinants: socioeconomic and political, structural, and intermediary (World Health Organization 2015). Socioeconomic and political determinants refer to the governance structures within different societies as well as social and public policies (e.g., labour, housing market, education, and social protection). Structural factors are mechanisms that generate stratification and social class divisions in the society as well as factors that define individual socioeconomic positions within hierarchies of power, prestige, and access to resources. Finally, intermediary factors are means through which structural mechanisms or determinants operate, such as age, religion, and marital status (World Health Organization 2015). With regard to malaria, socioeconomic factors include the state of the health system and the availability of health resources (de Savigny *et al.* 2004). As nongovernmental health providers are still common, particularly in rural regions, there can be difficulty regulating the quality and cost of care. Furthermore, the availability of antimalarial medications and the affordability of these impact malaria outcomes (de Savigny *et al.* 2004). Structural determinants related to malaria prevention and care-seeking include household economic status. In some cases, as much as 75% of malaria costs are borne by the household. This burden is greatest for the poorest households, which contributes to a continued cycle of poverty (de Savigny *et al.* 2004). Additionally, education and work status play a role as more highly educated families and families with steadier occupational status have better outcomes related to malaria (Taffa and Chepngeno 2005). Finally, intermediary determinants affecting health outcomes can include age, sex, ethnicity, and religion. Of utmost importance is caretakers' perception of the severity of illness and recognition of illness symptoms (Taffa and Chepngeno 2005). Health beliefs are argued to be as important as recognition of symptoms; in Ghana, Hill *et al.* (2003) found that only half of illnesses perceived as "severe" were taken to the hospital. Furthermore, some caretakers may play the "waiting game" to see if illness ameliorates on its own, especially if the cost of treatment is a burden (Taffa and Chepngeno 2005). This can delay access to life-saving treatment, especially in the case of malaria where prompt and accurate diagnosis is crucial for increasing the likelihood of successful treatment. These factors coalesce to influence health in a variety of ways as depicted in the demographic and epidemiological transitions. In this case study, we examined how these factors influenced care-seeking practices and case management for malaria among mothers in Nigeria with febrile children.

Methodology

Fieldwork was focused specifically in Lagos, Nigeria. Nigeria is the most populous country in Africa with an estimated population of 188 million. Nigeria is divided into 36 states and federal territories which are grouped into six geopolitical regions: North Central, North East, North West, South East, South, and South West. Hausa, Igbo, and Yoruba are the three major ethnic groups and languages spoken in the country. Lagos, where fieldwork was conducted, is the most populous state in Nigeria with over 17.5 million people and houses a large range of diverse sociocultural backgrounds (Afolabi *et al.* 2004, Lagos Bureau of Statistics 2015). Lagos is located in the rainforest and has two climactic seasons, the dry season and the wet season. The dry season occurs from November to March, while the rainy season lasts from April to October. The largest rainfalls occur from May to July. During this time, malaria transmission is particularly high. As a result, this location was ideal for understanding the multiple factors that influence child malaria diagnosis and treatment practices.

Traditional beliefs and customs are paramount in understanding Nigerian cultural and ethnic identities (Iwelunmor 2011). Beliefs in Nigeria can be categorised as social beliefs and religious beliefs (Okafor and Emeka 1998). Social beliefs refer to individuals' perspectives on human relationships, while religious beliefs focus on perspectives of supernatural forces and their interplay with human spirituality. Culture in Nigeria is expressed in a variety of ways such as dance, religion, rituals, ceremonies, art, food, and many others. Additionally, with over 444 languages spoken in Nigeria, language is described as one of the most important mechanisms of cultural expression (Okafor and Emeka 1998). Due to the rich and diverse culture in Nigeria, and the important role of culture in everyday life, it is possible that individual, sociocultural, and contextual factors may influence malaria case management in a variety of ways. Because Nigeria faces an extremely high burden of malaria, with one-quarter of all malaria cases in the World Health Organization African region occurring in Nigeria and contributing to 30% of child deaths, it is important to examine how these different factors influence effective diagnosis and treatment of malaria in children.

This study took place at a paediatric outpatient clinic, located in Ikeja, the capital of Lagos, a region endemic with malaria, in June and July 2010 during the rainy season in which malaria transmission peaks. The clinic was visited three to four times per week during the rainy season. The clinic was also chosen because it provides free services to everyone, and additional free testing and antimalarial medication (when available) to children under five years old (Iwelunmor *et al.* 2013). At the clinic, the first points of contact for caretakers were the nurses, who were in charge of leading patients to the clinic waiting room. The clinic was created with the purpose of providing high-quality and equitable healthcare services. In order to facilitate prompt diagnosis of malaria, healthcare services for children are free (including antimalarial medication for those diagnosed).

Structured interviews were conducted with mothers of febrile infants attending the paediatric clinic. All mothers were at least 18 years of age with children ages 2–12 years with reported febrile symptoms. Participants were informed of the study and its purposes in the waiting room of the clinic. Those who provided verbal and written consent were recruited to participate. Children were examined by physicians prior to being enrolled in the study. Interviews were conducted with 123 mothers and 135 children were enrolled in the study. Interviews were conducted in the consultation room at the clinic after each mother consulted with a healthcare provider about their febrile child. The in-depth interviews focused on perceptions of malaria, and treatment-seeking behaviours prior to bringing their child to the clinic. Mothers were asked to describe what events prompted them to bring their child to the clinic.

Structured interviews contained questions assessing individual (e.g., age, education), sociocultural (e.g., ethnicity, mothers' perceptions), and structural factors (e.g., access to medication, quality of healthcare). Sociocultural factors, or perceptions, included mothers' perceptions of children's symptoms, such as vomiting, diarrhoea, fever, coughing, runny nose, teething, or another condition. Mothers also reported whether they believed their child to have malaria or not. Physician diagnosis of these same symptoms was also recorded, as well as their diagnosis of malaria. The researchers examined whether there were sociocultural and demographic differences between mothers who correctly and incorrectly diagnosed their child with malaria when compared to a physician diagnosis. The researchers also examined the relationship between structural factors on mothers' malaria diagnosis and the accuracy of the diagnosis. We used this information to develop a more comprehensive understanding of what case management for malaria is in this context.

Results

Effects of maternal characteristics on diagnosis accuracy

Overall, we did not find sociocultural and demographic factors to influence mothers' accuracy of diagnosis. Structural factors, such as source of family water, distance to clinics, and quality of healthcare did not impact diagnosis accuracy either. We did find that mothers were more likely to diagnose their children with malaria if one of the reported symptoms was vomiting; however, the majority of these women did not correctly diagnose their child with malaria when compared to physician diagnosis. Mothers of children with diarrhoea listed among their symptoms were more likely to correctly presume that their child had malaria when compared to physician diagnosis. Additionally, mothers of children under age five were more likely to perceive that their child had malaria, but the accuracy of the diagnosis did not differ between mothers of children under age five versus mothers of children over age five.

Physician diagnosis

119 of 135 children (88%) were diagnosed as having malaria by a physician. Sixteen were diagnosed with another illness or with no illness. Only 35 out of 119 children (29%) diagnosed with malaria by a physician were correctly diagnosed by their mothers. There were no significant correlations between mother and physician diagnosis, which is consistent with the proportion of mothers correctly diagnosing their child with malaria. Age of the child was not significantly associated with physician diagnosis.

Case management of malaria in Nigeria

The purpose of this investigation was to explore multiple contextual factors that could influence malaria diagnosis to contribute to a better understanding of case management in Nigeria. Case management is defined as 1) access to accurate and reliable diagnosis of malaria based on parasitological confirmation of malaria parasites using microscopy or malaria rapid diagnostic tests (RDTs), and 2) prompt treatment of uncomplicated malaria at healthcare clinics using artemisinin combination therapies (World Health Organization 2015). Part of proper optimal case management for malaria is prompt detection, diagnosis, and treatment, which is dependent on appropriate symptom recognition. As we saw in this case study, the majority of mothers were incorrect in their perceptions of whether their child had malaria or not, as only 29% correctly diagnosed their child with malaria. Though many studies have found individual, sociocultural, and structural factors influence treatment-seeking practices and case management (Iwelunmor 2011), the relationships between individual and structural factors in this study were not robust; however we did find that sociocultural factors including perceptions of illness did play a role, as mothers with children experiencing vomiting and diarrhoea perceived the illness to be more severe. Despite this finding, the interplay between the three types of factors is still important to explore in order to improve malaria outcomes in the future. Indeed, emphasis on one determinant may be insufficient if efforts are not made to equally address the role they all play in influencing effective diagnosis and treatment of child malaria. Moreover, factors are inextricably linked with decision-making processes surrounding malaria management and control not only at the household level but also among physicians in healthcare settings who are also influenced by these determinants (Williams and Jones 2004). For example, it remains unclear which structural (or contextual) level determinants (such as free healthcare services or antimalarial drugs versus laboratory testing) play an active role in reducing malaria burden in sub-Saharan Africa. These questions have important implications for improving effective case management of child malaria management particularly among the poor who are often deterred from seeking care at most clinics (Iwelunmor 2011).

Moreover, there have been many advances in malaria research that hold promise for improving malaria control. For instance, there are continuing efforts to develop new medications, create new vaccines, increase distribution of insecticide-treated bed nets, and improve diagnostic efforts through greater distribution of rapid diagnostic tests (Crawley *et al.* 2010). While the increasing opportunities for malaria control are growing, the success of these strategies depends on understanding the contexts in which human behaviours occur, and carefully examining individual, sociocultural, and structural contexts. Thus, while the outcomes of this case primarily highlight the importance of perceptions of illness and illness severity, other contextual factors related to individual, groups, and healthcare systems cannot be ignored, as malaria occurs in a complex social, cultural, and behavioural context.

Discussion: Malaria in social context

Nigeria is considered by the World Health Organization to be a region of high malarial transmission, meaning there is more than one case per 1,000 population members. Despite the implementation of preventive measures such as insecticide-treated bed nets, indoor residual spraying, larval control, intermittent preventive treatment during pregnancy, and availability of free diagnostic tests, Nigeria is far from eradicating the burden of malaria. Surveillance policies and interventions are seriously lacking, leading to insufficient data to assess countrywide trends (World Health Organization 2014). The combined lack of adequate surveillance and community education not only can lead to misdiagnosis but also to under-diagnosis and underreporting of malaria in communities. As of 2015, despite the fact that progress was made in reducing cases of malaria and deaths due to malaria, the disease still remains a problem in many sub-Saharan African countries including Nigeria. In order for the objective to end preventable deaths of children under age five, as listed in Sustainable Development Goal 3, to be reached, effective local surveillance and education programmes are needed to improve the accuracy of at-home and clinical diagnoses, accuracy of reported cases of malaria, and prompt referral to healthcare providers if malaria is suspected in children.

This particular case study in Nigeria examined factors that influence mothers' perceptions of malaria in their children and whether they are correlated with physicians' diagnoses. The result from the analysis shows that there are no correlations between the mothers' and doctors' diagnosis of malaria, which is supported by the fact that only 29% percent of mothers who brought their child to the clinic correctly perceived their child to have malaria. The participants reported several symptoms that they believed were correlated with malaria, including diarrhoea, fever, and vomiting. Of all the symptoms mentioned, there were significant associations between the mothers' perception of malaria and reported symptoms of vomiting. Mothers who reported symptoms of vomiting were more likely to perceive that their child had malaria than mothers who did not report vomiting as a symptom. In addition, mothers reporting symptoms of diarrhoea in their children were more likely to correctly perceive malaria in their children. This is consistent with a wealth of literature describing nausea, vomiting, and diarrhoea as symptoms of malaria (Centers for Disease Control and Prevention 2014). Based on these results, if mothers report diarrhoea as a symptom in the child, a diagnosis of malaria is likely, underscoring the need to consult a health professional within 24 hours as most malaria deaths occur within 24 hours of symptom onset. However, not all symptoms that are associated with malaria were indicative of malaria diagnosis in this sample thus supporting that malaria is best diagnosed by a healthcare professional. Mothers with children under five years old were more likely to perceive that their child had malaria, possibly because malaria is one of the leading causes of death in children under age five; however, age of their child was not associated with whether the mothers' diagnosis was correct.

Similar to Isiguzo *et al.* (2014), our results highlight the importance of educating the public to seek help from a health professional when they believe their child to have symptoms associated with malaria, as inaccurate at-home diagnosis is prevalent across various social and demographic groups. In the context of the Sustainable Development Goal 3's objective of ending preventable deaths of newborns and children under five years of age, these results have implications for the recommendations of prompt and accurate diagnosis by a health professional. Though there is literature reporting discordance between physician diagnosis and the results of more accurate RDTs (Iwelunmor *et al.* 2013), the reality in many clinics is that the physicians' diagnosis is the best available indicator of malaria. Many clinics do not have funds to sustain regular use of RDTs; while this method would provide more accurate testing, the best available mechanism of diagnosis is the physician's opinion. The results of this case study support the need for parents of children with suspected malaria, or symptoms associated with malaria to consult a physician promptly in order to increase the child's chances of survival. Previous studies cite many barriers to prompt consultation with physicians, one of which is the inaccuracy of at-home diagnoses (Amexo *et al.* 2004). Community programmes underscoring the importance of physician consultation regardless of caretaker perceptions could potentially weaken this barrier. Furthermore, comprehensive and regular education for caretakers regarding malaria symptoms and developing a standard protocol of what to do in case malaria is suspected or detected could additionally help overcome the barrier of inaccuracy of at-home diagnoses. Another barrier to seeking out treatment of healthcare professionals is the large span of medicine sellers and healers, which are typically closer to community homes than healthcare providers (Goodman *et al.* 2007). This can lead to seeking ineffective or potentially harmful treatment in lieu of going to a healthcare professional. Due to the importance of traditional healers in African culture, there is a need for educational materials and programmes to be developed, tested, and implemented using culturally competent strategies. Especially as some religious groups in Nigeria do not seek biomedical healthcare (Bedford and Sharkey 2014), models combining faith-based and biomedical treatment are needed. It is important for all social, cultural, and religious groups to understand the importance of prompt and accurate diagnosis, however different forms of explanatory models may need to be used as barriers may vary between different communities.

Of note in this particular case study is that, despite the fact that we initially predicted care-seeking practices and perceptions of illness to differ based on social determinants of health, we, in fact, did not see that. This suggests that care-seeking practices in this sample did not differ significantly across different educational, occupational, and economic statuses, which further supports the need for widespread educational programmes to improve recognition of malaria symptoms and to encourage immediate consultation with a health professional. As this sample was limited to 135 mothers and their children, we cannot make sweeping generalisations to sub-Saharan Africa, or even Nigeria, as a whole; however, this finding is interesting and merits further investigation as to whether universal education and protocols for caretakers would be effective.

Broadly within this case study, we found a large discrepancy between mothers' perception of malaria in their children and physician diagnosis, indicating the need for community outreach and educational programmes targeting localised factors associated with perceptions of illness. The primary symptoms that mothers feel indicate malaria in their child were vomiting and diarrhoea, which is consistent with a large body of literature (Centers for Disease Control and Prevention 2014). Overall, these results, taken with the larger body of literature and the progress made by initiatives like the Millennium Development Goals agenda, support the need for countrywide education regarding malaria symptoms and the importance of health professional consultation. Although the 15-year span in which the Millennium Development Goals guided health initiatives worldwide resulted in a 37% decrease in malaria mortality worldwide, we are still far from achieving the goal of eliminating malaria-induced mortality. Estimates from United Nations for cause-specific mortality between 1970 and 2010 do indicate a steady decline, particularly in childhood, in all-cause mortality rates (including malaria). While these results did suggest that some of the ambitious goals set as part of the Sustainable Development Goals could feasibly be reached by 2030, the authors noted that these mortality decreases were markedly less rapid, if existent, in countries where the effects of HIV or political disturbance predominated (Norheim *et al.* 2015). As sub-Saharan Africa continues to have one of the most serious HIV epidemics in the world, with 2013 estimates suggesting 24.7 million individuals living with HIV were in sub-Saharan Africa (71% of the global burden), the region faces additional barriers to not only malaria reduction and eradication, but also management of other diseases (World Health Organization 2013). Furthermore, many countries in sub-Saharan Africa are facing political instability, including the threats of Boko Haram in Nigeria, the threats from Islamic extremists in West and East Africa, unrest in South Sudan, Central African Republic, and the Democratic Republic of the Congo, all of which have devastating socioeconomic and health impacts. This instability and unrest also contribute to other factors necessary for promoting good health including food security, clean water, electricity, and shelter (Maxwell 1999). Thus, when moving forward to address malaria prevention and rapid diagnosis and treatment, is it crucial to understand the larger context in which it occurs. Targeting malaria will require developing education, detection, and control methods that are appropriate to communities in need with the understanding of the variety of other factors that coalesce to influence the health of the region. Especially as sub-Saharan Africa is still in a demographic and epidemiological transition, dealing with high fertility rates as well as a dual burden of communicable and non-communicable diseases (Dalal *et al.* 2011), communities and healthcare infrastructures are working to address a variety of diseases, among which malaria is included, with limited resources and a fragile health system in which to operate. Therefore, working directly with caretakers to prevent malaria or intervene in a timely manner to facilitate diagnosis by a physician and early treatment provides a feasible and sustainable option as we move forward to address Sustainable Development Goal 3 of ensuring healthy lives for all and ending epidemics of malaria in sub-Saharan Africa.

Conclusion

In order to reach the Sustainable Development Goals, individual, social, and cultural factors need to be accounted for in future interventions, beginning with caretakers of young children, as this group shows the highest mortality from malaria. As noted in the demographic, epidemiologic and nutrition transition model, while the interplay of these factors is important, their relationship and influence change over time. Thus, as we move forward, it is important to recognise the shortcomings of past initiatives to reduce malaria transmission and death resulting from malaria and consider the changes that regions have undergone as a result of the epidemiologic and demographic transition as we modify these initiatives and programmes. Of particular importance is the development of educational programmes for caretakers, which could help improve the accuracy of mothers' and caretakers' perceptions of malaria, and which may prompt them to quickly visit a healthcare professional. Educational programmes should also emphasise the importance of rapid diagnosis by a physician in order to improve health outcomes. Once prompt and accurate diagnoses are able to be made from both the mother and physician perspectives, proper treatment and, eventually, control of malaria will be more feasible, reducing the death rate.

References

Afolabi, B.M., Brieger, W.R., and Salako, L.A., 2004. Management of childhood febrile illness prior to clinic attendance in urban Nigeria. *Journal of Health, Population, and Nutrition*, 22(1), 46–51.

Allen, L.K., Hatfield, J.M., DeVetten, G., Ho, J.C., and Manyama, M., 2011. Reducing malaria misdiagnosis: the importance of correctly interpreting Paracheck Pf® "faint test bands" in a low transmission area of Tanzania. *BMC Infectious Diseases*, 11(1), 308.

Amexo, M., Tolhurst, R., Barnish, G, and Bates, I., 2004. Malaria misdiagnosis: effects on the poor and vulnerable. *The Lancet*, 364(9448), 1896–1898.

Bedford, K.J.A. and Sharkey, A.B., 2014. Local barriers and solutions to improve care-seeking for childhood pneumonia, diarrhoea and malaria in Kenya, Nigeria and Niger: A qualitative study. *PLOS ONE*, 9(6), e100038.

Carneiro, I. *et al.*, 2010. Age-patterns of malaria vary with severity, transmission intensity and seasonality in sub-Saharan Africa: A systematic review and pooled analysis. *PLOS ONE*, 5(2), e8988.

Centers for Disease Control and Prevention, 2012. *Global health – Nigeria: Malaria* [Online]. Available from: http://www.cdc.gov/globalhealth/countries/nigeria/what/malaria.htm [Accessed: 5 June 2015].

Centers for Disease Control and Prevention, 2014. *Malaria* [Online]. Available from: http://www.cdc.gov/malaria/diagnosis_treatment/index.html [Accessed: 5 June 2015].

Crawley, J., Chu, C., Mtove, G., and Nosten, F., 2010. Malaria in children. *The Lancet*, 375(9724), 1468–1481.

Dalal, S. *et al.*, 2011. Non-communicable diseases in sub-Saharan Africa: What we know now. *International Journal of Epidemiology*, 40(4), 885–901.

Deressa, W., Hailemariam, D., and Ali, A., 2007. Economic costs of epidemic malaria to households in rural Ethiopia: Economic costs of malaria in Ethiopia. *Tropical Medicine and International Health*, 12(10), 1148–1156.

de Savigny, D. *et al.*, 2004. Care-seeking patterns for fatal malaria in Tanzania. *Malaria Journal*, 3(1).

Glik, D.C., Ward, W.B., Gordon, A., and Haba, F., 1989. Malaria treatment practices among mothers in Guinea. *Journal of Health and Social Behavior*, 30(4), 421–435.

Goodman, C., Kachur, S.P., Abdulla, S., Bloland, P., and Mills, A., 2007. Drug shop regulation and malaria treatment in Tanzania – why do shops break the rules, and does it matter? *Health Policy and Planning*, 22(6), 393–403.

Hill, Z., Kendall, C., Arthur, P., Kirkwood, B., & Adjei, E. (2003). Recognizing childhood illnesses and their traditional explanations: Exploring options for care seeking interventions in the context of the IMCI strategy in rural Ghana. *Tropical Medicine & International Health*, 8(7), 668–676.

Isiguzo, C. *et al.*, 2014. Presumptive treatment of malaria from formal and informal drug vendors in Nigeria. *PLOS ONE*, 9(10), e110361.

Iwelunmor, J., 2011. *Examining child malaria diagnosis and treatment practices at an outpatient clinic in southwest Nigeria*. Thesis (PhD). The Pennsylvania State University.

Iwelunmor, J., Belue, R., Nwosa, I., Adedokun, A., and Airhihenbuwa, C.O., 2013. Case-management of malaria in children attending an outpatient clinic in southwest Nigeria. *International Quarterly of Community Health Education*, 34(3), 255–267.

Krefis, A. *et al.*, 2010. Principal component analysis of socioeconomic factors and their association with malaria in children from the Ashanti Region, Ghana. *Malaria Journal*, 9(1), 201.

Lagos Bureau of Statistics, 2015. *Digest of statistics 2015*. Lagos, Nigeria: Lagos State Government.

Maxwell, D., 1999. The political economy of urban food security in sub-Saharan Africa. *World Development*, 27(11), 1939–1953.

Norheim, O.F. *et al.*, 2015. Avoiding 40% of the premature deaths in each country, 2010–30: Review of national mortality trends to help quantify the UN Sustainable Development Goal for health. *The Lancet*, 385(9964), 239–252.

Okafor, R. and Emeka, L., 1998. *Nigerian peoples and culture*. 1st ed. Enugu, Nigeria: New Generation Ventures.

Oresanya, O.B., Hoshen, M., and Sofola, O.T., 2008. Utilization of insecticide-treated nets by under-five children in Nigeria: Assessing progress towards the Abuja targets. *Malaria Journal*, 7(1), 145.

Peterson, I., Borrell, L.N., El-Sadr, W., and Teklehaimanot, A., 2009. Individual and household level factors associated with malaria incidence in a highland region of Ethiopia: A multilevel analysis. *The American Journal of Tropical Medicine and Hygiene*, 80(1), 103–111.

Popkin, B.M., 1994. The nutrition transition in low-income countries: An emerging crisis. *Nutrition Reviews*, 52(9), 285–298.

Popkin, B.M., 2002. An overview on the nutrition transition and its health implications: The Bellagio meeting. *Public Health Nutrition*, 5(1A), 93–103.

Rafael, M.E., Taylor, T., Magill, A., Lim, Y.M., Girosi, F., and Allan, R., 2006. Reducing the burden of childhood malaria in Africa: The role of improved. *Nature*, 444, 39–48.

Singh, S., Musa, J., Singh, R., and Ebere, U.V., 2014. Knowledge, attitude and practices on malaria among the rural communities in Aliero, Northern Nigeria. *Journal of Family Medicine and Primary Care*, 3(1), 39.

Snow, R.W., Trape, J.F., and Marsh, K., 2001. The past, present and future of childhood malaria mortality in Africa. *Trends in Parasitology*, 17(12), 593–597.

Taffa, N. and Chepngeno, G., 2005. Determinants of healthcare seeking for childhood illnesses in Nairobi slums. *Tropical Medicine & International Health*, 10(3), 240–245.

Talisuna, A.O. and Meya, D.N., 2007. Diagnosis and treatment of malaria. *BMJ*, 334(7590), 375–376.

United Nations, 2015. *Sustainable Development Goals: 17 goals to transform our world* [Online]. Available from: http://www.un.org/sustainabledevelopment/health/ [Accessed: 5 June 2015].

Uzochukwu, B.S., Obikeze, E.N., Onwujekwe, O.E., Onoka, C.A., and Griffiths, U.K., 2009. Cost-effectiveness analysis of rapid diagnostic test, microscopy and syndromic approach in the diagnosis of malaria in Nigeria: Implications for scaling-up deployment of ACT. *Malaria Journal*, 8(1), 265.

Williams, H.A. and Jones, C.O., 2004. A critical review of behavioral issues related to malaria control in sub-Saharan Africa: What contributions have social scientists made? *Social Science and Medicine*, 59(3), 501–523.

World Health Organization, 2013. *HIV/AIDS* [Online]. Available from: http://www.who.int/gho/hiv/en/ [Accessed: 5 June 2015].

World Health Organization, 2014. *World malaria report: Nigeria* [Online]. Available from: http://www.who.int/malaria/publications/country-profiles/profile_nga_en.pdf [Accessed: 5 June 2015].

World Health Organization, 2015. *Guidelines for the treatment of Malaria* [Online]. Available from: http://apps.who.int/iris/bitstream/10665/162441/1/9789241549127_eng.pdf [Accessed: 5 June 2015].

World Health Organization, 2016a. *Malaria* [Online]. Available from: http://www.who.int/mediacentre/factsheets/fs094/en/ [Accessed: 5 June 2015].

World Health Organization, 2016b. *Social determinants of health* [Online]. Available from: http://www.who.int/social_determinants/en/ [Accessed: 5 June 2015].

6 The impact of land use and land cover change on the spatial distribution of Buruli ulcer in southwest Ghana

Joseph R. Oppong and Warangkana Ruckthongsook

Introduction

Buruli ulcer (BU) is a neglected tropical disease that occurs in hot and humid climates and mostly affects people living in poor rural communities and poor hygiene environments (World Health Organization 2014, Johnson *et al.* 2015). It is a skin disease caused by *Mycobacterium ulcerans* (*M. ulcerans*), and the third most common disease of the mycobacterium family after tuberculosis and leprosy (World Health Organization 2007). Due to the low awareness of BU, it is difficult to diagnose and treat, and many people fail to recognise the symptoms in the early stage when treatment can be done without hospitalisation (World Health Organization 2007). Without medical treatment, *M. ulcerans* spreads to other organs through the lymphatic system or bloodstream in approximately three months (Debacker *et al.* 2004). Surgery is the only way to treat BU at that stage. Although the infected area is removed, the deformities and scars are permanent. Over 50% of BU patients suffer from functional limitations, social stigmatisation, and loss of livelihood after treatment (van der Werf *et al.* 2005). Despite considerable research effort, modes of transmission and hosts of the disease remain unknown (World Health Organization 2014).

BU is considered a non-communicable disease (Portaels *et al.* 2008) mostly attributed to human activity associated with water bodies, man-made environmental change, and degradation including excessive flooding during heavy rainfall, the damming of streams and rivers to create artificial lakes and wetlands, wetland modification, and deforestation practices that increase flooding (Merritt *et al.* 2005, Bratschi *et al.* 2013). Moreover, recent cross-sectional studies indicate that some environmental factors including stagnant water (Duker *et al.* 2006b, Pouillot *et al.* 2007, Brou *et al.* 2008,Wagner *et al.* 2008), wood/forest land cover (Brou *et al.* 2008), land elevation (Walsh *et al.* 2008, Wagner *et al.* 2008), seasonal variation (Duker *et al.* 2006b, Portaels *et al.* 2009), agriculture (Pouillot *et al.* 2007, Brou *et al.* 2008), and economic factors (Duker *et al.* 2006a) may be associated with BU. However, pinpointing the exact environmental factor in exposure has been a difficult challenge, because most patients are unable to recall where or when they possibly were exposed to *M. ulcerans* (van der Werf *et al.* 2005) since most BU infections are detected after a long period of time. Furthermore, due to severe underreporting, missed BU cases may indicate other risk factors or associations of BU in humans that have not yet been discovered.

In response to the alarming reports of widespread BU prevalence, the World Health Organization (WHO) established the Global Buruli Ulcer Initiative in 1998 to coordinate research efforts into diagnosis and prevention for this burgeoning public health problem. The urgent need is for interdisciplinary research to determine terrestrial–aquatic linkages of landscape disturbance and water quality, ecological conditions, and BU occurrence using Geographic Information System (GIS) and remote sensing technologies (Merritt *et al.* 2005). That is our goal in this study. Thus, this study does not fit neatly into the nutrition-related non-communicable diseases paradigm of Popkin (2002) but rather the third stage of the epidemiologic transition – the age of degenerative and man-made diseases (Omran 1971). In this study, using remote sensing and spatial analysis, we investigate the relationship between BU and environmental risk factors and how these change over time. The outcome of this geographic study will identify the possible locations that should be prioritised for targeting intervention.

Background

Buruli ulcer was first detected in Uganda in 1897 (World Health Organization 2007) but rapidly re-emerged in the 1980s (Walsh *et al.* 2008). This led the World Health Organization to establish the Global Buruli Ulcer Initiative in 1998 to promote research for BU intervention (World Health Organization 2002). Although most researchers agree that environmental and climate changes were the reasons for the endemic events in the 1980s (Walsh *et al.* 2008), the mode of transmission of this disease remains unclear. In addition, most BU studies focus on mode of exposure investigations through genetic experiments and surveys of reported BU cases. Few have applied health geography to study the environmental risk factors of BU. This study aims to fill this gap.

There are three stages of BU infection. At the first stage, BU infection appears in non-ulcerative forms such as nodule, papule, plaque, and oedematous (Debacker *et al.* 2004). This period of symptom development varies from several weeks to months before the infection forms ulcerate, which is the second stage of BU infection. During this stage, BU can be treated by injecting Bacille Calmette-Guérin vaccine for eight weeks to deactivate *M. ulcerans* (World Health Organization 2007). However, due to the mild symptoms in the early stages, patients usually are unaware of the infection and do not seek medical treatments. Without medical treatment, *M. ulcerans* spreads to other organs through the lymphatic system or bloodstream in the third stage which takes approximately three months (Debacker *et al.* 2004). Ten percent of the found BU cases have infected the bones (Portaels *et al.* 2009). As a result, the patients with BU in this stage face physical deformities and amputation. Up to now, the only way to treat BU in the last stage of infection is surgery. Even if the infected area is removed, the deformities and scars are permanent.

Geographic distribution

Although BU is not as deadly as tuberculosis, it has been reported in 33 countries around the globe, particularly in tropical and subtropical regions including Australia, China, and Japan (World Health Organization 2014). In Asia, BU has been found in tropical countries such as Indonesia, Malaysia, Papua New Guinea, and Sri Lanka (van der Werf *et al.* 2005). Latin American countries that have reported BU cases include Mexico, French Guyana, Peru (van der Werf *et al.* 2005), and Brazil (McGann *et al.* 2009). However, the majority of cases have been found in west and central Africa including Benin, Cameroon, Cote d'Ivoire, Democratic Republic of the Congo, and Ghana (World Health Organization 2014). The collective number of BU cases in Africa alone was estimated at 60,000 in 2008 (World Health Organization 2010) of which 20,000 were located in West Africa (Portaels *et al.* 2009).

Ghana is one of thirteen African countries confirming endemic BU, and the rate of new cases is increasing rapidly (van der Werf *et al.* 2005). Since 1993, Ghana has reported more than 11,000 cases (World Health Organization 2007). The highest prevalence rates occurred in south-west Ghana, ranging from 10.1 to 60 cases per 100,000 people, whereas Northern, Volta, and Upper West regions had the lowest prevalence rates at 7.0 to 10.0 per 100,000 people (World Health Organization 2002).

Environmental risk factors

Previous research suggests that many areas with high BU prevalence rates in tropical and subtropical countries were close to rivers, swamps, slow-flowing water, or stagnant water. In Ghana, BU cases were reported to be located near the Densu River, and there was an increasing rate of BU cases in wetlands (Duker *et al.* 2006b). In Liberia, BU prevalence rates were higher after a reservoir construction increased the number of swamp rice fields (Duker *et al.* 2006b). Similarly, the prevalence in Cameroon was also reported to be high among swamp rice farmers and patients living near the Nyong River (Pouillot *et al.* 2007). Debacker *et al.* (2006) and Brou *et al.* (2008) also proposed that environmental aquatic alteration might also pose a risk factor for BU. This was supported by the increased BU cases in Nigeria and Australia where flooding and damming in the areas preceded outbreaks (Pouillot *et al.* 2007). However, Wagner *et al.* (2008) and Brou *et al.* (2008) pointed out that BU cases in Benin and Cote d'Ivoire were not related to the locations of rivers.

Land elevation was suggested to be a BU risk factor by Wagner *et al.* (2008). From their study, the remote villages in low elevation located in drainage basins and near forest land cover in Benin showed high rates of BU. One possible explanation for higher prevalence rates in lower elevations may be that those areas contain more wet areas, and therefore, present a higher risk of human contact with *M. ulcerans* in water.

In addition, due to the higher rate of BU cases within a 20 kilometres radius around agricultural land use, Wagner *et al.* (2008) suggest that BU has an association with altered landscape. Also, some agriculture activities require a large quantity of water and therefore locate near rivers raising the risk of BU exposure (Duker *et al.* 2006a). Pouillot *et al.* (2007) noted that Cameroonian villagers living close to cocoa plantations showed higher risk of BU.

Walsh *et al.* (2008) proposed seasonal climate change as a risk factor for BU. For example, Duker *et al.* (2006b) indicate seasonal variations in BU cases. While BU incidence rates increased during the dry season in Papua New Guinea and Cameroon and were lower in Uganda during low rainfall months, the number of

BU cases in Ghana was higher in September and October, which were raining and flooding seasons (Portaels *et al.* 2009).

Proximity to forest vegetation is an environmental risk factor. Pouillot *et al.* (2007) found that humans living in close proximity to a forest were at risk of BU. Moreover, Brou *et al.* (2008) showed that BU prevalence rates in Cote d'Ivoire were higher in evergreen forest areas compared to sub-savannah and savannah areas.

Poverty and unclean living environment may be factors of BU risk. The related factors of poor housing conditions, malnutrition, lack of clean water supplies, low hygiene, and lack of modern health facilities reduce resistance in humans. Combined with ignorance of medical treatments and poor hygiene of certain cultural practices, these factors raise the risk of exposure to *M. ulcerans* in daily life (Duker *et al.* 2006b). Populations in remote areas, where poverty is high and education is low, mostly perceive BU disease as witchcraft and have improper treatment which causes the infection to become more severe (Pouillot *et al.* 2007). Moreover, due to difficulties of access to healthcare and cost of the treatment, those who would like to go to the hospital may not be able to afford the medical cost and, therefore, ignore the symptoms until the infection becomes untreatable.

Using remote sensing and spatial analysis, we aim to investigate the relationship between BU and environmental risk factors as well as how BU rates change over time. The objectives of this research are twofold: to investigate the effects of environmental factors on the spatial distribution pattern of BU in south-west Ghana and to explore the impact of environmental changes on the spatial distribution of BU over time. According to Yorke and Margai (2007), land use and land cover in Ghana changed significantly between 1990 and 2000. Since *M. ulcerans* is an environmental bacterium, it is possible that these changes may impact the spatial pattern of BU incidences in south-west Ghana between the 1990s and 2000s. Specifically, we address the following questions:

1 What is the spatial distribution of BU in south-west Ghana?
2 What are the environmental risk factors that explain the spatial distribution of BU in the study area?
3 Does land use/land cover change in the two periods (between the 1990s and 2000s) impact the spatial distribution of BU?

Methods

Study area

Ghana is located in West Africa and bounded by Togo (to the east), Cote d'Ivoire (to the west), Burkina Faso (to the north), and the Gulf of Guinea (to the south). The climate varies throughout the country due to seasonal and elevation differences. Ghana has two seasons – the rainy and dry seasons. The rainy season lasts from late April to October, and the dry season is between November and late March. Our study area (see Figure 6.1) covered six regions in south-west Ghana, including Ashanti, Brong Ahafo, Central, Eastern, Greater Accra, and Western regions, which contained 74 administrative districts and encompass 121,324.73 square kilometres with a resident population of 14,348,529 (Ghana Statistical Service 2009).

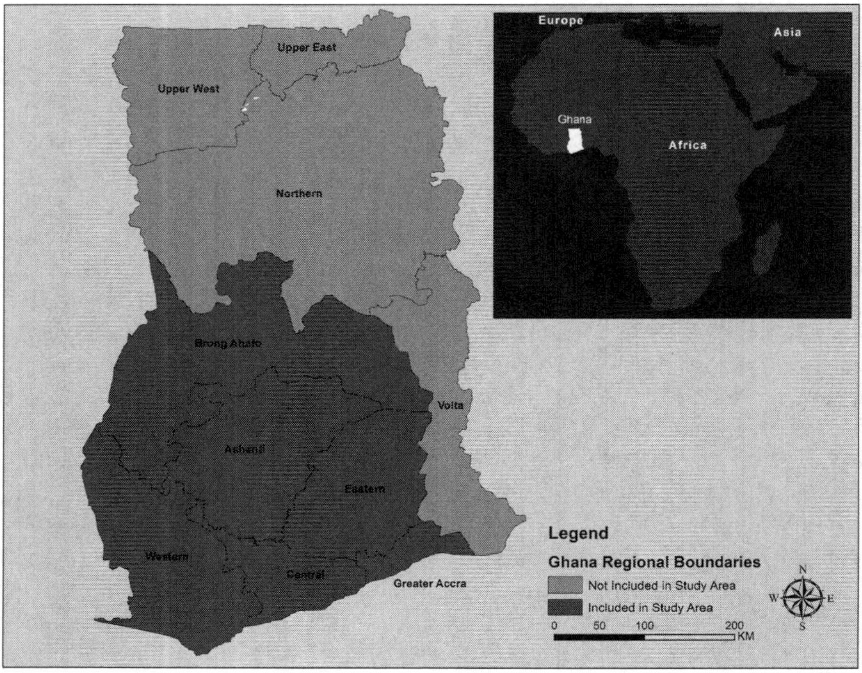

Figure 6.1 Map of study area.

Data

Population data were obtained from the Ghana 2000 census. Two BU datasets were used in this study: 1) the national search for cases of Buruli ulcer in Ghana conducted by the Centers for Disease Control (CDC) in v1999 comprising of 5,619 BU patients (Amofah *et al.* 2002) and 2) the Ghana Buruli Ulcer Control Program which contained 3,764 BU cases from 2005 to 2008 and was gathered from the healthcare centres. BU cases in each dataset were aggregated to the district level and joined to Ghana district shapefiles, obtained from Map Library (2007), in ArcGIS 10.2. All cartographic files are projected as Universal Transverse Mercator (UTM) Datum, WGS 84, Zone 30N.

Satellite image processing

Landsat Thematic Mapper (TM) and Landsat 7 Enhanced Thematic Mapper (ETM+) satellite images were acquired from the U.S. Geological Survey /Earth Resources Observation and Science Center in order to classify land use and land cover in the 1990s and 2000s respectively. To cover the study area, nine Landsat scenes – path 193 row 55 to 56, path 194 row 54 to 57, and path 195 row 54 to 56 – were downloaded. Each scene was selected with cloud cover less than 10% and in the dry season, between December and February. While the geometric

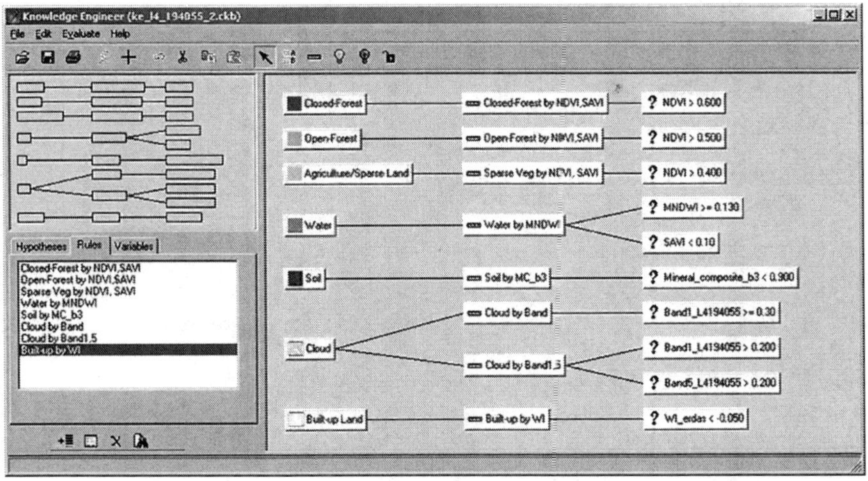

Figure 6.2 Classification rules and conditions for Landsat 4 path 194 row 55.

accuracy of Landsat 7 images exceeded the aforementioned requirements (Lee *et al.* 2004), Landsat TM images were calibrated using Chander and Markham's method (2003) to improve image accuracy.

To classify images, the knowledge-based classification method, expert classifier, was employed in ERDAS Imagine (2011) using the knowledge engineer tool. In order to perform this, hypothesis and classification rules were set as a hierarchical decision tree using seven indices – normalized difference vegetation index (Gillespie and MacDonald 2010), modification of normalized difference water index (Xu 2006), soil-adjusted vegetation index (Maselli *et al.* 2008), normalized difference built-up index (Zha *et al.* 2003), normalized difference blue band built-up index (Baraldi *et al.* 2006), wetness index (Huang *et al.* 2002), and mineral composite (Dogan 2009). Each scene was classified as water, soil, built-up land, agriculture/sparse land, closed-forest (dense forest), and open-forest (woodland). Due to different reflectance, each scene had its own threshold values and rules. Figure 6.2, for instance, illustrates a hierarchical decision tree of Landsat 4 scene 194/55.

Then post-classification comparison change detection technique was implemented in ArcMap 10.2 to evaluate land use and land cover changes. This technique reduced the atmospheric sensor influence and environmental differences between images that were acquired from multi-temporal images (Lu *et al.* 2003) and was straightforward subtraction of the 2000s- from the 1990s- classified images.

Spatial and statistical analyses

Spatial distribution of Buruli ulcer

We used the spatially adaptive filters method (Tiwari and Rushton 2005) to produce spatially continuous representations of BU rates in south-west Ghana. First, grid points were generated based on geographic units and control population (Yiannakoulias 2007).

At each grid point, a series of overlapping filters were placed, and then a disease rate was computed by dividing the number of BU cases by the population that falls within the same filter. These processes were implemented using Web-based Disease Mapping and Analysis Program (WebDMAP) developed by Tiwari (2008).

Influence of environmental factors on the spatial distribution of Buruli ulcer

For each district, as a spatial unit, we computed the percentages of land area classified as closed-forest, open-forest, agriculture/sparse land, water bodies, soil, and built-up land during the image classification and joined these to the BU rates. Spearman's rank correlation was employed to determine the relationship between environmental factors and BU rates. Consequently, Poisson and zero-inflated Poisson (ZIP) regression models, which are appropriate for analysing rare events, were implemented to generate the BU prediction models. The model of each dataset was selected based on Akaike's Information Criterion (AIC) values in which the lowest AIC values indicated the most preferred model (Akaike 1973). The general form of these regressions is illustrated in equation (1). The number of BU cases and the percentages of land use and land cover types were dependent and predictor variables respectively. To avoid overestimation, multicollinearity between independent variables was determined before fitting the models.

$$\log_e\left(Y_j\right) = \beta_0 + \beta_{ij}X_{ij} + \beta_{i+1,j}X_{i+1,j} + \ldots + \beta_{nj}X_{nj} \quad (1)$$

Where, Y_j = the number of BU cases in district j

β_0 = the intercept

β_{ij} = the coefficient of covariate X_{ij}

X_{ij} = the percentage of land use and land cover i in district j

n = the total number of predictors

When, I = land use and land cover type (closed-forest, open-forest, agriculture/grassland, water, soil, and built-up land)

J = administrative district in Ghana

Evaluating the spatial distribution of Buruli ulcer over time

The Kolmogorov-Smirnov test (KS test) was used to determine whether the changes in land use and land cover affected the distribution of Buruli ulcer or not. We excluded the districts with no reported BU cases before performing the analysis to avoid misinterpretation.

Results and discussion

Spatial distribution of Buruli ulcer

In 1999, 11 of 74 districts in the study area reported no cases of BU. The incidence rates of BU in the study area in 1999 range from 0 to 433.98 per 100,000 people. The national rate of BU in Ghana was 20.7 per 100,000 people (Amofah *et al.* 2002) compared to the average rate in the study area in 1999 of 42.83 per 100,000 people. The incidence rate from the study area is considered relatively high. Five districts have extremely high rates, exceeding 165 per 100,000 people, including Amansie West (434.11, the darkest colour south of Kumasi), Upper Denkyira (350.41, around the darkest colour south of Kumasi), Asutifi (209.50, the darkest colour south of Sunyani), Ga (201.65, the small darkest colour north of Accra), and Afigya-Sekyere (165.62, the darkest colour in north of Kumasi) (Figure 6.3A). It clearly illustrates that the higher rates occur in the south-west Kumasi area whereas the north-east portion of the study area has much lower rates. Notably, the high incidence rates of BU in 1999 occurred away from the urban areas.

The incidence rates of BU from 2005 to 2008 range from 0 to 134.63 per 100,000 people (Figure 6.3B), and the average incidence rate is 20.08 per 100,000 people. During this period, the higher rates occurred in the central part of the study area, and the lower rates were in the eastern study area similar to the pattern of BU in 1999 (Figure 6.3A).

Influence of environmental risk factors

The results show that the percentages of land use and land cover types of two time periods in the study area are slightly different (see Appendix A). In the early 2000s, the average percentages of closed-forest, open-forest, and agriculture/sparse land decreased from the early 1990s, whereas soil land cover and built-up land increased. These results suggest that the expansion of urban areas and/or built-up land probably led to deforestation and decreased agriculture/sparse land in the study area. While the mean of land use and land cover types are not statistically different between the two datasets, the locations of each land use and land cover type differs geographically (Figure 6.4).

Relationship between Buruli ulcer and environmental factors

Figure 6.4A illustrates that closed-forest mostly covers the western part of the study area along the border of Ghana and Cote d'Ivoire toward the south of Kumasi. Soil mainly covers the northern part of the study area and around the water bodies. Built-up land only occurs in regional capitals and cities. In contrast, the areas that used to be closed-forest along the border of Ghana and Cote d'Ivoire in the early 1990s are changed to open-forest in the early 2000s (Figure 6.4B). Closed-forest in this period mostly occurs around Koforidua. Agriculture/sparse land at the northern part of the study area and the north of Accra are changed to soil. Also, built-up land is expanded around the regional capital cities.

Considering the spatial distribution of BU incidence in 1999 (Figure 6.3A) and land use and land cover map (Figure 6.4A) together, the high BU endemic areas have the same distribution pattern as closed-forest. On the other hand, lower BU rates mostly appear in northern and eastern portions of the study area where soil and agriculture/sparse land dominate as land cover. Likewise, the results from Spearman's rank correlation confirm that the incidence of

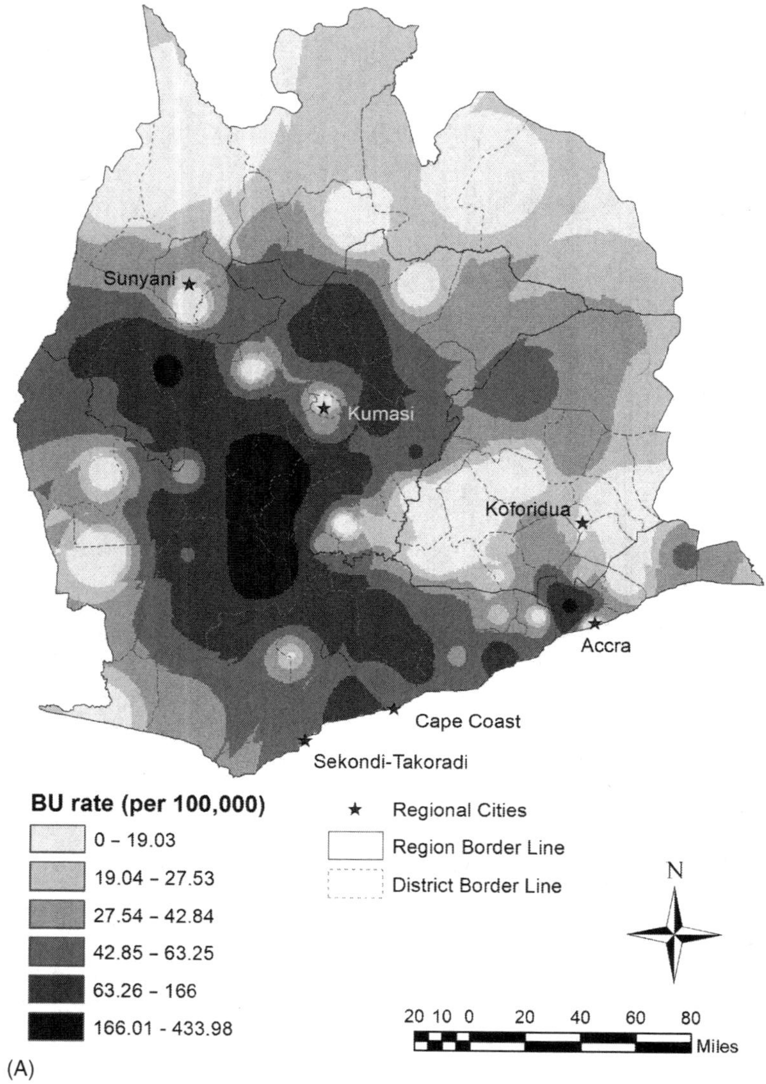

Figure 6.3 Geographic distribution of BU incidence (A) in 1999 and (B) between 2005 and 2008.

BU in 1999 has a significantly positive correlation with closed-forest areas ($r = 0.382$, $p < 0.001$) but inversely correlates to agriculture/sparse land and soil areas ($r = -0.319$, $p < 0.006$ and $r = -0.384$, $p < 0.001$ respectively) (Table 6.1). Considering the incidence of BU between 2005 and 2008 (Figure 6.3B) and land use and land cover in the early 2000s (Figure 6.4B), the high BU endemic areas partly occur in both closed-forest and open-forest land covers. Though they do not correspond to the distribution of closed-forest, the incidence rate of BU between

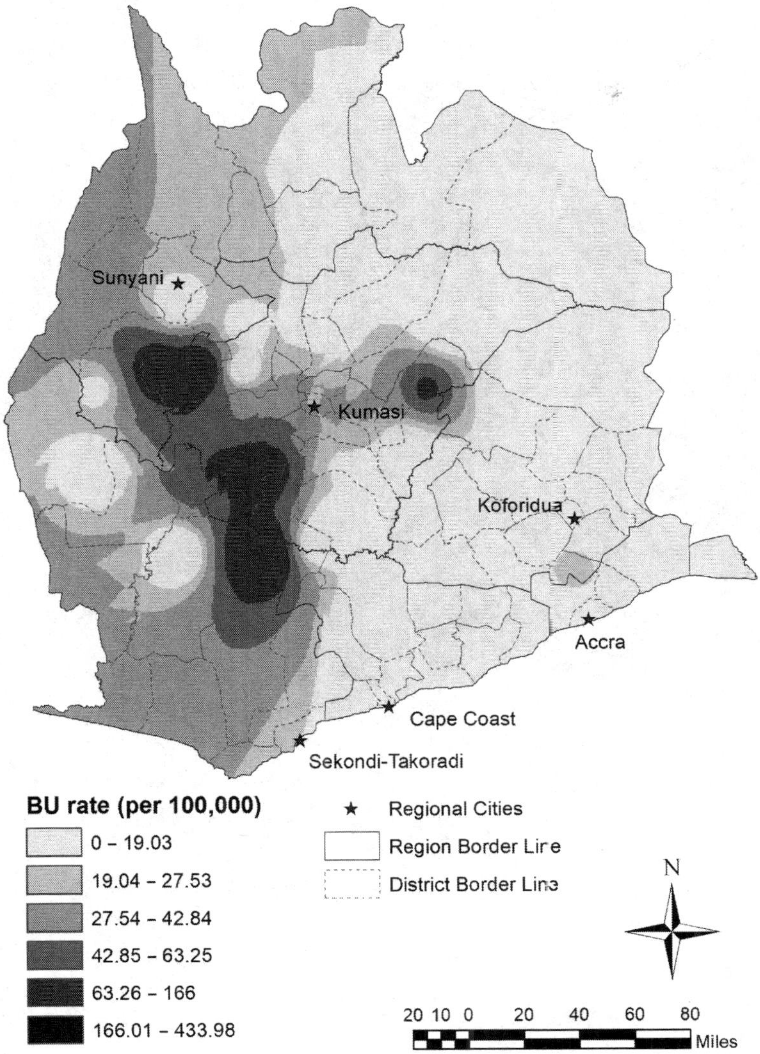

(B)

Figure 6.3 Continued.

2005 and 2008 statistically correlates to closed-forest and open-forest land covers ($r = 0.351, p < 0.002$ for both land covers) and yet inversely correlates to agriculture/sparse and built-up land covers ($r = -0.372, p < 0.001$ and $r = -0.391, p < 0.001$ respectively) (Table 6.1). These results also confirm the previous studies from Duker *et al.* (2006a), Pouillot *et al.* (2007), and Brou *et al.* (2008). However, the relationship between water bodies and the geographic distribution of BU in the study area cannot be determined due to the limitations of data and satellite imagery resolution.

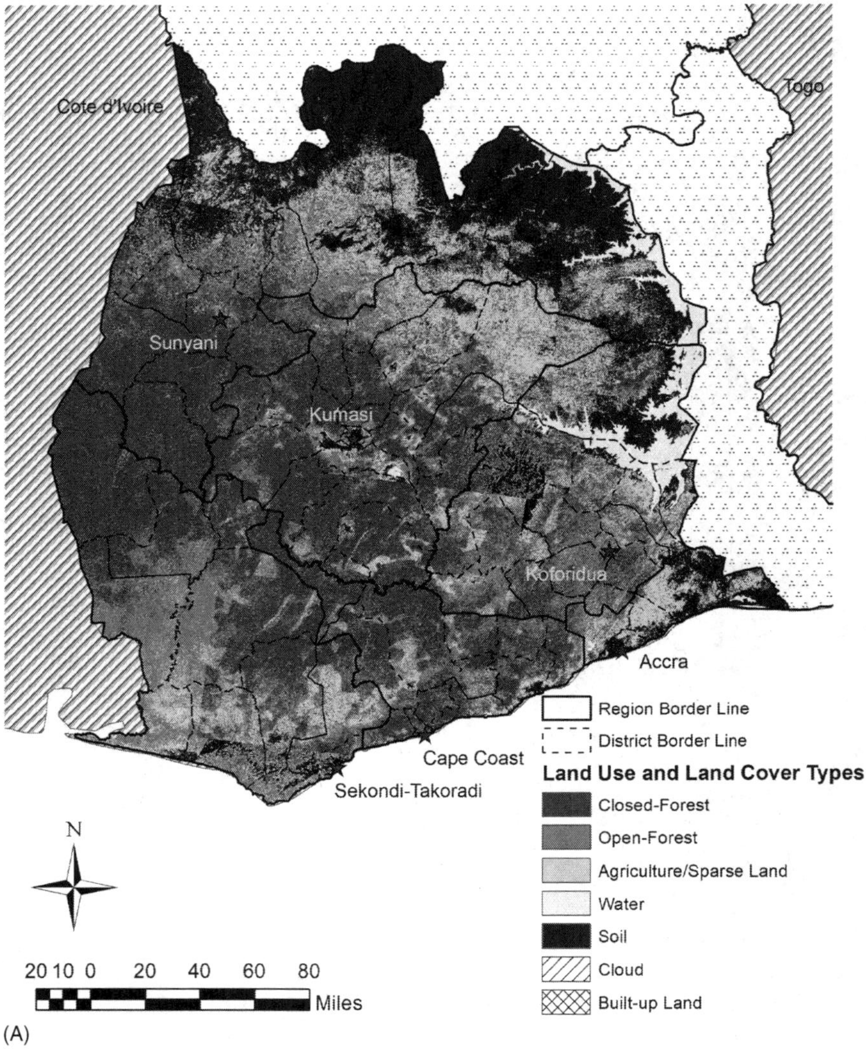

Figure 6.4 Land use and land cover classification in the early (A) 1990s and (B) 2000s.

Buruli ulcer prediction

For the BU_{1999} model, multicollinearity exists for three predictor variables – agriculture/sparse land, closed-forest, and soil (see Appendix B). Therefore, eight different models were run for each regression (see Appendix C for some candidate models). The results show that model ZIP42, which derives from ZIP regression, is the most preferred model for predicting BU cases in 1999 (equation (2)). It predicts that 14% of all districts have no cases, which is almost identical to the BU dataset in 1999 which showed 15%.

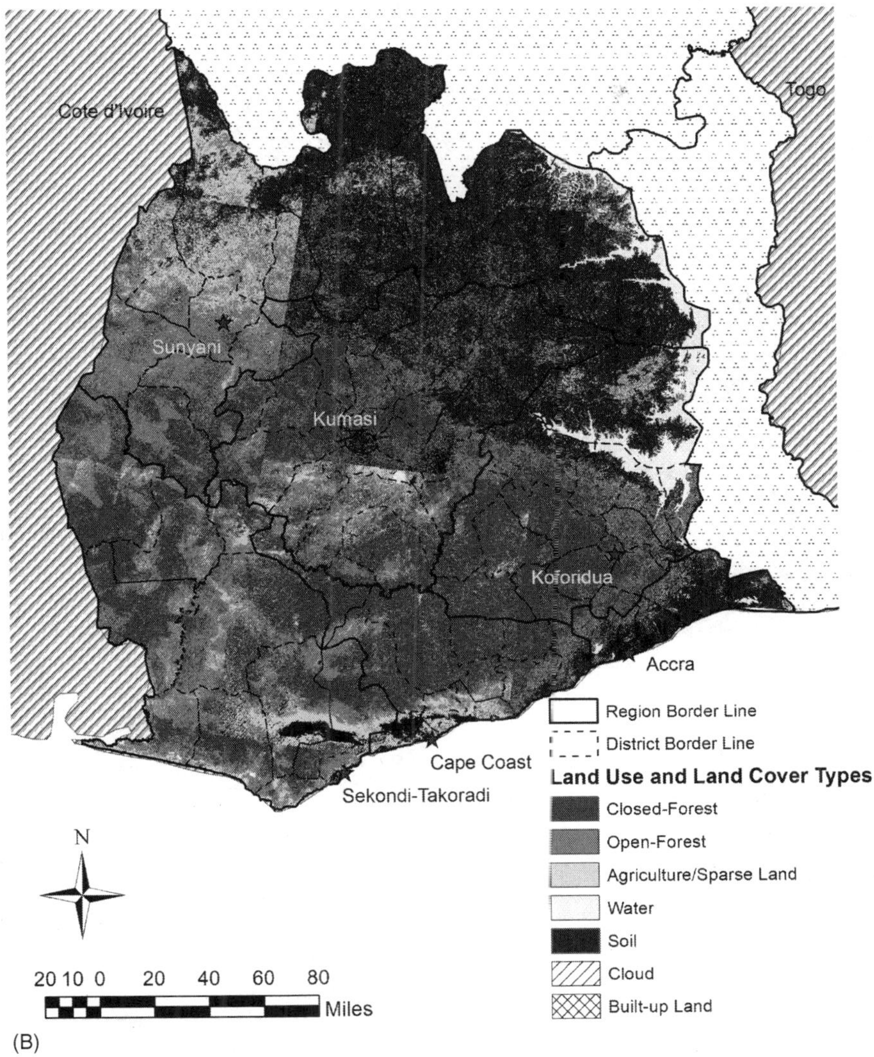

Figure 6.4 Continued.

$$\begin{aligned} BU_{1999} = \exp\big(&5.2871 - 0.02602(\%\,Open - forest) \\ &+ 0.01890(\%\,Agriculture/sparse\ land) \\ &- 0.00810(\%\,Water\,body) - 0.06791(\%\,Soil) \\ &- 0.07436(\%\,Built - up\,land)\big) \end{aligned} \tag{2}$$

For the BU_{2000s} model, there is no multicollinearity among predictor variables. Thus, all variables are included in the model. Like the BU_{1999} model, the most preferred model of BU_{2000s} is derived from ZIP (equation (3)). The ZIP model also predicts that 50% of all districts have zero cases, which is the same as the BU dataset between 2005 and 2008.

$$\begin{aligned} BU_{2000s} = \exp\big(&38.1316 - 0.3538(\%\,Closed - forest) \\ &- 0.3437(\%\,Open - forest) \\ &- 0.3572(\%\,Agriculture/sparse\,land) \\ &- 0.4397(\%\,Water\,body) \\ &- 0.3577(\%\,Soil) - 0.3510(\%\,Built - up\,land)\big) \end{aligned} \tag{3}$$

Though equation (2) and (3) precisely predict the percentage of districts with no cases, they overestimate the BU cases. Therefore, other factors such as human activities should be included in order to generate a more precise prediction model.

Influence of land use and land cover change on the spatial distribution of Buruli ulcer

Land use and land cover changes mostly occur along the coast of the study area and around the regional capital cities (Figure 6.5A). These can imply that the economic growth and infrastructure developments are mainly in urban areas. However, Figure 6.5A cannot be used to determine the BU rates in each changed area because there are many small groups of changed areas (isolated pixels) throughout the image. The best possible technique to solve the problem shown in Figure 6.5A is the use of zonal statistics to investigate the majority area (change or no change) in each district. The result shows that 57 of 74 districts are identified as land use and land cover changes (Figure 6.5B). The rest of districts, which have less change, are labelled as "no change".

Overlaying the spatial distribution of BU incidence on the changing area map clearly shows that the high endemic areas of both time periods occur in the changed areas (Figure 6.6A and 6B). The KS test is used to determine the impact of land use and land cover change on the geographic distribution of BU.

Table 6.1 Spearman's rank correlation among BU Rates and the percentage of land use and land cover types

Land use and land cover types	*Incidence of BU in 1999 with 1990s classified image*		*Incidence of BU from 2005 to 2008 with 2000s classified image*	
	r	P-*value*	r	P-*value*
Closed-forest	0.382**	0.001	0.351**	0.002
Open-forest	0.090	0.447	0.351**	0.002
Agriculture/sparse land	–0.319**	0.006	–0.372**	0.001
Water bodies	0.018	0.878	–0.155	0.187
Soil	–0.384**	0.001	–0.218	0.062
Built-up land	–0.152	0.195	–0.391**	0.001

** Correlation is significant at the 0.01 level (2-tailted).

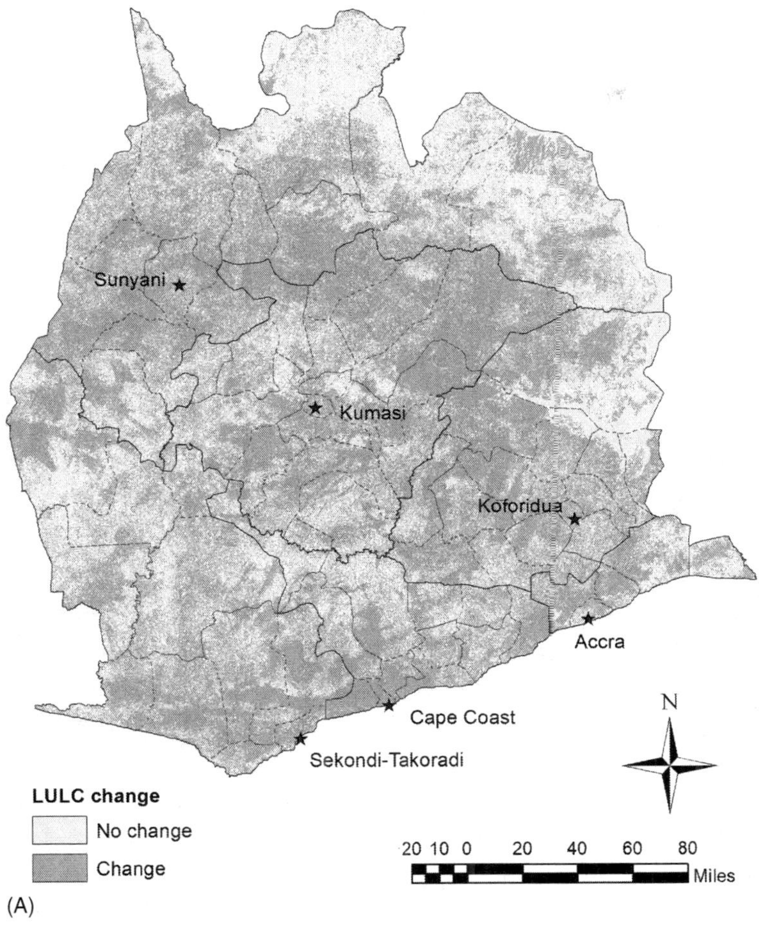

Figure 6.5 Land use and land cover change between the 1990s and 2000s: (A) changed area from the classified images and (B) majority of changing area at the district level.

Only 27 of 57 districts are included in the KS test analysis. The result confirms that there is a significant difference between the incidence of BU in 1999 and the incidence of BU between 2005 and 2008 ($p = 0.023$). Thus, changes of land use and land cover influence the spatial pattern of BU.

Limitations

There are several limitations that are worth mentioning. First, because there is no periodic BU dataset from the same source, it is difficult to investigate the trend of BU distribution. In addition, the satellite imagery of the study area had a high percentage of cloud cover and it was not possible to obtain the remotely sensed images in the same month and year. Moreover, the coarse resolution (30m by 30m) of the Landsat imagery precludes the detection of water bodies with a surface area less than

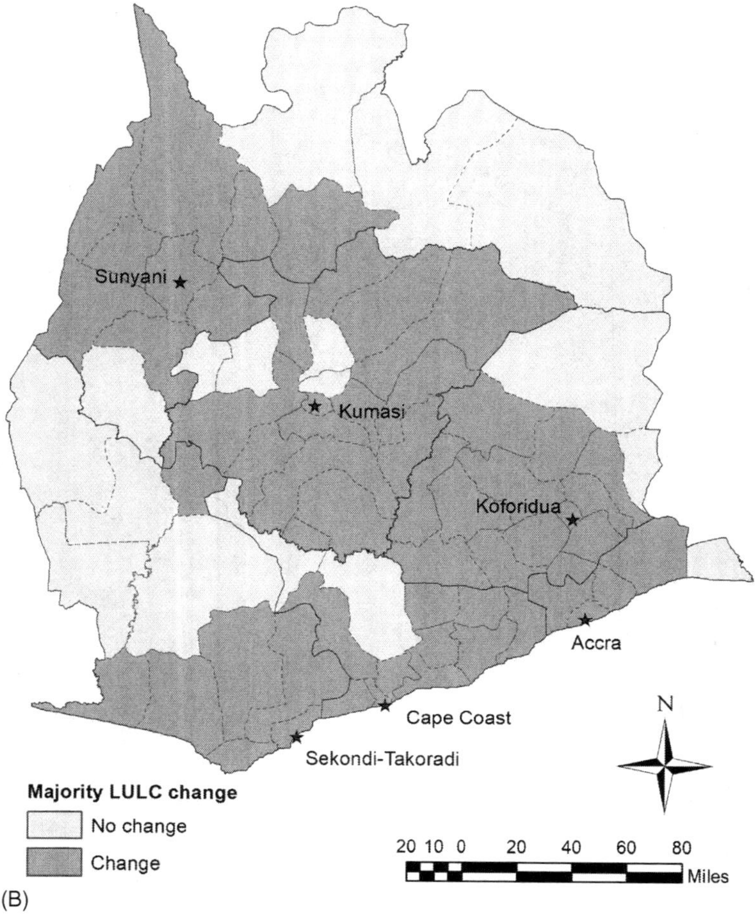

(B)

Figure 6.5 Continued.

900 square metres (roughly equivalent to a football field). Therefore, small water and/or stagnant water that would be a major environmental risk factor to facilitate BU cannot be detected, and only six types of land use and land cover can be classified from the Landsat. Due to lack of high resolution images, ground truth data, and aerial photographs, the study cannot measure the accuracy assessment of the classified images, which is an important process of classifying remotely sensed images. Finally, in order to distinguish between soil and built-up land, many indices are applied together because of the similarity background reflectance of soil and built-up land.

Conclusion

Although the BU incidence rate in 1999 (42.83 per 100,000 people) in the study area is considered relatively higher than what is observed between 2005 and 2008, the spatial distribution patterns are similar. The highest endemic areas of both

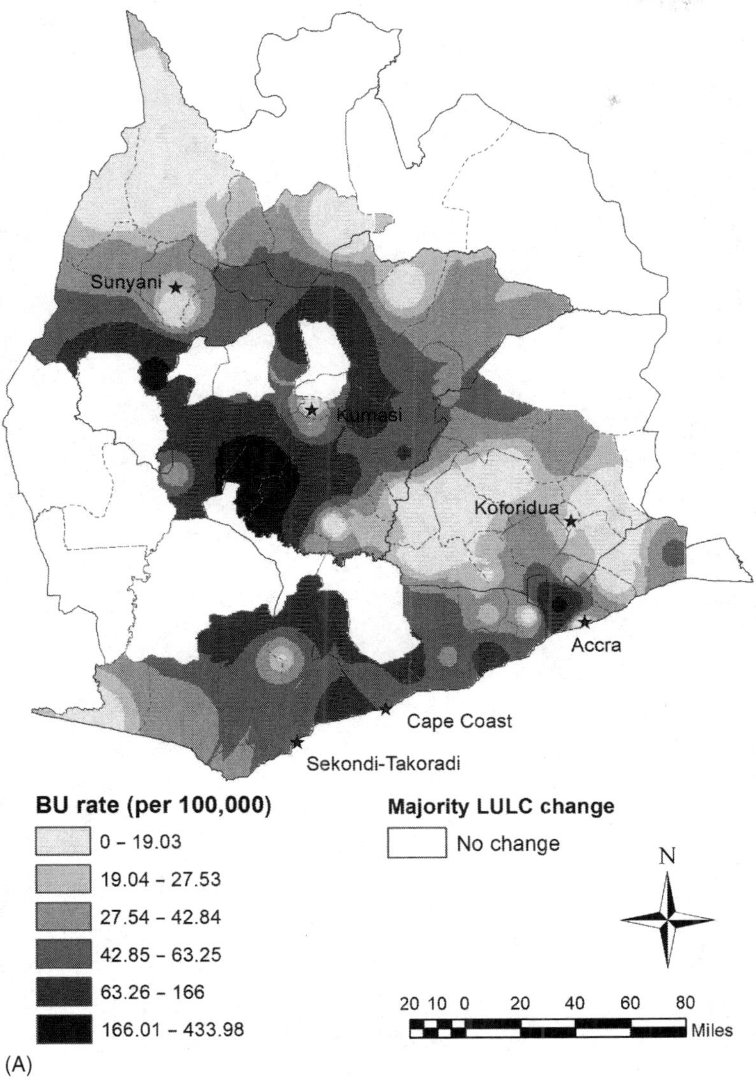

Figure 6.6 The overlay of spatial distribution of BU and the change of land use and land cover maps: (A) BU incidence in 1999 on the change map; (B) BU incidence between 2005 and 2008 on the change map.

periods occur in the southern portion of Sunyani and Kumasi and seem to expand to the western portion of the study area along the border of Ghana and Cote d'Ivoire. In contrast, the north-east portion of the study area appears to be low endemic areas during both periods. Also, BU rates around the regional capital cities are lower than other areas. The Spearman's rank correlation results indicate that closed-forest and open-forest are the most important environmental factors that explain the spatial distribution pattern of BU. The results also suggest that BU mostly occurs in rural areas, in the large forest, rather than in urban areas, which have better infrastructure.

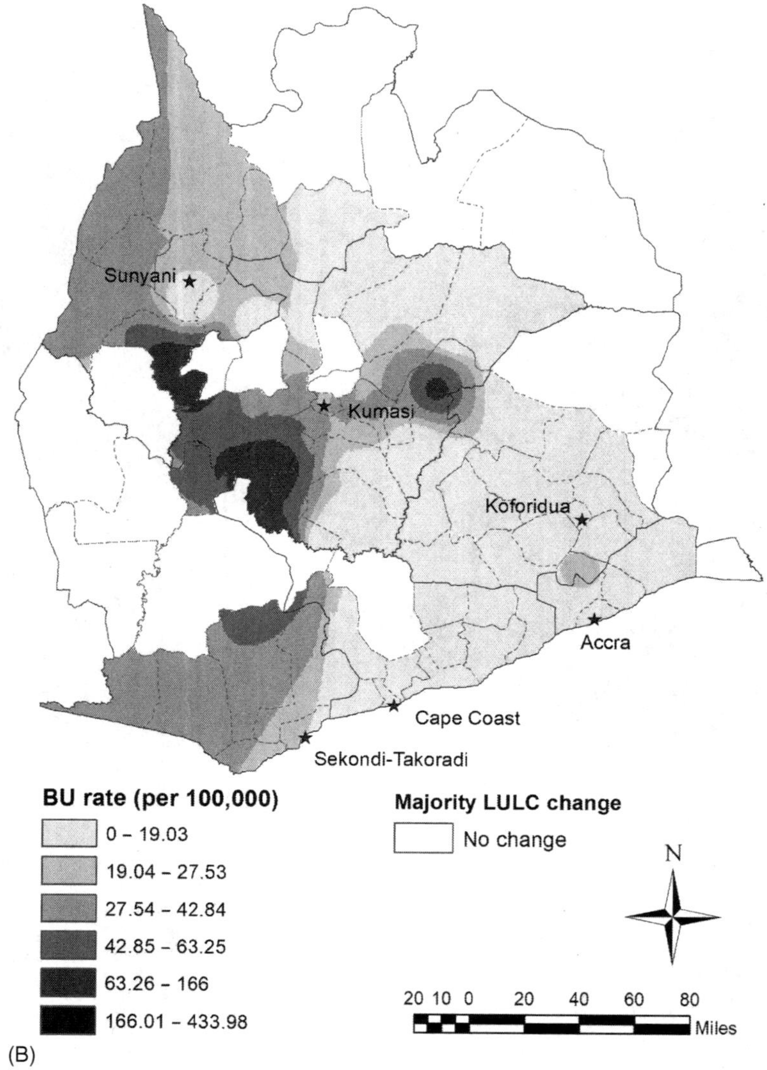

Figure 6.6 Continued.

While the ZIP model is the most appropriate, the resulting models overestimate BU cases. Further, other factors such as human activities and socioeconomic status may need to be included in order to generate a more precise prediction model.

The study found that land use and land cover change statistically affects the spatial distribution of BU; however, further research is also needed to investigate the changes of each land use and land cover type in order to determine how those changes affect the BU rates. Unlike previous studies, this study could not determine the relationship between water bodies and the spatial distribution of BU due to the limitation of imagery resolution. To confirm this finding, higher resolution satellite images are recommended for further study to classify land use and land cover and detect more small water bodies, particularly in the closed-forest areas.

References

Akaike, H., 1973. Information theory and an extension of the maximum likelihood principle. *In* B.N. Petrov and F. Csaki., eds. *Second International Symposium on Information Theory*, Budapest: Academiai Kiado, 267–81.

Amofah, G. *et al.*, 2002. Buruli ulcer in Ghana: Results of a national case search. *Emerging Infectious Disease*, (8)2, 167–70.

Baraldi, A., Puzzolo, V., Blonda, P., Bruzzone, L., and Tarantino, C., 2006. Automatic spectral rule-based preliminary mapping of calibrated Landsat TM and ETM+ images. *IEEE Transactions on Geoscience and Remote Sensing*, 44(9), 2563–86.

Bratschi, M.W. *et al.*, 2013. Geographic distribution, age pattern and sites of lesions in a cohort of Buruli ulcer patients from the Mape Basin of Cameroon. *PLOS Neglected Tropical Diseases*, 7(6), e2252.

Brou, T., Broutin, H., Elguero, E., Asse, H., and Guegan, J-F., 2008. Landscape diversity related to Buruli ulcer disease in Cote d'Ivoire. *PLOS Neglected Tropical Diseases*, 2(7), e271.

Chander, G. and Markham, B., 2003. Revised landsat-5 TM radiometric calibration procedures and post-calibration dynamic range. *IEEE Geoscience and Remote Sensing*, 41, 2674–77.

Debacker, M. *et al.*, 2004. *Mycobacterium ulcerans* disease (Buruli ulcer) in rural hospital. Southern Benin, 1997–2001. *Emerging Infectious Diseases*, 10(8), 1391–98.

Debacker, M. *et al.*, 2006. Risk factors for Buruli ulcer, Benin. *Emerging Infectious Diseases*, 12(9), 1325–31.

Dogan, H.M., 2009. Mineral composite assessment of Kelkit River Basin in Turkey by means of remote sensing. *Journal of Earth System Science*, 118(6), 701–10.

Duker, A.A., Portaels, F., and Hale, M., 2006a. Pathways of *Mycobacterium ulcerans* infection: A review. *Environment International*, 32(4), 567–73.

Duker, A.A., Stein, A., and Hale, M., 2006b. A statistical model for spatial patterns of Buruli ulcer in the Amansie West district, Ghana. *International Journal of Applied Earth Observation and Geoinformation*, 8(2), 126–36.

Ghana Statistical Service, 2009. CountrySTAT Ghana [Online]. Available from: http://countrystat.org/index.asp?ctry=GHA&t=1&lang=1 [Accessed 15 October 2010].

Gillespie, T.W. and MacDonald, G.M., 2010. Chapter 9 vegetation. *In* B. Gomez and J.P. Jones III, eds. *Research methods in geography*. Singapore: Willey-Blackwell, 137–54.

Huang, C., Yang, W.L., Homer, C., and Zylstra, G., 2002. Derivation of a tasseled cap transformation based on Landsat 7 at-satellite reflectance. *International Journal of Remote Sensing*, 23(8), 1741–48.

Johnson, R.C. *et al.*, 2015. Assessment of water, sanitation, and hygiene practices and associated factors in a Buruli ulcer endemic district in Benin (West Africa). *BMC Public Health*, 15, 801.

Lee, D.S., Storey, J.C., Choate, M.J., and Hayes, R.W., 2004. Four years of Landsat-7 on-orbit geometric calibration and performance. *IEEE Transactions on Geoscience and Remote Sensing*, 42(12), 2786–95.

Lu, D.P., Mausel, P., Brondizio, E., and Moran, E., 2003. Change detection techniques. *International Journal of Remote Sensing*, (25)12, 2365–407.

Maselli, F., Gardin,L., and Bottai, L., 2008. Automatic mapping of soil texture through the integration of ground, satellite and ancillary data. *International Journal of Remote Sensing*, (29)19, 5555–69.

McGann, H. *et al.*, 2009. Buruli ulcer in United Kingdom tourist returning from Latin America. *Emerg Infect Dis*, 15(11), 1827–29.

Merritt, R.W., Benbow, M.E., and Small, P.L.C., 2005. Unraveling and emerging disease associated with disturbed aquatic environments: The case of Buruli ulcer. *Frontiers in Ecology and the Environment*, 3(6), 323–31.

Omran, A.R. 1971. The epidemiologic transition: A theory of the epidemiology of population change. *Millbank Mem Fund Q*, 49(4), 509–37.

Popkin, B.M. 2002. An overview on the nutrition transition and its health implications: The Bellagio meeting. *Public Health Nutrition*, 5(1A), 93–103.

Portaels, F. *et al.*, 2008. First cultivation and characterization of *Mycobacterium ulcerans* from the environment. *PLOS Neglected Tropical Diseases*, 2(3), e178.

Portaels, F., Silva, M.T., and Meyers, W.M., 2009. Buruli ulcer. *Clinics in Dermatology*, 27(3), 291–305.

Pouillot, R. *et al.*, 2007. Risk factors for Buruli ulcer: A case control study in Cameroon. *PLOS Neglected Tropical Diseases*, 1(3), e101.

Tiwari, C., 2008. A spatial analysis system for environmental health surveillance. Unpublished doctoral dissertation. The University of Iowa.

Tiwari, C. and Rushton, G., 2005. Using spatially adaptive filters to map late stage colorectal cancer incidence in Iowa. *In* P. Fisher, ed. *Developments in Spatial Data Handling*. Berlin: Springer-Verlag, 665–76.

van der Werf, T.S. *et al.*, 2005. *Mycobacterium ulcerans* disease. *Bulletin of the World Health Organization*, 83(10), 785–91.

Wagner, T., Benbow, M.E., O Brenden, T., Qi, J., and Johnson, R.C., 2008. Buruli ulcer disease prevalence in Benin, West Africa: Associations with land use/cover and the identification of disease clusters. *International Journal of Health Geographies*, 7(25).

Walsh, D.S., Portaels, F., and Meyers, W.M., 2008. Buruli ulcer (*Mycobacterium ulcerans* infection). *Transactions of the Royal Society of Tropical Medicine and Hygiene*, 102(10), 969–78.

World Health Organization, 2002. Buruli ulcer: *Mycobacterium ulcerans* infection: An overview of reported cases globally. *The Weekly Epidemiological Record*, 20, 194–200.

World Health Organization, 2007. Buruli ulcer disease (*Mycobacterium ulcerans* infection) [Online]. Available from: http://www.who.int/mediacentre/factsheets/fs199/en/ [Accessed 6 October 2010].

World Health Organization, 2010. Neglected tropical diseases: Frequently asked questions [Online]. Available from: htp://www.who.int/neglected_diseases/faq/en [Accessed 6 October 2010].

World Health Organization, 2014. Buruli ulcer (*Mycobacterium ulcerans* infection) [Online]. Available from: http://www.who.int/mediacentre/factsheets/fs199/en/ [Accessed 10 August 2015].

Xu, H., 2006. Modification of normalised difference water index (MNDWI) to enhance open water features in remotely sensed imagery. *International Journal of Remote Sensing*, 14, 3025–33.

Yiannakoulias, N., 2007. Using quad trees to generate grid points for application in geographic disease surveillance. *Advance in Disease Surveillance*, 3(2), 1–9.

Yorke, C. and Margai, F.M., 2007. Monitoring land use change in the Densu river basin, Ghana using GIS and remote sensing methods. *African Geographical Review*, 26(1), 87–110.

Zha, Y., Gao, J., and Ni, S., 2003. Use of normalized difference built-up index in automatically mapping urban areas from TM imagery. *International Journal of Remote Sensing*, 24(3), 583–94.

Appendix A

Table 6.2 Descriptive statistics of land use and land cover types in the study area

LULC types	*Early 1990s*		*Early 2000s*	
	Average (%)	*S.D.*	*Average (%)*	*S.D.*
Closed-forest	37.09	27.39	32.92	27.51
Open-forest	30.80	16.19	29.49	18.18
Agriculture/sparse land	17.73	14.51	11.15	12.99
Water bodies	2.19	6.24	2.04	5.74
Soil	9.95	17.32	19.12	26.64
Built-up land	1.12	3.75	2.80	6.93
Cloud cover	1.05	3.03	1.63	5.41
Unclassified	0.07	0.20	0.87	1.61

Appendix B

Table 6.3 Correlation coefficient of land use and land ccver types (predictor variables) in the 1990s

LULC 1990s	*Closed-forest*	*Open-forest*	*Agriculture*	*Water*	*Soil*	*Built-up*
Open-forest	–0.024	1				
Agriculture	–0.896*	0.033	1			
Water	–0.450	–0.244	–0.457	1		
Soil	–0.773*	–0.355	–0.791*	0.542	1	
Built-up	–0.364	–0.224	–0.282	0.236	0.453	1

* Correlation coefficients greater than 0.700 or less than -0.700 indicate multicollinearity between those two variables

Appendix C

Table 6.4 BU_{1999} prediction model from Poisson regression and zero–inflated Poisson regression (ZIP)

Parameters		*Poisson regression*					*Zero–inflated Poisson regression (ZIP)*				
		P41	*P42*	*P43*	*P44*	*P45*	*ZIP41*	*ZIP42*	*ZIP43*	*ZIP44*	*ZIP45*
Goodness of fit	AIC	9089.38	9088.27	9105.23	9155.67	10388.88	7773.3	7771.9	7775.9	7847.6	8876.9
Inflation probability	Estimate	–	–	–	–	–	**0.1390**	**0.1395**	**0.1381**	**0.1473**	**0.1487**
	P–value	–	–	–	–	–	**0.001**	**0.001**	**0.001**	**<0.001**	**<0.001**
Intercept (b_0)	Estimate	**4.4760**	**5.0846**	**7.0940**	**–0.9587**	**4.5800**	5.7925	**5.2871**	**7.1706**	–0.1292	**4.8465**
	P–value	**<0.001**	**<0.001**	**<0.001**	**<0.001**	**<0.001**	0.4501	**<0.001**	**<0.001**	0.5771	**<0.001**
Closed–forest percentage (b_1)	Estimate	0.0061	–	**–0.0200**	**0.0600**	–	**–0.00503**	–	**–0.01873**	**0.05377**	–
	P–value	0.3496	–	**<0.001**	**<0.001**	–	**<0.001**	–	**<0.001**	**<0.001**	–
Open–forest percentage (b_2)	Estimate	**–0.0181**	**–0.0243**	**–0.0447**	**0.0379**	**–0.0074**	**–0.0312**	**–0.02602**	**–0.04528**	**0.02983**	**–0.0115**
	P–value	**0.0074**	**<0.001**	**<0.001**	**<0.001**	**<0.001**	**0.0394**	**<0.001**	**<0.001**	**<0.001**	**<0.001**
Agriculture/ sparse land percentage (b_3)	Estimate	**0.0264**	**0.0205**	–	**0.0782**	–	0.01393	**0.01890**	–	**0.07074**	–
	P–value	**<0.001**	**<0.001**	–	**<0.001**	–	0.0747	**<0.001**	–	**<0.001**	–
Water bodies percentage (b_4)	Estimate	0.0033	–0.0023	**–0.0206**	**0.0511**	**–0.0309**	**–0.0129**	**–0.00810**	**–0.02582**	**0.04146**	**–0.0269**
	P–value	0.6314	0.5019	**<0.001**	**<0.001**	**<0.001**	**<0.001**	**0.018**	**<0.001**	**<0.001**	**<0.001**
Soil percentage (b_5)	Estimate	**–0.0696**	**–0.0768**	**–0.0995**	–	–	**–0.0737**	**–0.06791**	**–0.08910**	–	–
	P–value	**<0.001**	**<0.001**	**<0.001**	–	–	**<0.001**	**<0.001**	**<0.001**	–	–
Built–up land percentage (b_6)	Estimate	**–0.0521**	**–0.0579**	**–0.0765**	–0.0073	**–0.1488**	**–0.0786**	**–0.07436**	**–0.08980**	**–0.0334**	**–0.1254**
	P–value	**<0.001**	**<0.001**	**<0.001**	0.521	**<0.001**	**<0.001**	**<0.001**	**<0.001**	**0.008**	**<0.001**

AIC: Akaike's Information Criterion.
Set 1: All predictor variables.
Set 2: Exclude closed–forest variable.
Set 3: Exclude agriculture/sparse land variable.
Set 4: Exclude soil variable.
Set 5: Exclude closed–forest, agriculture/sparse land, and soil variables.

7 Perilous outcomes

The intersection of culture, maternal health, and HIV/AIDS on Malawian women in the face of an international development consensus

Linda L. Semu

Introduction

Could the Millennium Development Goals (MDGs) have functioned as a catalyst for a fundamental realignment of the political economy of development at the global level that reflects the emancipatory aspirations contained therein (Saith 2006)? The MDGs provided measurable targets and indicators of progress for governments, UN agencies, international financial institutions, and civil society (Antrobus 2005, Barton 2005, Saith 2006, Fried *et al.* 2012). However, the MDGs also had some strategic limitations. They set a minimalist agenda and represented a technocratic effort to solve a systemic political issue (Barton 2005). The MDGs also relied on a top-down strategy for eradicating poverty with a focus on implementation in the global south without creating similar accountability mechanisms for the countries of the global north (Barton 2005). Furthermore, treating each goal as a separate issue ignores the ways they are interrelated, especially when it comes to women's empowerment (Antrobus 2005). All of the goals are significantly related to the position of women and the conditions under which they live (Antrobus 2005). For example, three of the eight MDG targets are health-related: reduction of child mortality; improving maternal health; and combating HIV, malaria, and other diseases (Goals 4, 5, and 6 respectively) (Fowkes *et al.* 2016). The world has since transitioned to the Sustainable Development Goals (SDGs) comprised of 17 aspirational goals and 169 targets. In particular, SDG 3 "was designed to address the unfinished business of MDG 5a which called for a 75% reduction in maternal mortality rate" (Ogu *et al.* 2016). Although sub-Saharan Africa had registered a 49% decrease in the maternal mortality rate (MMR) by 2015, this was short of the stated global target (Ogu *et al.* 2016).

In order for the SDGs to overcome the MDG's failure to meet the maternal health goals, Fowkes *et al.* (2016) advocate for an integrated approach that addresses the interrelated health issues of maternal mortality, HIV, TB, malaria, and other preventable infectious diseases and nutritional deficiencies among pregnant women and children (Fowkes *et al.* 2016). Ogu *et al.* (2016) further argue that health inequalities are exacerbated by social, cultural, and economic conditions as well as policies that limit women's access to evidence-based clinical and preventive interventions for reducing maternal mortality. Fowkes *et al.* and Ogu *et al.*'s views help situate this chapter within Popkin's (2002) theoretical framework that focuses on demographic, epidemiological, and nutritional transitions. In particular, the demographic transition from high mortality to reduced mortality and the epidemiological transition from high prevalence infectious disease to receding pestilence and poor environmental conditions both lead to a focus on family planning, infectious disease control, and policy initiatives and behaviour change (Popkin 2002) – the stage which Malawi faces. For Malawi, maternal health (maternal mortality and morbidity or MMM) and HIV/AIDS are indicative of interlocking sociocultural, structural, and systemic processes that drive gender disempowerment and poverty, gender inequality, and gender-based violence. Furthermore, cultural norms that perpetuate gender inequality by sanctioning behaviours and practices that reinforce women's subordinate position produce and reproduce interrelated HIV/AIDS and maternal health outcomes. It follows, therefore, that no single goal can be achieved without progress on all goals, particularly the cross-cutting goal of gender equality and empowerment (Fried *et al.* 2012). Moser and Moser (2005) go further and identify three critical aspects for the achievement of progress: gender equality, gender mainstreaming, and gender empowerment. Gender equality is the recognition that women and men have different needs and priorities (Moser and Moser 2005). Gender mainstreaming comprises the assessment of implications on women and men of any planned action at all levels (Moser and Moser 2005). Gender empowerment is about promoting women's participation in decision-making processes, having their voices heard, and their issues put on the agenda (Moser and Moser 2005).

Although the MDGs came to an end in 2015, the gender equality, mainstreaming, and empowerment goals were not attained for Malawian women. The National Statistics Office carried out Malawi's MDG Endline Survey in 2014, a nationally representative survey of 28,479 households with a 98.8% response rate, with participants drawn from the Northern (12.2%), Central (39.5%), and Southern (48.3%) regions. 86% of participants lived in rural areas. Major findings show:

- The total fertility rate is 5.0 births per woman.
- The maternal mortality ratio is 574 deaths per 100,000 live births.
- 44.7% of the women have been seen at least four times by an antenatal care provider during their pregnancy.
- 31.3% of women ages 20–24 had had at least one live birth by age 18.
- There is some age-mixing among sexual partners with 9% of women ages 15–24 having been with men ten or more years older than them.

(National Statistics Office 2014)

These results corroborate the Malawi Government which in 2010 identified the MDGs of education, improvement of maternal health, and promotion of gender equality and women's empowerment as unlikely to be attained by 2015 (Malawi Government 2010b). At the time, the primary school net enrolment ratio was

83%, and the ratio of girls to boys was 1.03. However, the ratio of girls to boys dropped to 0.79 by secondary school, constricting the pool of girls available for tertiary levels of education and the opportunities that thereby accrue (Malawi Government 2010b). These perilous outcomes form the backdrop to the interlocking system of gender, HIV, and MMM highlighting the need for a holistic approach to development under the SDGs.

HIV, maternal mortality, and morbidity

The majority of the world's women are mothers, especially in developing countries where childlessness is rare, forming a connection between women, motherhood, and development. Family and parenting are shaped by gender and entrenched through structural inequality. Thus, women's development is not about women *per se* but is about gender as a societal issue (Reid and Shams 2013). Women's empowerment and gender equality are not only essential for current human development but also for improving human development across future generations. Addressing issues related to motherhood is, therefore, fundamental to reducing structural barriers so as to allow women more opportunities in the public sphere and transform gender relationships in the domestic sphere (Reid and Shams 2013). Unravelling structural and systemic drivers of HIV and MMM helps to illuminate barriers to the attainment of gender role transformation and equality. In Malawi, these drivers operate within a context where culture, women's caregiving role, and economic disempowerment shape gender and power dynamics (Macintyre *et al.* 2013).

In 2014, there were 36.9 million people living with HIV globally (UNAIDS 2015). That year, there were 2 million new infections and 1.2 million AIDS-related deaths (UNAIDS 2015). The scenario within the sub-Saharan Africa (SSA) region in 2014 was even dire; 1.4 million new infections occurred, of which 190,000 were children. In total, the region suffered 790,000 AIDS-related deaths. 25.8 million HIV-positive people were women, comprising more than half of the population living with HIV. Although the rate of new infections declined by 39% between the years 2000 and 2014, SSA represented 70% of all global new infections (UNAIDS 2015).

The heavy burden of HIV has implications on MMM; an estimated 1.5 million women living with HIV in low- and middle-income countries were pregnant (Joint UN Program on HIV/AIDS 2012). Although preventing reproductive-age women from acquiring HIV and avoiding unintended pregnancies are both critical, 13–21% of women living with HIV who knew their status had unmet for family planning needs. Clearly, MMM and HIV are connected both in outcomes and solutions, especially in sub-Saharan Africa where HIV is the leading cause of maternal death (Fried *et al.* 2012). Hence, prevention of unintended pregnancy and access to contraceptives were identified as the two most important HIV-related prevention efforts (Joint UN Program on HIV/AIDS 2012).

Malawi faces a heterosexual HIV epidemic with one of the highest HIV prevalence rates in the world. An escalated infection pattern was observed early in the epidemic where a sample of pregnant women attending antenatal clinics in the city of Blantyre showed a rise in HIV incidence from 2.6% in 1986 to more than 30% in 1998, decreasing only slightly to 28.5% in 2001 (World Health Organization/UNAIDS 2005). By 2005, approximately 14.1% of the country's adult population aged 15–49 were living with HIV/AIDS, giving Malawi one of the highest adult prevalence rates in the world (US Department of State 2009). Women comprise 59% of the 850,000 adults over age 15 living with HIV in Malawi (UNAIDS 2013). The 2010 Malawi Demographic and Health Survey (MDHS), which included HIV testing of over 7,000 women aged 15–49 and over 6,800 men aged 15–54 found a 10.6% prevalence of HIV within the 15–49 age group (12.9% and 8.1% for women and men respectively). It also found that HIV prevalence is two times higher in urban areas than in rural areas: 17.4% versus 8.9% (National Statistics Office Malawi and ICF Macro 2011). Although the figures show a reduction from the 2005 data, this should not obscure the gendered processes that underscore women's vulnerability to HIV infection and, therefore, requires a deliberate engagement with the gender context of infection (Anderson 2012). For example, the largest disparity is among 15- to 19-year-olds: 3.7% for women versus 0.4% for men (National Statistics Office Malawi and ICF Macro 2011). These figures reflect young women's vulnerability to poverty and the persistence of sociocultural attitudes that restrict women's control over their sexual and reproductive freedom (White 2007). Not surprisingly, the 2004 MDHS found that among the youth between the ages of 15 and 19, more young women (36%) were married, divorced, or widowed compared to young men (3%) (National Statistics Office Malawi and ORC Macro 2005).

Women are at the epicentre of Malawi's HIV epidemic, with the highest risk occurring among young women, mostly through heterosexual transmission (Anderson 2012, Kumwenda *et al.* 2008). Efforts to mainstream gender into HIV/AIDS programmes have largely failed to address structural and systemic drivers that create gender disparities, underscoring the need for transformative planning. It is therefore imperative that development practitioners should find innovative ways to address HIV/AIDS and gender relations (Tiessen 2005). These should comprise integrated efforts that are critical towards the reduction of HIV, MMM, and, by extension, the attainment of reproductive and sexual health. The normal expected outcome of pregnancy is a healthy delivery, yet we should remain cognizant of other common outcomes in SSA: miscarriage, stillbirth, infant death, unsafe abortion, or long-term morbidity. This makes sexual health an imperative, yet the high incidence of HIV has undermined the capacities of health, social, and community institutions in addressing MMM and other development-related issues (Fried *et al.* 2012). Integrated responses that take into account common sociocultural, structural, and systemic drivers that create HIV and MMM outcomes are, therefore, the most effective in addressing and overcoming both challenges.

The need for integrated responses notwithstanding, HIV and MMM have traditionally been treated separately (including under the MDGs) with each focusing on addressing its own immediate needs such as acquiring adequate antiretrovirals (ARVs) or providing obstetric care. They have also followed strategies that are programmatically contradictory and are contradictory within the context of women's circumstances. For example, the tendency by HIV/AIDS programmes to discourage women from childbearing assumes that they have the power to make decisions about sexual and reproductive health. In addition, the antenatal programmes for the prevention of mother-to-child transmission may discourage women diagnosed with HIV from breastfeeding. This does not take into account the fact that pervasive perceptions and expectations about motherhood may force women to reveal their HIV-positive status to their extended families were they to not breastfeed, hence running the risk of negative repercussions including rejection, blame, and violence. On the other hand, when these women attend postnatal care clinics for their children, they may be admonished for not breastfeeding. Hence, the women find themselves in a quagmire: non-integrated programmes have failed to strengthen overall health delivery systems and have also overlooked the development of effective responses to structural and systemic problems that determine women's health outcomes (Fried *et al.* 2012). Although driven by good intentions, the MDGs focused more on establishing technocratic, quantifiable goals than on making the oft-messy, qualitative, and cross-cutting gender equality and empowerment objective a central tenet of all goals. The Malawi scenario echoes Fried *et al.*'s (2012) call for an integrated approach, anchored in gender equality, as a way to wrestle with pressing health needs while addressing sociocultural and structural drivers of HIV and MMM.

Overlapping structural drivers of HIV and MMM

Gendered structures of power that underscore women's vulnerability to MMM and HIV are not always fully understood and as a result, responses often fail to effectively engage the gender context (Anderson 2012). This explains why the overarching goal of gender empowerment and transformation was not mainstreamed in all the MDG goals. Similarly, HIV and MMM ought to be tackled in an integrated manner because they occur due to overlapping sociocultural, structural, and systemic drivers. In addition, HIV and MMM are linked to the larger gender equality agenda in which 18% of global MMM is attributable to HIV, especially in SSA where HIV rates are high among women and girls. In generalised epidemics, early sexual debut, early marriage, and sexual violence are significantly associated with an increased risk of women acquiring HIV infection. Improving women's social and economic status would significantly moderate their risk of acquiring HIV infection by reducing their dependence on male partners and boosting their decision-making power (Joint United Nations Program on HIV/AIDS 2012). The overlapping structural drivers include economic disempowerment (feminised poverty); sociocultural practices that produce uneven power dynamics; gender inequality; convoluted perceptions about women, reproduction, and their bodies; and gender-based violence. The system-related drivers create inadequate access to and unavailability of health and other development-related services (Fried *et al.* 2012). This section describes these processes in an in-depth manner so as to show their juxtaposition on HIV, MMM, and gender equality and transformation.

Feminised poverty: The economic situation of women in Malawi

Access to the primary determinants of human development – income, education, and health – are critical for women's empowerment as these have powerful multiplicative effects for maximising their current status and the development of future generations due to women's primary responsibility in childrearing (Reid and Shams 2013). Malawian women are generally worse off than their male counterparts in almost all spheres of health, social, and economic life. As one of the world's poorest countries, Malawi ranks low on several international development and gender indicators. As an agriculturally-based economy, Malawi's uniqueness lies in its overwhelming reliance on land for food provision. However, only 15–30% of households produce enough to last all year. The poorest households rarely grow enough to last more than a month or two and the 30% that are above the ultra-poverty line are likely to experience transitory food insecurity. At less than 0.8ha, the average land-holding is insufficient for adequate household production, hence the acute poverty and food insecurity. Reliance on rain-fed agriculture, extensive soil degradation, and unaffordable inputs (especially fertilisers) have all contributed to a decline in maize production (Harrigan 2008). In 2013, the World Bank classified Malawi as a low-income country where 51% of its 16.4 million people lived in poverty, with an average life expectancy of 55 years, and gross national income (GNI) of $270 (World Bank 2017). Poverty in Malawi is particularly feminised as seen in international measures such as the Human Development Index, Gender Inequality Index, and the Global Gender Gap that all rank Malawi low on their measures (Organization for Economic Co-operation and Development 2012). The 2010 Gender Gap Index by the World Economic Forum (WEF) ranked Malawi at 68 out of 134 countries, with a score of 168. The index shows that Malawian women fare worse in educational attainment, economic participation and opportunity, and political empowerment (World Economic Forum 2010). Similarly, the Social Institutions and Gender Index (SIGI) that measures underlying discrimination against women by evaluating early marriage, discriminatory inheritance practices, violence against women, son bias, restrictions on access to public space, and restricted access to productive resources ranks Malawi at 38 out of 86 countries with an SIGI value of 0.217098. This is indicative of high inequality on a scale ranging from 0 (inequality) to 1 (equality) (Organization for Economic Co-operation and Development 2012).

Regarding control over resources, both the matrilineal and patrilineal systems in Malawi discriminate against women who are often viewed as inferior and on the receiving end of men's decisions (White 2007). Women access land through male heads of households. Although access is determined by husbands and their patrilineage in the patrilineal system and by maternal uncles in the matrilineal system, men are considered heads of household in both systems. Hence, they make key decisions related to land use. In a rather convoluted way, Malawi's National Land Policy allows for the name of the "head" of a family to be registered as the proprietor of family land, resulting in men's (as husbands and brothers) names being recorded, a policy that disenfranchises women and youth's land rights (Peters and Kambewa 2007). This limitation of women's access extends to other areas of property ownership due to customary practices. These limitations occur despite Malawi's constitutional provision (chapter IV) that grants a broad coverage of human rights and specifically the chapter's non-discrimination clause (section 24) that grants women same rights as men in civil law, including the right to enter into contracts and to acquire and maintain rights in property (Malawi Government 2006a). The situation of Malawian women is typical of SSA where the level of poverty and hunger deepened despite the MDGs' call for reduction of the same. Although rural women substantially contribute to rural livelihoods and household food security, their attempt to uplift themselves out of poverty is contingent upon engendered access to markets; supportive institutions, policies, and programmes; and land for cultivation (Thamaga-Chitja 2012). The World Economic Forum thus notes how customary law usurps rights through gender-linked stereotypes and roles in customary practices where women are seen as owning less valuable property (such as kitchen utensils), while higher value property such as land or cars belongs to men (World Economic Forum 2010). These same attitudes spill over to access to credit where mainstream banks' collateral requirements pose as barriers to women who are less likely to own "hard property" (World Economic Forum 2010). There is thus income disparity, with men earning double women's income (Malawi Government 2006b).

Malawi's Gender and Development Index (MGDI) is a composite index of both quantitative assessment of gender equality (in the social, economic, and political spheres) and qualitative evaluation of government's performance in implementation of specific treaties, declarations, and resolutions. The index shows Malawi falling short in fulfilling obligations to women in education, employment, and leadership at all levels of society (Malawi Government 2006b). As Fried *et al.* (2012) argue, this disempowerment limits women's access to the resources, information, social networks, and tools to protect themselves and reduce their dependence on male partners. Disempowerment also inhibits their ability to negotiate sex and thereby increases their vulnerability to HIV, unwanted pregnancies, and MMM (Fried *et al.* 2012). Women's economic dependence within formal systems does not imply they are not working. On the contrary, they make significant contributions to household and national economies. While rural Malawian men work 8-hour days on average, women have a "double shift," working 16 hours a day on average, although they often do not control the proceeds of their labour (Saur *et al.* 2005). These labour burdens weaken their immunity, increasing vulnerability to HIV and/or progression to AIDS. Childbearing in such a scenario adds to the disease and death burden borne by women, hence the high rates of MMM.

Sociocultural practices: Power dynamics, gender inequality, and women's bodies

Malawian women's unequal status is a conundrum borne out of structural and interlocking factors related to poverty, gender discrimination, and vulnerability to HIV/AIDS. Furthermore, cultural norms that perpetuate gender inequality within communities, organisations, and institutions are produced and reproduced on a daily basis through the sanctioning of norms, behaviours, and practices that reinforce women's subordinate position in society (Macintyre 2013, Tiessen 2005). Malawi's strategy for HIV prevention has centred on behaviour change and abstinence. This strategy does not acknowledge structural constraints and the gender context that hinders negotiation of voluntary and protected sex. The poorest and most marginalised women have inadequate access to sexual, reproductive, and maternal health services and are more likely to have limited information on HIV transmission (Fried *et al.* 2012). Among 15- to 49-year-olds in Malawi, only 50% of women have access to the media (i.e., reading a newspaper or magazine, listening to the radio, or watching television at least once a week), compared to 73% of their male counterparts (National Statistics Office 2014). The importance of the media in accessing information on HIV and MMM cannot be overstated. Women's vulnerability to infection within the private sphere has public health implications. Hence, women's bodies and their embodied experiences of power should be at the centre of analysis (Anderson 2012). Anderson's 2012 qualitative study of various stakeholders (policy makers, organisations engaged in decision-making and implementation, and representatives of HIV/AIDS organisations in Malawi) highlights the following gendered constructions of women's bodies that amplify their risk of HIV infection:

- Women's reproductive roles as a "vessel for life"
- Male control over the female body and surrender of the female body for male sexual pleasure
- The utilisation of the female body for cultural sexual practices
- The stigma of men being at risk from "infectious women," coupled with the myth that virgin bodies are infection free and can cure HIV.

(Anderson 2012: 270)

Women's reproductive role is central to their value, status, and identity. Within the matrilineal system, the woman is looked upon as the root of the lineage (Phiri 1997) where her ability to reproduce enhances the power and prosperity of her ethnic group, especially that of her uncle and brother who act as guardians of the offspring. Once a woman begins childbearing (*kuchembeza)*, she is no longer called by her name but either by that of her first born or the number of children she has, for example:*achemberekamodzi, achemberekawiri* (mother of one, mother of two, etc.). Malawian women are under pressure to have many children such that among 20- to 24-year-olds, 31.3% have had at least one live birth by age 18. In addition, 10.3% are married by age 15, compared to 1.5% males, and at least 50% of the women are married by age 18, compared to 9% males (National Statistics Office 2014). Women's culturally-expected role in reproduction is not in tandem with the HIV abstinence strategy. Delayed sexual debut and marriage facilitates attainment of financial self-reliance, independence from male partners, and more autonomous sexual decision-making leading to better maternal health, sexual, and reproductive health outcomes. Conversely, early onset of sexual activity increases the likelihood of MMM and HIV infection (Joint United Nations Program on HIV/AIDS 2012).

Cultural norms and economic disparities have resulted in the tendency for age asymmetry in sexual relationships whereby 9% of women between the ages of 15 and 24 have been with men ten or more years older than them (National Statistics Office 2014). Compounding this is the fact that the virgin body is desired by older men who prefer sex with younger girls due to the perception that the girls are not infected as well as the popular myth that sexual intercourse with virgins can cure HIV (Anderson 2012). Poverty also pushes girls to go for older men (sugar daddies) because they are viewed as being financially secure and can enable them to access the three Cs: cell phone, car, and cash. Some women turn to sex work as a means to support their children due to impoverishment from social status positions, like widowhood or divorcée, that strip them of most privileges. These sexual liaisons occur despite prevailing social norms surrounding women's sexual activity that are characterised by a culture of silence when it comes to sexual reproductive health for women and girls (Anderson 2012, Rankin *et al.* 2005). The economic disparities and dependence have led to intense competition for men among women. For the most part, men prefer having loose relationships with multiple partners, thereby avoiding formal duties that are incidental to marriage (Saur *et al.* 2005). Among men aged 15–49, 10.7% report having had sexual intercourse with more than one sexual partner in the past year (National Statistics Office 2014). In addition, married men (20%) are just as likely as unmarried men (21%) to pay for sex (National Statistics Office and ORC Macro 2001). While some married women are driven towards marriage due to poverty and the desire for companionship, most realise that this does not guarantee security and protection from HIV due to men's infidelity and the fact that polygynous marriage is legal in Malawi (Mkandawire-Valhmu *et al.* 2013). Married women live with the lingering threat that their husband may take on additional wives at any time (Saur *et al.* 2005). Hence, 14% of married women aged 15–49 are in polygynous unions (National Statistics Office 2014). Although marriage is an important risk factor for HIV infection among women, most feel they have no option but to get married due to their dependent economic situation.

Compared to their husbands, most Malawian women are less educated, have been socialised to obey their husbands, and are on the receiving end in sexual relationships where men control the sexual reproductive health decisions of the female body (Anderson 2012). Since men are the main breadwinners, wives are obligated to be sexually available to their husbands and to provide offspring, irrespective of HIV status. Women are unable to negotiate for safe sex even if the husband is HIV-positive or has other STIs (National Statistics Office and ORC Macro 2005). The transmission of HIV is directly linked to decisions men make in matters of sex: men's control over condom usage deprives women the ability to negotiate for protected sex. A woman's request for condom use can instigate gender-based violence as the suggestion might lead to accusations of infidelity by their partner. Consequently, women have no guaranteed protection from HIV transmission since their sexual and reproductive health choices are overtaken by sociocultural expectations and their subordinate status in society (Kathewera-Banda *et al.* 2005). There is also stigma around women accessing condoms, so it is the men who purchase, carry, and decide on condom usage. For Malawian women, being seen with a male condom carries the risk of stigmatisation and being labelled as a prostitute (Macintyre *et al.* 2013). As observed by Forster's ethnographic study, condom use goes against norms of male virility whereby among Malawian men:

> casual sexual activity is an essential expression of masculine enjoyment; sexual promiscuity is linked with geographical mobility and labour migration that facilitates sexual activity away from wives, neighbours, and kin who might otherwise report. There is also widespread view that "skin to skin" contact during sex is more appropriate. Commonly used analogies are that using a condom is like eating a sweet with the wrapper on or like having a shower with one's raincoat on. Furthermore, the desire for children is strongly emphasised culturally, hence some reject condoms because they prevent conception.
>
> (Forster 2001)

Although men control matters of sexuality and decisions about family size, contraceptive and condom use, most HIV programmes target women through HIV diagnoses made during antenatal care (Mkandawire-Valhmu *et al.* 2013). In fact, 91% of women between ages 15 and 49 who have had a live birth in the last two years report getting antenatal care during their pregnancy through which they were offered and accepted an HIV test and were given results (National Statistics Office 2014). Therefore, women are often the first to know, are blamed for infecting their husbands, and are seen as using their infected bodies to seduce and transmit HIV to men (Rankin *et al.* 2005, Anderson 2012). Most prevention of mother-to-child transmission of HIV (PMTCT) programmes demand partner disclosure which has, in most instances, resulted in women's abandonment by their husbands (Njunga and Blystad 2010, Joint United Nations Program on HIV/AIDS 2012, Mkandawire-Valhmu 2013). As a result, the mothers not only have to deal with the fear of transmitting the virus to their infants but also the loss of income and the social disgrace of a ruined family. A study of rural mothers enrolled in a PMTCT programme in Malawi's Chiradzulu district found a pattern of family disruption to such an extent that surrounding communities referred to the programme as "*pulojekitiyothetsamabanja*" [family disruption programme] (Njunga and Blystad 2010). In addition, the culturally-mandated postpartum abstinence period that lasts between 6 and 12 months makes women vulnerable to HIV infection since men are likely to engage in sexual activity with other women as they (the men) are viewed as not being capable of physically surviving for long periods without sex (Mkandawire-Valhmu *et al.* 2013). Hence, despite high levels of HIV awareness among Malawians (99% for both women and men) (National Statistics Office 2014), infection rates have not abated due to cultural practices.

Finally, the "sexual contract" implies a universal "masculine sex-right" to women's bodies for cultural sexual purposes including sexual cleansing (Anderson 2012). In November 2016, a 45-year old HIV-positive man was convicted after admitting to having slept with 104 women and girls in sexual cleansing rituals (Agence France-Presse 2016). According to the Malawi Human Rights Commission (MHRC), many cultural practices related to marriage, rites of passage, pregnancy, funerals, and other matters invariably involve sex. For example, widow cleansing is practised to appease a departed husband's spirit by hiring a *"fisi"* (hyena) to perform the cleansing (MHRC 2006), which in itself is a risk factor for HIV. Similarly, initiation practices are closely related to sexual behaviour since it is during such procedures that sexual instruction is provided, and sexual activity is expected to follow immediately after initiation where an older and sexually experienced "*fisi*" (hyena) is hired to perform a sexual act with the initiates (Forster 2001). Age asymmetry between the older man and the young girls increases the likelihood of HIV infection, especially since the men who act as "hyena" tend to hire out their services professionally, hence increasing the number of partners and HIV risk. Though on a limited scale, there is an emerging trend where some cultural practices that involve sexual acts for completion are being moderated through condoms and HIV-testing The repurposing of some traditional herbs known for healing ailments to symbolise sexual acts has also been reported. Although by no means implying a complete abandonment of associated traditions, this represents a renegotiation of cultural practices and meanings associated with particular rites of passage (Banda and Kunkeyani 2015). Cultural renegotiation should, therefore, be a critical part of a comprehensive HIV-prevention effort comprising women's empowerment and gender transformation.

Gender-based violence

Gender-based violence (GBV), including sexual violence, is a global phenomenon and is directly related to HIV because intimate partner violence increases the risk of HIV infection in women and girls. Broad, socially-transformative programmes that promote gender equality and discourage gender-based violence are a critical step towards curtailing HIV and attaining the health-related SDGs. GBV is an explicit manifestation of discrimination. Women who experience physical and/or sexual violence are less likely to negotiate condom use. Moreover, the unequal power between men and women and age asymmetry in pairings heighten the risk of abuse and vulnerability to HIV (Fried *et al.* 2012). GBV is deeply entrenched in Malawian society. Apart from the physical form, it manifests structurally through the exclusion of women from acquisition and control over resources such as land, jobs, education, credit, and other goods and services on the basis of their gender (Saur *et al.* 2005). Generally, physical violence in Malawi is accepted as a tool for resolving conflict. The use of corporal punishment in prisons, schools, and sometimes hospitals is common such that beating is a normative, accepted, and condoned way of interaction. This leads to internalisation of the idea that "educational beating" is a necessary tool to shape a person into a responsible adult or wife (Saur *et al.* 2005). Not surprisingly, 72% of children between the ages 1 and 14 have experienced psychological aggression or physical punishment (National Statistics Office 2014). This is pronounced among girls, 44% of whom report being touched on the breasts, butt, or genitals without permission (Mellish *et al.* 2015). Such socialisation normalises violence so that GBV in adulthood becomes acceptable, creating a pathway towards vulnerability. Gender inequality and the fear of violence makes it difficult for women living with HIV to access health services and stay healthy during pregnancy (Fried *et al.* 2012) leading to high rates of MMM. There is an acceptance that a husband is justified in beating his wife if she goes out without telling him, neglects the children, argues with him, refuses to have sex with him, or burns the food (National Statistics Office and ORC Macro 2005). Studies on the incidence of "intimate partner violence" in Malawi show this to be deeply embedded with women being subjected to various forms of violence: emotional, physical (being pushed, shaken, slapped, or punched), severe physical (being strangled, burned, threatened with a knife, gun, or with another weapon), and sexual (Bazargan-Hejazi *et al.* 2013). This has created the perfect storm for women's disempowerment and vulnerability to HIV and MMM.

Perpetrators of sexual violence are much more than the individuals who undertake such acts but also comprise institutions (family, workplaces, religious institutions, or public services) and structures – including the state – that are often complicit in the creation and/or protection of social, legal, or political–economic systems that enable, and often encourage, the prevalence of sexual violence in society (Kathewera-Banda *et al.* 2005). For example, linkages exist between the construction of sex and sexuality; the transaction of sex within the local economy and fish industry; and the influence of cultural practices on women's vulnerability to HIV transmission. A study in Nkhota-kota, a major fishing district in Malawi's central region found that unequal gender and power relations perpetuate violence against women, increase women's vulnerability to HIV infection, and infringe on their sexual and reproductive health rights. Meanwhile, women are often powerless to protect themselves from infection, to negotiate the terms of sex, or to leave a high risk and/or violent relationship. The study concludes that sex and sexuality are perceived and constructed in ways that rob women of their bodily autonomy, thereby making them more vulnerable to the intersecting experiences of sexual violence and HIV infection (Kathewera-Banda *et al.* 2005). The abuse is an even more urgent matter in cases where women living with HIV/AIDS experience abuse from their husbands/partners, families, and friends in the form of blame, humiliation, abandonment, and hopelessness that often leads to a deep sense of despair that prompts suicidal ideation in some cases (Chilemba *et al.* 2014). HIV-positive women report:

- Husbands/partners refusing to consistently use condoms
- Repeated incidents of physical assault and emotional pain
- Being forced to blame themselves for having HIV/AIDS through accusations of infidelity

- Husbands/partners being at the forefront, leading other people to humiliate them by showing the women's medication to other people
- Being abandoned or chased away by their husbands/partners immediately after diagnosis
- Friends and family members humiliating them through gossip and rejection.

(Chilemba *et al.* 2014)

GBV has a significant impact on Malawi's domestic product since most victims are unable to work during periods of victimisation. It is estimated that GBV costs 29 billion Malawi Kwacha per year (2013 exchange rate of US$1=MK325; Mellish *et al.* 2015). The 2006 Domestic Violence Act covers spousal relationships and includes "relations between family members" or financially dependent relations and was followed by the adoption of the National Strategy to combat GBV 2008–2013. However, poor dissemination and lack of budgetary support have affected implementation (Mellish *et al.* 2015). For example, the act considers lack of consent within intimate relationships (including, but not limited to, marriage) as a form of sexual coercion and, therefore, a form of sexual violence. However, some Malawians do not believe that forced sex can occur within the confines of marriage since women have a marital obligation to engage in sex with their husbands/partners. They believe that rape only occurs if a woman is forced to have sex by someone who is not her partner (Kathewera-Banda *et al.* 2005). This calls for involving men in programmes addressing sexual and reproductive health and GBV as an effective strategy to challenge harmful gender norms (Joint United Nations Program on HIV/AIDS 2012).

Systemic drivers: Institutional incapacity and unavailability of health and other services

The impact of systemic drivers that have contributed to the lack of access to high-quality effective health services and other development resources are manifested in HIV and MMM outcomes (Fried *et al.* 2012). The current state of healthcare in Malawi and other countries in SSA can be traced to the Structural Adjustment Programs where governments had to disinvest from social programmes and in the process transferred the social welfare burden to households. As a result, healthcare was transformed from a basic human right into a commodity. The Malawi government's expenditure on health constitutes only 9.7% of total government expenditure, which is far below the African Heads of State resolution to allocate 15% of the national budget to health (Zere *et al.* 2007). Families have increasingly drawn upon women to fill in the gap: a crumbling public healthcare system and poverty, compounded by HIV/AIDS, have exposed Malawian women to multiple burdens made more onerous by sexual practices and cultural beliefs. This has had devastating consequences on women as they have added the caregiving role for critically ill family members on top of their other caring and household duties (Rankin *et al.* 2005) in addition to their ill health due to HIV and MMM. It is not surprising that the women are "sick and tired, yet they continue to shoulder the burden of care which health policy formulated without acknowledging the real costs of exploiting women's labour as a substitute for genuine state-led health provision" (Burger 2004, i). Malawian women provide 87% of caregiving labour. This burden on women's time, labour, and resources has been added without changing traditional gender roles that would redistribute responsibilities (Burger 2004). The burden of caregiving has been extended to all generations of women in Malawi including elderly grandmothers who are increasingly called upon to take care of children orphaned by AIDS. As of 2013, Malawi had 790,000 AIDS orphans (UNAIDS 2013). Faced with the task of providing for children's basic needs, older women are often too weak for the physical labour required for subsistence farming. This is compounded by small land parcels such that inadequate yields are the norm. Hence, many older women rely on "*ganyu*" (piecework) in exchange for food and cash (Littrell *et al.* 2012).

Conclusion

Although many African countries underwent democratisation in the 1990s, this cannot be viewed as a transformative process when the majority of its population (women) are not able to experience the manifestation of such change through social and developmental dividends (Okonofua 2005). Forty-seven percent of deaths in pregnancy and childbirth occur in SSA where for every woman who dies due to obstetric complication, 30–50 women suffer morbidity and disability (Simwaka *et al.* 2005). High MMM reflects women's low status and gender inequality that leads to poor nutrition, ill health, multiple and unwanted high-risk pregnancies, illiteracy and lack of education, little or no control over resources, and limited autonomy to make decisions about reproductive health (Simwaka *et al.* 2005). These processes act as barriers to gender transformation and are manifested in women's vulnerability to HIV and MMM. Therefore, safeguarding women's rights and promoting gender equality is central to achieving all the MDGs (Fried *et al.* 2012). Although quite ambitious and wide-reaching, the implementation of MDGs focused on isolated objectives. The SDGs provide a new iteration of a global development agenda with an integrated approach that should facilitate tackling the root cause of gender disempowerment that fuels women's vulnerability to HIV and MMM. While commendable, ratifying conventions and making declarations about gender mainstreaming at international and national forums (Mbilizi 2013, Prugl and Lustgarten 2006, FAO 2011, Walthouse 2014) is insufficient on its own. Neither are constitutional provisions on gender equality, promotion of women's rights, and prohibition of gender discrimination (Malawi Government 2006a) if these instruments are used in isolation and the fundamental processes that disempower women remain untouched. Without an integrated approach, improvement in maternal health, promotion of gender equality, and women's empowerment will remain elusive for Malawian women (Malawi Government 2010a), and perilous outcomes with respect to HIV and MMM will remain the norm.

References

Agence, F.-P., 2016. Court convicts "hyena" who had sex with over 100 women [Online]. *The Guardian*. Available from https://www.theguardian.com/world/2016/nov/18/court-convicts-malawi-hyena-who-had-sex-with-100-women [Accessed 10 July 2017].

Anderson, E.L., 2012. Infectious women: Gendered bodies and HIV in Malawi. *International Feminist Journal of Politics*, 14(2), 267–287.

Antrobus, P., 2005. Critiquing MDGs from a Caribbean perspective 1. *Gender and Development*, 13(1), 94–104.

Banda, F. and Kunkeyani, T.E., 2015. Renegotiating cultural practices as a result of HIV in the eastern region of Malawi. *Culture, Health and Sexuality*, 17(1), 34–47

Barton, C., 2005. Where to for women's movements and the MDGs? *Gender and Development*, 13(1), 25–34.

Bazargan-Hejazi, S., Medeiros, S., Mohammadi, R., Lin, J., and Dalal, K., 2013. Patterns of intimate partner violence: A study of female victims in Malawi. *Journal of Injury and Violence Research*, 5(1), 38–50.

Burger, E., 2004. "*Ndadwala ndi Ndatopa*" (I am sick and I am tired): Women's caregiving and the real cost of health policy reform in Malawi Master's Thesis. Saint Mary's University.

Chilemba, W., van Wyk, N.C., and Leech, R., 2014. Development of guidelines for the assessment of abuse in women living with HIV/AIDS in Malawi. *African Journal for Physical, Health Education, Recreation and Dance*, 20(3:2), 1189–1201.

Food and Agricultural Organization of the United Nations (FAO), 2011. *Gender inequalities in rural employment in Malawi: Policy context*. FAO: Gender, Equity and Rural Employment Division.

Forster, P., 2001. AIDS in Malawi: Contemporary discourse and cultural continuities. *African Studies*, 60(2), 245–261.

Fowkes, F.J.L., Draper, B.L., Hellard, M., and Stoove, M., 2016. Achieving development goals for HIV, tuberculosis and malaria in sub-Saharan Africa through integrated antenatal care: barriers and challenges. *BMC Medicine*, 14, 202.

Fried, S., Harrison, B., Starcevich, K., Whitaker, C., and O'Konek, T., 2012. Integrating interventions on maternal mortality and morbidity and HIV: A human rights-based framework and approach. *Health and Human Rights*, 14(2), 21–33.

Harrigan, J., 2008. Food security, poverty and the Malawian starter pack: Fresh start or false start? *Food Policy*, 33(3), 237–249.

Joint United Nations Program on HIV/AIDS, 2012. *Together we will end AIDS*. Geneva: UNAIDS.

Kathewera-Banda, M. *et al.*, 2005. Sexual violence and women's vulnerability to HIV transmission in Malawi: A rights issue. *International Science Journal*, 57(186), 649–660.

Kumwenda, N.I. *et al.*, 2008. HIV-1 incidence among women of reproductive age in Malawi. *International Journal of STD and AIDS*, 19(5), 339–341.

Littrell, M., Murphy, L., Kumwenda, M., and Macintyre, K., 2012. Gogo care and protection of vulnerable children in rural Malawi: Changing responsibilities, capacity to provide, and implications for well-being in the era of HIV and AIDS. *Journal of Cross-Cultural Gerontology*, 27(4), 335–355.

Macintyre, L.M. *et al.*, 2013. Socially disempowered women as the key to addressing change in Malawi: How do they do it? *Health Care for Women International*, 34(2), 103–121.

Malawi Government, 2006a. *Constitution of the republic of Malawi.*

Malawi Government, 2006b. *Malawi growth and development strategy. From poverty to prosperity: 2006–2011.*

Malawi Government, 2010a. *Gender and development index, 2010.* Lilongwe: Ministry of Gender, Children and Community Development and the National Statistical Office.

Malawi Government, 2010b. *2010 Malawi millennium development goals report.* Lilongwe: Ministry of Development Planning and Cooperation.

Malawi Government, 2012. *Malawi growth and development Strategy II.* International Monetary Fund.

Malawi Human Rights Commission (MHRC), 2006. *Cultural practices and their impact on the enjoyment of human rights, particularly the rights of women and children in Malawi.* Lilongwe: MHRC.

Mbilizi, M.A., 2013. When a woman becomes president: Implications for gender policy and planning in Malawi. *Journal of International Women's Studies*, 14(3), 148–162.

Mellish, M., Settergren, S., and Sapuwa, H., 2015. *Gender-based violence in Malawi: A literature review to inform the national response.* Washington, DC: Futures Group, Health Policy Project.

Mkandawire-Valhmu, L. *et al.*, 2013. Marriage as a risk factor for HIV: learning from the experiences of HIV-infected women in Malawi. *Global Public Health*, 8(2), 187–201.

Moser, C. and Moser, A., 2005. Gender mainstreaming since Beijing: A review of success and limitations in international institutions. *Gender and Development*, 13(2), 11–22.

National Statistics Office Malawi and ORC Macro, US, 2001. *Malawi demographic health survey (MDHS), 2000.* Maryland: NSO and ICF Macro.

National Statistics Office Malawi and ORC Macro, US, 2005. *Malawi demographic health survey (MDHS), 2004.* Maryland: NSO and ICF Macro.

National Statistics Office Malawi and ICF Macro, US, 2011. *2010 Malawi demographic health survey (MDHS).* Maryland: NSO and ICF Macro.

National Statistics Office, 2014. *Malawi MDG endline survey* 2014, *key findings.* Zomba, Malawi: NSO.

Njunga, J. and Blystad, A., 2010. "The divorce program": Gendered experiences of HIV-positive mothers enrolled in PMTCT programs – the case of rural Malawi. *International Breastfeeding Journal*, 5, 14–19.

Ogu, R.N., Agholor, K.N., and Okonofua, F.E., 2016. Engendering the attainment of the SDG-3 in Africa: Overcoming the sociocultural factors contributing to maternal mortality. *African Journal of Reproductive Health (Special Edition),* 20(3), 62–74.

Okonofua, F.E., 2005. Achieving millennium development goals in Africa: How realistic? *African Journal of Reproductive Health*, 9(3), 7–14.

Organization for Economic Co-operation and Development, 2012. Social institutions and gender index (SIGI) [Online]. Available from http://www.genderindex.org/#_ftnref3/ [Accessed 10 July 2017].

Peters, P. and Kambewa, D., 2007. Whose security? Deepening social conflict over 'customary' land in the shadow of land tenure reform in Malawi. *Journal of Modern African Studies*, 45(3), 447–472.

Phiri, I.A., 1997. *Women, Presbyterianism and patriarchy: Religious experience of Chewa women in central Malawi.* Blantyre, Malawi: Christian Literature Association of Malawi.

Popkin, B.M., 2002. An overview on the nutrition transition and its health impacts. *Public Nutrition*, 5(1A), 93–103.

Prugl, E. and Lustgarten, A., 2006. Mainstreaming gender in international organization. In Jaquette, J.S. and Summerfield, G., eds. *Women and gender equity in development theory and practice: Institutions, resources and mobilization.* London: Duke University Press.

Rankin, S.H., Lindgren, T., Rankin, W.W., and Ng'oma, J., 2005. Donkey work: Women, religion and HIV/AIDS in Malawi. *Health Care for Women International*, 26(1), 4–16.

Reid, J. A. and Shams, T., 2013. Gender and Multigenerational Global Human Development, *Sociology Compass*, 7(8), 612–629.

Saith, A., 2006. From universal values to millennium development goals: Lost in translation. *Development and Change*, 37(6), 1167–1999.

Saur, M., Semu, L., and Hauya Ndau, S., 2005. *Nkhanza: Listening to people's voices: A study of gender-based violence, Nkhanza in three districts of Malawi.* Zomba: Montfort Media.

Simwaka, B.N., Theobald, S., Amekudzi, Y.P., and Tolhurst, R., 2005. Meeting millennium development goals 3 and 5. *BMJ*, 331, 708.

Thamaga-Chitja, J., 2012. How has the rural farming woman progressed since the setting up of the Millennium Development Goals for eradication of poverty and hunger? *Agenda: Empowering Women for Gender Equity*, 26(1), 67–80.

Tiessen, R., 2005. Mainstreaming gender in HIV/AIDS programs: Ongoing challenges and new opportunities in Malawi. *Journal of International Women's Studies*, 7(1), 8–25.

UNAIDS, 2013. *Malawi: HIV and AIDS estimates (2013).*

UNAIDS, 2015. Factsheet: 2014 Statistics [Online]. Geneva: UNAIDS. Available from http://www.unaids.org/en/resources/documents/2015/20150714_factsheet [Accessed 10 July 2017].

United States Department of State, 2009. U.S. president's emergency plan for emergency AIDS relief (PEPFAR) [Online]. Available from http://2006-2009.pepfar.gov/press/75919.htm [Accessed 10 July 2017].

Walthouse, E., 2014. Malawi's 50-50 Campaign [Online]. *BORGEN Magazine.* Available from http://www.borgenmagazine.com/malawis-50-50-campaign/ [Accessed 10 July 2017].

White, S., 2007. *Malawi: Country gender profile.* Japan International Cooperation Agency.

World Bank, 2017. Online data: Malawi [Online]. http://data.worldbank.org/country/malawi [Accessed 10 July 2017].

World Economic Forum, 2010. The Global Gender Gap Report 2010 [Online]. Available from http://www3.weforum.org/docs/WEF_GenderGap_Report_2010.pdf [Accessed 10 July 2017].

World Health Organization, 2005. *Malawi: HIV fact sheet, summary country profile for HIV/AIDS scale-up.*

Zere, E., Moeti, M., Kirigia, J., Mwase, T., and Kataika, E., 2007. Equity in health and healthcare in Malawi: an analysis of trends. *BMC Public Health*, 7(78).

Part III

Noncommunicable/ degenerative disease complex

8 Physical activity, nutrition, and hypertension in sub-Saharan Africa

The case of Ghana and South Africa

Eric Y. Tenkorang

Introduction

Hypertension, often referred to as high blood pressure, is a chronic condition that affects people from different ethnicities, religions, and socioeconomic backgrounds (Van de Vijver *et al.* 2013, Tenkorang *et al.* 2015b). Estimates by the World Health Organization (WHO) indicate that approximately 1 billion people aged 25 and above were diagnosed with hypertension in 2008 (World Health Organization 2013). Thus, hypertension has become a major cause of mortality leading to about 7.5 million deaths globally (World Health Organization 2013). It is also a major risk factor and often considered a co-morbid condition for other cardiovascular ailments such as stroke, diabetes, and coronary heart diseases. For instance, it is estimated that hypertension accounts for 51% of deaths due to stroke and 45% of ischemic heart diseases (World Health Organization 2013). While developed countries have a high prevalence of hypertension globally, low- to middle-income countries contribute substantially to global prevalence. In fact, recent estimates show that developing countries contribute more than 40% to the global prevalence of hypertension compared to 35% from the developed world (World Health Organization 2013). This means hypertension is not only associated with affluent or wealthier societies as was perceived in the past (Van de Vijver *et al.* 2013).

Similar to other developing countries, hypertension is highly prevalent in sub-Saharan Africa. Approximately, 80 million people live with hypertension in sub-Saharan Africa, and this is projected to increase to 150 million by 2025 (Van de Vijver *et al.* 2013). It is widely documented that such figures may be underestimated given that the majority of hypertensive cases go undiagnosed and undetected (Tenkorang *et al.* 2015b,Ataklte *et al.* 2015, Guwatudde *et al.* 2015, Echouffo-Tcheugui *et al.* 2015). The increasing prevalence of hypertension and other chronic diseases has been discussed in the context of Africa's changing socioeconomic, demographic, and epidemiologic transitions. For instance, Africa's contribution to the global economy has increased over the past few years as several countries have achieved sustained economic growth (African Development Bank 2012). The changing socioeconomic circumstances have had implications for the region's fertility and mortality transitions. Although relatively high compared to other parts of the world, mortality (mainly infant and child deaths) and fertility rates have reduced over the past several years (Bercher *et al.* 2004, Penfold *et al.* 2013). The changing mortality and fertility schedules have implications for Africa's population age structure. With declining fertility, some countries in the sub-Saharan African region have been projected to experience population ageing even faster than what was experienced by North America and Europe (Bongaarts 2010, Bongaarts and Casterline 2013, Shapiro 2015). Thus, the increase in chronic diseases, including hypertension, is often interpreted as a natural consequence of the changing age structure accompanying it.

It is important to acknowledge, however, that Africa's demographic or epidemiologic transition is unique, especially as declines in infant and child deaths have not completely given way to chronic and non-communicable diseases. Popularly called the 'double burden of disease', it is widely documented that several countries in the sub-Saharan African region have infectious/communicable diseases that cause mortality in infants and children coexisting with non-communicable/chronic diseases, including hypertension. Also, the socioeconomic improvements preceding these demographic transitions have led to changes in lifestyle and dietary habits within several populations in this region of the world.

The extant literature, mostly from the West, has documented lifestyle factors such as unhealthy diet, physical inactivity, tobacco use, etc., as risk factors for hypertension. However, the evidence of a link between these potential risk factors and hypertension has remained inconclusive in sub-Saharan Africa. Using data from Ghana and South Africa, I contribute to this debate by examining links between physical activity, nutrition, and hypertension in sub-Saharan Africa.

Theoretical and empirical considerations

As the study focuses on nutrition and physical activity, we employ Popkin's (2002) nutrition transition model to explore the risks of living with hypertension in Ghana and South Africa. Similar to the demographic and epidemiological transition models, the nutrition transition model examines the historical development of nutrition and physical activity as nations experience socioeconomic development. According to Popkin (2002), the initial phase of the nutrition transition was characterised by higher levels of famine and acute scarcity of food. At this stage of the life course, diets were high in carbohydrates but low in saturated fat. Populations were mostly nomadic, meaning physical activity levels were very high and obesity rates low. As nations experienced socioeconomic development, nutrition and levels of physical activity changed. The consumption of fruits, vegetables, and animal protein increased, while diets rich in staples and carbohydrates reduced. Activity patterns began shifting as inactivity and leisure increased (Popkin 2002). However, this stage was transitional and ushered in the post-transitional stage characterised by diets high in saturated fat, cholesterol, sugar, and other refined carbohydrates. With decreased physical activity, sedentary lifestyles became more common and obesity rates also increased. As a result, chronic and non-communicable diseases, including hypertension, diabetes, stroke, and other cardiovascular diseases became more prevalent.

High-income societies have completed the nutrition transition, but low- and middle-income countries, including those in sub-Saharan Africa, are yet to complete theirs. It is important to acknowledge that the transition in sub-Saharan African countries is delayed mainly due to political instability, economic stagnation, and the re-emergence of infectious and communicable diseases such as HIV/AIDS, cholera, Ebola, etc. (Popkin 2002).

Ghana and South Africa are at different phases of the development process, and as a result, different stages of their epidemiologic transitions. Ghana has enjoyed relative political stability for over two decades, and it is believed this stability has aided its economic transformation (Boafo-Arthur 2008). For instance, with a GDP of $37.86 billion dollars, Ghana is ranked a lower middle-income country. The discovery of oil in commercial quantities has added to its economic fortunes (Tenkorang *et al.* 2015a). Ghana's economic transformation reflects in its ongoing demographic and epidemiologic transitions, leading to the sudden emergence of chronic diseases as major causes of morbidity and mortality.

South Africa, on the other hand, is an upper middle-income country and arguably the wealthiest in sub-Saharan Africa. Similar to other countries in Africa, South Africa is known to have experienced rapid and complex health transitions (Kahn 2011). A dissection of the South African population paints a more complex picture of its epidemiologic transition especially as this differs by race and ethnicity. It is widely documented that the epidemiologic transition of South Africa has been delayed mainly due to the HIV/AIDS scourge (Oni and Mayosi 2016, Pillay-van Wyk *et al.* 2016).

While much has been written on the demographic and epidemiologic transition in sub-Saharan Africa, not a lot is known about its nutrition transition. Popkin (2002) attributes this lacuna to the lack of data on physical activity and nutritional measures on national surveys in sub-Saharan Africa. Meanwhile, studies on nutrition, physical activity, and hypertension have been inconclusive. While some studies discover links between fruit and vegetable intake and the risks of living with hypertension (Alonso *et al.* 2004, Utsugi *et al.* 2008, McCall *et al.* 2009), others find limited or no evidence of a relationship between these variables (Nunez-Cordoba 2009, Wang *et al.* 2012, Li *et al.* 2016). It is established, however, that fruit and vegetable intake is quite low among African populations compared to populations elsewhere (Peltzer and Phaswana-Mafuya 2012). Similar to nutrition, levels of physical activity have been found to be low among Africans (Assah *et al.* 2011, Sibai *et al.* 2013). Therefore, it is not too surprising that both childhood and adult obesity rates have been increasing for several African countries in the past few years (Adeboye *et al.* 2012, Scott *et al.* 2012, Muthuri *et al.* 2014, Gebremedhin 2015). Meanwhile, physical activity and obesity are identified as significant correlates of living with hypertension. We contribute to this body of literature by using data from the WHO's Study on Global Ageing and Adult Health (SAGE) for both Ghana and South Africa.

Data and methods

Data used for this study came from Wave 1 of SAGE collected in 2007/2008 by the World Health Organization (see www.who.int/healthinfo/sage/en/). The WHO's SAGE provides nationally-representative longitudinal data on the health and wellbeing of adult populations aged 50 years and above for six countries including Ghana and South Africa. It is important to note, however, that the survey also included smaller samples of young adults aged 18–49 years. The SAGE employed a multi-stage random sampling procedure to collect data from the six participating countries. First, primary sampling units were identified and stratified by administrative regions and localities. Enumeration areas were then selected and survey respondents further sampled from each stratum. For Ghana and South Africa, approximately 5,563 and 4,227 respondents were interviewed respectively. However, the analytical sample for this study was limited to respondents whose three systolic and diastolic values were available (Ghana = 5,071 and South Africa = 3,908). Ethical clearance was obtained from the WHO and the local ethical authorities for each participating country.

Measures

The dependent variable, hypertension, was measured through physical examinations using a Boso Medistar Wrist BP Monitor Model S (Minicuci *et al.* 2014). The Boso Medistar Wrist BP Monitor has been validated by the European Society of Hypertension as an efficient monitoring device that generates accurate measurements when the arm is well-positioned in relation to the heart (Basu and Millett 2013, Lloyd-Sherlock 2014). Data enumerators were trained to take physical or biometric measurements to ensure they were accurate and reliable. Before taking their measurements, respondents were asked to remain seated with their arms positioned at the level of their heart. Measurements for hypertension were taken on respondents' wrists after three deep slow breaths with a minute each in between measurements (Minicuci *et al.* 2014). For this study, a respondent was considered hypertensive if the average of the three measurements was greater than or equal to 140 mmHg for (systolic BP) and/or greater than or equal to 90 mmHg for (diastolic BP). It has been demonstrated that using systolic cut-points in the range of 130–180mm Hg and taking measurements two times or more may significantly reduce misclassification errors resulting from choosing specific systolic cut points (Friedman-Gerlicz and Lilly 2009).

Given the framework employed, we used measures of nutrition, physical activity, and obesity as focal independent variables. As the data used are cross-sectional, we were unable to capture changes in these predictors but were able to estimate if these measures were associated with the risks of becoming hypertensive. Two important measures of nutrition were used. These included questions that asked respondents: "How many servings of fruit (banana, mango, apple, orange, papaya, tangerine, grapefruit, peach, pear, etc.) do you eat on a typical day?" and "How many servings of vegetables (tomato, cauliflower, cucumber, peas, corn, lettuce, squash, bean, etc.) do you eat in a typical day?" The WHO recommends a minimum of five servings of fruits and vegetables per day, thus we recoded these variables to reflect this. Respondents' level of physical activity was measured with three dummy-coded variables. These included questions that asked respondents: "Does your work involve vigorous activity that causes large increases in breathing or heart rate (like heavy lifting, digging, or chopping wood) for at least 10 minutes continuously?", "Do you walk or use a bicycle (pedal cycle) for at least 10 minutes continuously to get to and from places?", and "Do you do any vigorous-intensity sports, fitness, or recreational (leisure) activities that cause large increases in breathing or heart rate (like running or football) for at least 10 minutes continuously?" Obesity rates within the samples were measured with a derived Body Mass Index variable created from anthropometric measures (height and weight of respondents). This was categorised into 'underweight', 'normal weight', 'overweight', and 'obese'. Socioeconomic variables (including education, income quintile, and occupation) and demographic variables (age, ethnicity/race, and religious affiliation) were used as statistical controls. In addition, we controlled for respondents living with other co-morbidities, such as diabetes and angina (severe pain in the chest).

Data analysis

The outcome variables used are dichotomous, requiring that binary logit models to be employed to estimate the relationship between nutrition, physical activity, obesity, and the risks of living with hypertension. Given the documented differences in the rates of hypertension for males and females across the world, the models are gender-specific. Three separate models were estimated: Model 1 includes nutrition, Model 2 adds variables on physical activity, and Model 3 adds obesity. In all these models, socioeconomic, demographic, and variables capturing other co-morbidities were controlled. Results are explained in odds ratios.

Results

Descriptive results for both Ghana and South Africa are presented in Tables 8.1 and 8.2. Results show a high prevalence of hypertension in both samples although higher in South Africa compared to Ghana. Gender differences are observed in prevalence for both countries. Males have higher prevalence compared to females in both populations, although the gender difference is larger for South Africa than Ghana. The WHO recommends at least five servings of fruits and vegetables daily. It is clear from the descriptive findings that the majority of Ghanaians and South Africans do not consume the recommended number of vegetables and fruits. On physical activity, we observe that Ghanaians compared to South Africans are engaged in work that involves vigorous activity causing large increases in breathing and, overall, are more physically active than their South African counterparts. Physical activity was higher among male respondents compared to females in both countries. It is, thus, not too surprising that the majority of South Africans are overweight or obese compared to their Ghanaian counterparts. However, obesity is higher among the female population for both countries when compared to the male population.

Table 8.1 Percentage distribution of selected dependent and independent variables (SAGE 2008)

	Ghana		South Africa	
	Male	*Female*	*Male*	*Female*
Variables				
Has hypertension?				
No	56.4	57.6	46.4	53.7
Yes	43.6	42.4	53.6	46.3
Nutrition				
Servings of fruits on a typical day?				
Less than 5 servings	90.5	90.0	80.5	76.9
5 or more servings	9.50	10.0	2.8	3.9
Not stated	–	–	16.7	19.1
Servings of vegetables on a typical day?				
Less than 5 servings	98.9	99.4	87.5	84.7
5 or more servings	1.1	0.6	6.0	4.7
Not stated	–	–	6.5	10.6
Physical activity				
Vigorous activity that causes heart to beat?				
No	40.8	57.6	79.4	87.9
Yes	59.2	42.4	20.6	12.1
Walk or bike for at least 10 minutes?				
No	18.8	27.1	52.7	66.6
Yes	81.2	72.9	47.3	33.4
Vigorous intensity sports, fitness, recreation?				
No	90.8	97.4	90.9	95.0
Yes	9.2	2.6	9.1	5.0
Body Mass Index				
Normal weight	65.0	48.2	38.3	29.6
Underweight	9.4	8.8	5.7	2.3
Overweight	17.2	26.4	27.9	26.6
Obese	8.4	16.6	28.1	41.3
Socioeconomic Controls				
Educational background				
No education	21.4	36.7	7.6	9.4
Primary education	32.4	34.4	31.0	25.5
Secondary education	40.4	26.3	54.2	57.7
Higher education	5.8	2.6	7.2	7.3
Wealth quintile				
Poorest	13.4	16.9	18.7	19.0
Poorer	17.7	18.7	22.7	16.6
Middle	19.3	18.7	17.7	22.9
Richer	21.6	23.2	22.7	16.4
Richest	27.9	22.5	18.2	25.1

(*continued*)

Table 8.1 Continued

	Ghana		*South Africa*	
	Male	*Female*	*Male*	*Female*
Employment status				
No	15.0	18.3	31.1	38.3
Yes	85.0	81.7	43.8	36.6
Not stated	–	–	25.1	25.1
Demographic Controls				
Average age of respondent	45.12	43.89	41.55	42.05
Marital status				
Married	84.3	58.8	53.6	29.5
Never married	8.6	9.1	27.2	34.5
Divorced/Widowed/Separated	7.1	32.1	19.2	36.1
Residence				
Rural	55.8	52.1	31.3	30.1
Urban	44.2	47.9	68.7	69.9
Co-morbidities				
Has diabetes				
No	98.6	97.1	97.7	95.9
Yes	1.4	2.9	2.3	4.1
Has stroke				
No	98.0	99.2	99.1	98.1
Yes	2.0	0.8	0.9	1.9

Note: Ethnicity not included here because this is different for both countries.

Table 8.2 examines the gross effects of nutrition, physical activity, and Body Mass Index of Ghanaians and South Africans on the likelihood of living with hypertension. Socioeconomic, demographic, and variables capturing co-morbidities are also examined. Results on the effects of nutrition on hypertension are counter-intuitive. Female and male respondents in South Africa who ate the recommended five servings of fruits and vegetables daily had higher odds of becoming hypertensive. On physical exercise, we observed that Ghanaian females who walked for at least 10 minutes continuously had lower odds of becoming hypertensive. Respondents' Body Mass Index was significantly associated with hypertension. Specifically, Ghanaians and South Africans who were obese had higher risks of becoming hypertensive compared to those with normal weights.

Tables 8.3 and 8.4 present results from multivariate models of hypertension for Ghana and South Africa, respectively. In all, I estimated three models for males and females and for the two countries. In Ghana, it was observed that having the recommended servings of fruits daily was associated with higher odds of living with hypertension for both males and females. Consuming the recommended servings of vegetables was, however, associated with lower odds of living with hypertension for females. For both male and female Ghanaians, engaging in physical

Table 8.2 Bivariate models of hypertension among Ghanaian and South African males and females (SAGE 2008)

	Ghana		South Africa	
	Male	*Female*	*Male*	*Female*
Nutrition				
Servings of fruits on a typical day?				
Less than 5 servings	1.00	1.00	1.00	1.00
5 or more servings	1.07 (.274)	2.13 (.631)***	.422 (.406)	2.17 (1.79)
Servings of vegetables on a typical day?				
Less than 5 servings	1.00	1.00	1.00	1.00
5 or more servings	1.20 (.750)	.543 (.322)	6.99 (5.07)***	4.08 (3.06)
Physical activity				
Vigorous activity that causes heart to beat?				
No	1.00	1.00	1.00	1.00
Yes	.822 (.141)	.790 (.149)	1.56 (.789)	1.13 (.179)
Walk or bike for at least 10 minutes?				
No	1.00	1.00	1.00	1.00
Yes	.816 (.179)	.585 (.125)***	1.40 (.592)	1.42 (.502)
Vigorous intensity sports, fitness, recreation?				
No	1.00	1.00	1.00	1.00
Yes	.647 (.218)	.709 (.482)	.760 (.432)	.784 (.809)
Body Mass Index				
Normal weight	1.00	1.00	1.00	1.00
Underweight	.875 (.239)	.955 (.875)	1.86 (1.21)	1.55 (.933)
Overweight	1.60 (.374)**	1.49 (.349)	2.13 (1.17)	1.61 (.812)
Obese	4.09 (1.48)***	2.54 (.709)***	3.26 (1.56)***	5.05 (2.45)***
Socioeconomic Controls				
Educational background				
No education	1.00	1.00	1.00	1.00
Primary education	.803 (.167)	.862 (.186)	.268 (.140)***	.820 (.296)
Secondary education	.923 (.183)	.641 (.156)	.284 (.145)***	.219 (.079)***
Higher education	.457 (.196)	.478 (.306)	.559 (.338)	.254 (.165)**
Wealth quintile				
Poorest	1.00	1.00	1.00	1.00
Poorer	.965 (.257)	1.07 (.325)	1.83 (1.11)	.761 (.429)
Middle	1.38 (.368)	2.11 (.625)***	4.59 (2.86)**	.746 (.460)
Richer	1.16 (.305)	1.92 (.559)**	1.59 (.950)	.632 (.300)
Richest	1.28 (.329)	1.34 (.399)	1.58 (.883)	.541 (.291)
Employment status				
No	1.00	1.00	1.00	1.00
Yes	.902 (.201)	.785 (.164)	.754 (.339)	.377 (.165)**
Demographic Controls				
Age of respondent	1.02 (.010)***	1.03 (.010)***	1.04 (.013)***	1.07 (.012)***
Marital status				
Married	1.00	1.00	1.00	1.00

(*continued*)

Table 8.2 Continued

	Ghana		*South Africa*	
	Male	*Female*	*Male*	*Female*
Never married	.744 (.252)	1.24 (.468)	.794 (.401)	.319 (.131)***
Divorced/Widowed/ Separated	1.73 (.427)**	1.84 (.352)***	.756 (.435)	.465 (.199)
Residence				
Rural	1.00	1.00	1.00	1.00
Urban	.978 (.168)	1.82 (.340)***	.832 (.340)	.524 (.187)
Co-morbidities				
Has diabetes				
No	1.00	1.00	1.00	1.00
Yes	.792 (.404)	2.68 (1.57)	1.01 (.573)	2.51 (1.28)
Has stroke				
No	1.00	1.00	1.00	1.00
Yes	1.28 (.860)	5.31 (1.87)***	4.70 (2.35)***	7.46 (4.14)***

Note: **p < .05; ***p < .01; ethnicity not included here because this is different for both countries.

activity was associated with lower odds of living with hypertension. Respondents' Body Mass Index emerged as the most important predictor of hypertension judging from improvements in pseudo r-squares and reductions in -2 log likelihood ratios from Models 2 and 3 in Table 8.3.

Results are quite different for South Africa. For male South Africans, having five or more servings of fruits was associated with lower odds of living with hypertension. However, consuming the recommended servings of vegetables was rather associated with higher odds of living with hypertension, especially for male South Africans. Also engaging in physical activity was associated with higher risks of living with hypertension. Similar to Ghana, respondents' Body Mass Index appeared to be the most important predictor of hypertension given significant improvement in pseudo r-square and reductions in –2 log likelihood ratios. This was, however, more the case for female than male South Africans.

Although not the focus of this chapter, it is important to highlight the gender and country-specific differences in the influences of the socioeconomic characteristics of respondents on their risks of living with hypertension. For instance, in Ghana, we did not see any effects of income on hypertension for men. For women, however, higher income was associated with higher odds of living with hypertension. Higher education was protective of hypertension for both male and female Ghanaians. In South Africa, higher income was protective of the risks of living with hypertension for women but not for men. South African men and women with primary and secondary education were significantly less likely to be hypertensive compared to those with no education. It was also observed that while men with employment were more likely to be hypertensive compared to the unemployed, women with employment were rather less likely to be hypertensive.

Table 8.3 Multivariate models of hypertension among Ghanaian males and females (SAGE 2008)

	Male			*Female*		
	Model 1	*Model 2*	*Model 3*	*Model 1*	*Model 2*	*Model 3*
Nutrition						
Servings of fruits on a typical day?						
Less than 5 servings	1.00	1.00	1.00	1.00	1.00	1.00
5 or more servings	1.48 (.153)***	1.61 (.156)***	1.42 (.161)**	3.00 (.165)***	3.13 (.167)***	3.23 (.173)***
Servings of vegetables on a typical day?						
Less than 5 servings	1.00	1.00	1.00	1.00	1.00	1.00
5 or more servings	.490 (.470)	.526 (.472)	.652 (.483)	.167 (.617)***	.205 (.627)***	.135 (.677)***
Physical activity						
Vigorous activity that causes heart to beat?						
No		1.00	1.00		1.00	1.00
Yes		.812 (.100)**	.845 (.103)		1.07 (.102)	1.14 (.105)
Walk or bike for at least 10 minutes?						
No		1.00	1.00		1.00	1.00
Yes		.891 (.113)	1.12 (.120)		.735 (.106)***	.736 (.110)***
Vigorous intensity sports, fitness, recreation?						
No		1.00	1.00		1.00	1.00
Yes		.625 (.165)***	.492 (.176)***		.493 (.308)**	.434 (.319)***
Body Mass Index						
Normal weight			1.00			1.00
Underweight			.736 (.156)**			.903 (.180)
Overweight			2.06 (.125)***			1.43 (.121)***
Obese			4.28 (.186)***			3.04 (.141)***

Socioeconomic Controls						
Educational background						
No education	1.00	1.00	1.00	1.00	1.00	1.00
Primary education	.659 (.149)***	.661 (.150)***	.612 (.153)***	.803 (.123)	.755 (.125)**	.726 (.138)***
Secondary education	.703 (.151)**	.743 (.152)**	.729 (.155)**	.396 (.142)***	.380 (.144)***	.349 (.148)***
Higher education	.316 (.245)***	.293 (.247)***	.282 (.255)***	.323 (.311)***	.385 (.316)***	.401 (.320)***
Wealth quintile						
Poorest	1.00	1.00	1.00	1.00	1.00	1.00
Poorer	.830 (.163)	.847 (.164)	.786 (.167)	1.12 (.162)	1.13 (.163)	1.13 (.166)
Middle	1.27 (.159)	1.24 (.158)	1.11 (.161)	2.56 (.157)***	2.57 (.158)***	2.28 (.163)***
Richer	.970 (.164)	.930 (.165)	.823 (.169)	2.44 (.159)***	2.45 (.160)***	2.04 (.167)***
Richest	1.32 (.168)	1.30 (.170)	1.01 (.175)	1.73 (.170)***	1.77 (.172)***	1.38 (.180)
Employment status						
No	1.00	1.00	1.00	1.00	1.00	1.00
Yes	.857 (.132)	.950 (.135)	1.03 (.140)	1.17 (.121)	1.18 (.122)	1.13 (.130)
Demographic controls						
Age of respondent	1.02 (.004)***	1.01 (.004)***	1.02 (.004)***	1.02 (.004)***	1.02 (.004)***	1.02 (.004)***
Marital status						
Married	1.00	1.00	1.00	1.00	1.00	1.00
Never married	1.24 (.189)	1.35 (.193)	1.36 (.200)	2.22 (.181)***	2.58 (.194)***	3.30 (.199)***
Divorced/Widowed/Separated	1.19 (.182)	1.18 (.182)	1.38 (.187)	1.46 (.110)***	1.47 (.111)***	1.39 (.114)***
Residence						
Rural	1.00	1.00	1.00	1.00	1.00	1.00
Urban	.963 (.100)	.940 (.105)	.887 (.107)	1.92 (.101)***	1.88 (.103)***	1.66 (.107)***

(*continued*)

Table 8.3 Continued

	Male			*Female*		
	Model 1	*Model 2*	*Model 3*	*Model 1*	*Model 2*	*Model 3*
Ethnicity						
Akan	1.00	1.00	1.00	1.00	1.00	1.00
Ewe	1.28 (.213)	1.22 (.214)	.945 (.221)	.922 (.229)	.877 (.566)	.857 (.239)
Ga Adangbe	.609 (.149)***	.592 (.150)***	.629 (.157)***	1.10 (.144)	1.06 (.146)	.919 (.150)
Northern tribes	.589 (.137)***	.609 (.138)***	.605 (.141)***	1.38 (.138)**	1.39 (.138)**	1.28 (.141)
Other	.429 (.142)	.446 (.143)***	.406 (.148)***	.409 (.149)***	.405 (.150)***	.377 (.154)***
Co-morbidities						
Has diabetes						
No	1.00	1.00	1.00	1.00	1.00	1.00
Yes	.798 (.392)	.816 (.394)	.786 (.413)	1.75 (.291)	1.64 (.292)	1.93 (.300)**
Has stroke						
No	1.00	1.00	1.00	1.00	1.00	1.00
Yes	.797 (.349)	.827 (.353)	.887 (.373)	4.37 (.627)**	4.33 (.628)**	5.09 (.712)**
Pseudo R-squared	**.069**	**.076**	**.126**	**.181**	**.188**	**.223**
–2 Log-likelihood ratio	**3079.59**	**3065.27**	**2942.35**	**2849.98**	**2835.87**	**2730.29**

Note: $^{**}p < .05$; $^{***}p < .01$; Odds ratios reported and standard errors are in brackets.

Table 8.4 Multivariate models of hypertension among South African males and females (SAGE 2008)

	Male			*Female*		
	Model 1	*Model 2*	*Model 3*	*Model 1*	*Model 2*	*Model 3*
Nutrition						
Servings of fruits on a typical day?						
Less than 5 servings	1.00	1.00	1.00	1.00	1.00	1.00
5 or more servings	.332 (.414)***	.312 (.419)***	.336 (.404)***	.403 (.365)***	.377 (.366)***	.569 (.379)
Not stated	1.39 (.180)	1.39 (.184)	1.38 (.187)	1.34 (.166)	1.33 (.166)	2.781 (.195)***
Servings of vegetables on a typical day?						
Less than 5 servings	1.00	1.00	1.00	1.00	1.00	1.00
5 or more servings	7.34 (.327)***	8.81 (.331)***	2.23 (.385)**	2.64 (.367)***	2.60 (.366)***	1.04 (.390)
Not stated	.765 (.281)	.652 (.297)	.650 (.299)	2.23 (.219)***	2.23 (.221)***	.924 (.264)
Physical activity						
Vigorous activity that causes heart to beat?						
No		1.00	1.00		1.00	1.00
Yes		1.68 (.165)***	1.47 (.170)**		1.46 (.182)**	1.61 (.193)***
Walk or bike for at least 10 minutes?						
No		1.00	1.00		1.00	1.00
Yes		1.65 (.125)***	1.90 (.127)***		1.09 (.130)	1.13 (.139)
Vigorous intensity sports, fitness, recreation?						
No		1.00	1.00		1.00	1.00
Yes		.767 (.227)	.836 (.236)		.916 (.257)	1.51 (.146)
Body Mass Index						
Normal weight			1.00			1.00
Underweight			1.80 (.265)**			.455 (.395)**
Overweight			1.67 (.155)***			1.50 (.161)***
Obese			3.08 (.165)***			5.19 (.154)***

(continued)

Table 8.4 Continued

	Male			*Female*		
	Model 1	*Model 2*	*Model 3*	*Model 1*	*Model 2*	*Model 3*
Socioeconomic Controls						
Educational background						
No education	1.00	1.00	1.00	1.00	1.00	1.00
Primary education	.413 (.257)***	.452 (.265)***	.461 (.274)***	1.29 (.231)	1.30 (.232)	1.03 (.249)
Secondary education	.486 (.267)***	.485 (.276)***	.405 (.289)***	.893 (.231)	.848 (.233)	.468 (.255)***
Higher education	1.25 (.331)	1.14 (.342)	.971 (.363)	1.50 (.316)	1.49 (.319)	1.27 (.345)
Wealth quintile						
Poorest	1.00	1.00	1.00	1.00	1.00	1.00
Poorer	2.08 (.171)***	1.99 (.175)***	1.90 (.182)***	.912 (.193)	.855 (.195)	.736 (.208)
Middle	7.07 (.200)***	7.91 (.210)***	6.32 (.222)***	1.11 (.185)	1.16 (.186)	.943 (.199)
Richer	1.33 (.178)	1.47 (.189)**	1.45 (.199)	.630 (.195)**	.635 (.197)**	.437 (.216)***
Richest	1.32 (.215)	1.59 (.225)**	1.45 (.237)	.513 (.229)***	.526 (.231)***	.292 (.251)***
Employment status						
No	1.00	1.00	1.00	1.00	1.00	1.00
Yes	1.67 (.148)***	1.65 (.155)***	1.81 (.160)***	.672 (.137)***	.670 (.141)***	.802 (.152)
Not stated	.633 (.168)***	.644 (.171)***	.611 (.117)	.764 (.149)	.768 (.150)	.863 (.165)
Demographic Controls						
Age of respondent	1.05 (.006)***	1.05 (.006)***	1.04 (.006)***	1.07 (.005)***	1.07 (.005)***	1.05 (.006)***
Marital status						
Married	1.00	1.00	1.00	1.00	1.00	1.00
Never married	1.79 (.166)***	1.76 (.170)***	1.72 (.175)***	.486 (.155)***	.470 (.157)***	.669 (.170)**
Divorced/Widowed/Separated	1.26 (.154)	1.24 (.159)	1.26 (.163)	.265 (.150)***	.250 (.154)***	.295 (.162)***

Residence						
Rural	1.00	1.00	1.00	1.00	1.00	1.00
Urban	.542 (.134)***	.591 (.136)***	.550 (.137)***	.599 (.138)***	.623 (.141)***	.891 (.155)
Ethnicity						
African	1.00	1.00	1.00	1.00	1.00	1.00
White	1.05 (.266)	1.33 (.267)	1.16 (.277)	.748 (.295)	.696 (.297)	1.01 (.306)
Coloured	6.96 (.282)***	8.32 (.291)***	7.61 (.306)***	.652 (.183)**	.618 (.187)***	.724 (.203)
Indian	6.41 (.434)***	7.97 (.443)***	8.65 (.457)***	1.89 (.262)**	1.85 (.268)**	2.69 (.300)***
Others	1.03 (.152)	1.15 (.146)	.850 (.159)	.892 (.175)	.914 (.178)	.691 (.155)
Co-morbidities						
Has diabetes						
No	1.00	1.00	1.00	1.00	1.00	1.00
Yes	.276 (.367)***	.256 (.374)***	.312 (.371)	1.24 (.277)	1.25 (.278)	.797 (.292)
Has stroke						
No	1.00	1.00	1.00	1.00	1.00	1.00
Yes	2.30 (.809)	2.62 (.827)	4.47 (.932)	3.33 (.507)**	3.44 (.509)**	5.17 (.556)***
Pseudo R-squared	**.304**	**.321**	**.329**	**.365**	**.368**	**.447**
–2 log-likelihood ratio	**2131.4**	**2089.78**	**2012.91**	**2.167.16**	**2160.32**	**1913.71**

Note: **p < .05; ***p < .01; Odds ratios reported and standard errors are in brackets.

Discussion

With prevalence of over 40%, the analyses demonstrate that hypertension is a major public health threat for Ghana and South Africa. Most important is that the prevalence estimated in this study reflects what exists across the African continent, although some heterogeneity or differences may be observed in several other countries. In their article titled "Burden of undiagnosed hypertension in sub-Saharan Africa", Akakite *et al.* (2015) estimated the pooled prevalence of hypertension in sub-Saharan as 30%. These estimates clearly suggest changes in the morbidity profile for a region of the world where communicable diseases such as cholera, tuberculosis, diphtheria, etc. were common and accounted for the majority of deaths within its populations.

However, the changing morbidity and mortality profile in Africa may be understood as part of the broader socioeconomic and demographic transitions occurring within these populations. For instance, over the years, infant and child mortality have declined steadily (Amouzou and Hill 2004, Kanmiki *et al.* 2014). Combined with declines in fertility rates, life expectancy has continued to increase, albeit slowly. Thus, the sudden emergence of chronic conditions, such as hypertension, may reflect the beginning of an ageing process within these populations (Stuckler 2008, Ezeh *et al.* 2015). Similarly, the changing socioeconomic circumstances and increasing urbanisation have led to drastic lifestyle changes reflected mostly in diets and physical activity levels. It is documented that increasingly Africans have abandoned their traditional diets which are rich in fibre, vegetables, and other nutritious foods for imported and processed foods that are rich in saturated fat and sugar (Vorster *et al.* 2011, Abrahams *et al.* 2011). The data showed that very few respondents from both Ghana and South Africa consumed the recommended servings of fruits and vegetables often considered healthy and protective of these chronic conditions. This is consistent with the findings of Peltzer and Phaswanya-Mafuya (2012) who also reported inadequate fruit and vegetable intake in the South African population using the same data.

A growing middle class, concentrated mostly in white-collar occupations and with increased access to computers and automobiles, has also meant limited physical activity within these populations. It is, thus, not surprising that Ghanaians and South Africans rarely engaged in physical exercise as witnessed in this analysis. The combined effects of unhealthy diet and limited physical activity might explain the high rates of obesity in these populations, especially in South Africa.

Guided by Popkin's framework, we examined the effects of nutrition and physical activity on the risks of living with hypertension for both Ghana and South Africa. Results were mixed with some consistent with theoretical expectations and others contrary to such expectations. For instance, we found evidence of higher odds for South Africans who ate the recommended number of servings of vegetables, but lower odds for those who ate the recommended number of fruits daily. In Ghana, however, consuming the recommended servings of vegetables was associated with lower odds for females, while consuming the recommended number of fruits was linked to lower odds of living with hypertension for both males and females. These inconsistent findings are not particularly surprising as previous studies have shown such inconsistencies between fruit and vegetable consumption and hypertension (see Fan *et al.* 2010, Wang *et al.* 2012, Li *et al.* 2016). Li and colleagues (2016) note that fruits and vegetables have specific micronutrients, including minerals, vitamins, and folic acid, that benefit the functioning of the endothelium, whose dysfunction has been found to increase the risks of hypertension. This might explain the lower odds of hypertension found in respondents with sufficient and recommended servings of fruits and vegetable consumption. For those who consumed the recommended servings of fruits and vegetables and were hypertensive, it is possible that the decision to consume higher levels of fruits and vegetables was taken retroactively – that is, after they had been diagnosed with hypertension. However, as the data are cross-sectional, we are unable to show the temporal sequencing between fruit/vegetable consumption and when respondents were diagnosed with hypertension. Longitudinal data with proper time order may help deal with the endogenous relationship between these two variables.

Similar to nutrition, physical activity was associated with the risks of hypertension, but the direction of the association was not clear. For instance, both male and female Ghanaian respondents who engaged in physical exercise had lower odds of living with hypertension. This was not the case in South Africa where engaging in physical activity increased the odds of living with hypertension. Physical activity is known to have psychological benefits including dealing with stress, a known risk factor for hypertension (Tsioufis *et al.* 2010, Teh *et al.* 2015, Borjesson *et al.* 2016). Thus, it is expected that engaging in moderate to vigorous physical activity should help deal with emotions that trigger stress-related responses with implications for high blood pressure. Similar to the explanation on nutrition, we believe the counter-intuitive relationship between physical activity and hypertension may be a result of retroactive decisions taken by respondents after they had been diagnosed with the condition. Again, the limitations of the data mean we are unable to clearly demonstrate that physical activity may have occurred after diagnosis. It is important to note, however, that a cross-classification analysis of the nutrition and physical activity variables and the Body Mass Index of respondents showed that a higher percentage of the obese and overweight in Ghana and, especially, in South Africa were more likely to eat healthily and engage in physical activity, compared to respondents with normal weight. Meanwhile, the obese and overweight were significantly more likely to be hypertensive. These results taken together may be indicative that respondents' knowledge of their diagnosis influenced their decision to be physically active. The extant literature is conclusive on the links between Body Mass Index and the risks of hypertension. Generally, the finding is that the overweight and obese have increased risks of living with hypertension (Landsberg *et al.* 2013, Wang *et al.* 2014, Hall 2015) and the results of this study corroborate these documented and established findings.

Although not the focus of this chapter, the findings demonstrated gender and country-specific differences in the effects of socioeconomic variables on hypertension. In Ghana, for instance, higher education was protective of both males and females against hypertension. Similarly, education was protective of South African males and females, but mostly for those with primary and secondary education. Formal education is a necessary, albeit insufficient, condition for health promotion and disease prevention. It equips individuals with knowledge of the causes and symptoms of disease, making it crucial for avoiding chronic ailments (Dalstra *et al.* 2005, Choi *et al.* 2011). Income, on the other hand, provides access to resources including information and healthcare. Yet these results showed that higher income was a significant risk factor for Ghanaian women and South African men but not South African women. Most important was that Body Mass Index attenuated and completely mediated the risks of wealthy Ghanaian men and South African women. These findings are consistent with the work of Cois and Erhlich (2014) who, using South African data, reported Body Mass Index as a strong mediator of the effects of socioeconomic status on blood pressure. Furthermore, these findings corroborate the conclusion that in developing and transitional countries such as Ghana and South Africa, the relationship between socioeconomic status and blood pressure reflects a high prevalence of obesity, especially among those with higher income (Colhoun *et al.* 1998). The gender dynamics suggest that obesity rates are higher among South African men and wealthy Ghanaian women respectively, placing them at increased risks of hypertension.

The findings suggest that policymakers pay specific attention to nutrition, physical activity, and obesity rates within these populations and other similar countries in sub-Saharan Africa. As the results showed, the majority of respondents did not consume the recommended servings of fruits and vegetables and rarely engaged in physical activity. It was, thus, not surprising that obesity rates were high in these samples. It is important for policymakers to target populations in sub-Saharan Africa with messages of healthy eating and the importance of engaging in physical activity. Although the evidence is not very conclusive, the analysis suggested that the decision to engage in healthy eating and lifestyle may be occurring retroactively, mostly after diagnosis. This means health promotion strategies should target individuals earlier in the life course and also focus on early detection. As there are gender differences in hypertension risks, it will be important to consider gender-specific interventions. For instance, as obesity and hypertension risks were high for Ghanaian women, it will be important to target this sub-sample with relevant health promotion messages that can reduce their risks.

References

Abrahams, Z., Mchiza, Z., and Steyn, N., 2011. Diet and mortality rates in Sub-Saharan Africa: Stages in the nutrition transition. *BMC Public Health*, 11(1).

Adeboye, B., Bermano, G., and Rolland, C., 2012. Obesity and its health impact in Africa: A systematic review. *Cardiovascular Journal of Africa*, 23(9), 512–521.

African Development Bank, 2012. *Republic of Ghana country strategy paper, 2012–2016.* Tunis-Belvedere, Tunisia.

Alonso, A. *et al.*, 2004. Fruit and vegetable consumption is inversely associated with blood pressure in a Mediterranean population with a high vegetable-fat intake: the

Seguimiento Universidad de Navarra (SUN) Study. *British Journal of Nutrition*, 92(2), 311.

Amouzou, A. and Hill, K., 2004. Child mortality and socio-economic status in sub-Saharan Africa. *Etude de la Population Africaine*, 19(1), 1–12.

Assah, F., Ekelund, U., Brage, S., Mbanya, J.C., and Wareham, N.J., 2011. Urbanization, physical activity, and metabolic health in sub-Saharan Africa. *Diabetes Care*, 34(2), 491–496.

Ataklte, F. *et al.*, 2015. Burden of undiagnosed hypertension in sub-Saharan Africa: A systematic review and meta-analysis. *Hypertension*, 65(2), 291–298.

Basu, S. and Millet, C., 2013. Social epidemiology of hypertension in middle-income countries: determinants of prevalence, diagnosis, treatment, and control in the WHO SAGE study. *Hypertension*, 62(1), 18–26.

Becher, H. *et al.*, 2004. Risk factors of infant and child mortality in rural Burkina Faso. *Bulletin of the World Health Organization*, 82(4).

Boafo-Arthur, K., 2008. *Democracy and stability in West Africa*. Uppsala: Department of Peace and Conflict Research, Uppsala University.

Bongaarts, J., 2010. *The causes of educational differences in fertility in sub-Saharan Africa*. New York: The Population Council.

Bongaarts, J. and Casterline, J., 2013. Fertility transition Is sub-Saharan Africa different? *Population Development Review*, 38(Suppl 1), 153–168.

Börjesson, M., Onerup, A., Lundgvist, S., and Dahlöf, B., 2016. Physical activity and exercise lower blood pressure in individuals with hypertension: narrative review of 27 RCTs. *British Journal of Sports Medicine*, 50(6), 356–361.

Choi, A. *et al.*, 2011. Association of educational attainment with chronic disease and mortality: The kidney early evaluation program (KEEP). *American Journal of Kidney Diseases*, 58(2), 228–234.

Cois, A. and Ehrlich, R., 2014. Analysing the socioeconomic determinants of hypertension in South Africa: a structural equation modelling approach. *BMC Public Health*, 14(1).

Colhoun, H., Hemingway, H., and Poulter, N., 1998. Socio-economic status and blood pressure: an overview analysis. *Journal of Human Hypertension*, 12(2), 91–110.

Dalstra, J., 2005. Socioeconomic differences in the prevalence of common chronic diseases: an overview of eight European countries. *International Journal of Epidemiology*, 34(2), 316–326.

Echouffo-Tcheugui, J., Kengne, A., Erqou, S., and Cooper, R., 2015. High blood pressure in sub-Saharan Africa: the urgent imperative for prevention and control. *The Journal of Clinical Hypertension*, 17(10), 751–755.

Ezeh, O.K., Agho, K.E., Dibley, M.J., Hall, J.J., and Page, A.N., 2015. Risk factors for postneonatal, infant, child and under-5 mortality in Nigeria: a pooled cross-sectional analysis. *BMJ Open*, 5(3), e006779–e006779.

Fan, A., Mallawaarachchi, D.S., Gilbertz, D., Li, Y., and Mokdad, A.H., 2010. Lifestyle behaviors and receipt of preventive health care services among hypertensive Americans aged 45 years or older in 2007. *Preventive Medicine*, 50(3), 138–142.

Friedman-Gerlicz, C. and Lilly, I., 2009. Misclassification rates in hypertension diagnosis due to measurement errors. *SIAM Undergraduate Research Online*, 2(2), 46–57.

Gebremedhin, S., 2015. Prevalence and differentials of overweight and obesity in preschool children in Sub-Saharan Africa. *BMJ Open*, 5(12), e009005.

Guwatudde, D. *et al.*, 2015. The burden of hypertension in sub-Saharan Africa: a four-country cross sectional study. *BMC Public Health*, 15(1).

Hall, J.E., do Carmo, J.M., da Silva, A.A., Wang, Z., and Hall, M.E., 2015. Obesity-induced hypertension: interaction of neurohumoral and renal mechanisms. *Circulation Research*, 116(6), 91–1006.

Kahn, K., 2011. Population health in South Africa: dynamics over the past two decades. *Journal of Public Health Policy*, 32(S1), S30–S36.

Kanmiki, E. *et al.*, 2014. Socio-economic and demographic determinants of under-five mortality in rural northern Ghana. *BMC International Health Human Rights*, 14(1).

Landsberg, L. *et al.*, 2013. Obesity-related hypertension: pathogenesis, cardiovascular risk, and treatment-A position paper of the Obesity Society and the American Society of Hypertension. *Obesity*, 21(1), 8–24.

Li, B., Li, F., Wang, L., and Zhang, D., 2016. Fruit and vegetables consumption and risk of hypertension: a meta-analysis. *Journal of Clinical Hypertension*, 18(5), 468–476.

Lloyd-Sherlock, P., Beard, J., Minicuci, N., Ebrahim, S., and Chatterji, S., 2014. Hypertension among older adults in low- and middle-income countries: prevalence, awareness and control. *International Journal of Epidemiology*, 43(1), 116–128.

McCall, D. *et al.*, 2009. Dietary intake of fruits and vegetables improves microvascular function in hypertensive subjects in a dose-dependent manner. *Circulation*, 119(16), 2153–2160.

Minicuci, N. *et al.*, 2014. Sociodemographic and socioeconomic patterns of chronic non-communicable disease among the older adult population in Ghana. *Global Health Action*, 7(0).

Muthuri, S. *et al.*, 2014. Evidence of an overweight/obesity transition among school-aged children and youth in sub-Saharan Africa: a systematic review. *PLoS ONE*, 9(3), e92846.

Nuñez-Cordoba, J.M., Alonso, A., Beunza, J.J., Palma, S., Gomez-Gracia, E., and Martinez-Gonzalez, M.A., 2009. Role of vegetables and fruits in Mediterranean diets to prevent hypertension. *European Journal of Clinical Nutrition*, 63(5), 605–612.

Oni, T. and Mayosi, B., 2016. Mortality trends in South Africa: progress in the shadow of HIV/AIDS and apartheid. *The Lancet Global Health*, 4(9), e588–e589.

Peltzer, K. and Phaswana-Mafuya, N., 2012. Fruit and vegetable intake and associated factors in older adults in South Africa. *Global Health Action*, 5(0).

Penfold, S., Willey, B., and Schellenberg, J., 2013. Newborn care behaviours and neonatal survival: evidence from sub-Saharan Africa. *Tropical Medicine and International Health*, 18(11), 1294–1316.

Pillay-van Wyk, V. *et al.*, 2016. Mortality trends and differentials in South Africa from 1997 to 2012: second National Burden of Disease Study. *The Lancet Global Health*, 4(9), e642–e653.

Popkin, B., 2002. Part II. What is unique about the experience in lower- and middle-income less-industrialized countries compared with the very-high income industrialized countries? *Public Health Nutrition*, 5(1a).

Scott, A., Ejikeme, C., Clottey, E., and Thomas, J., 2012. Obesity in sub-Saharan Africa: development of an ecological theoretical framework. *Health Promotion International*, 28(1), 4–16.

Shapiro, D., 2015. Accelerating fertility decline in sub-Saharan Africa. *Population Horizons*, 12(1).

Sibai, A.M., Costanian, C., Tohme, R., Assaad, S., and Hwalla, N., 2013. Physical activity in adults with and without diabetes: from the 'high-risk' approach to the 'population-based' approach of prevention. *BMC Public Health*, 13(1).

Stuckler, D., 2008. Population causes and consequences of leading chronic diseases: a comparative analysis of prevailing explanations. *The Milbank Quarterly*, 86(2), 273–326.

Teh, C. *et al.*, 2015. Association of physical activity with blood pressure and blood glucose among Malaysian adults: a population-based study. *BMC Public Health*, 15(1).

Tenkorang, E., Kuuire, V., Luginaah, I., and Banchani, E., 2015a. Examining risk factors for hypertension in Ghana: evidence from the Study on Global Ageing and Adult Health. *Global Health Promotion*, 24(1), 14–26.

Tenkorang, E.Y., Sedziafa, P., Sano,Y., Kuuire, V., and Banchani, E., 2015b. Validity of self-report data in hypertension research: findings from the Study on Global Ageing and Adult Health. *The Journal of Clinical Hypertension*, 17(12), 977–984.

Tsioufis, C. *et al.*, 2010. Relation between physical activity and blood pressure levels in young Greek adolescents: the Leontio Lyceum study. *The European Journal of Public Health*, 21(1), 63–68.

Utsugi, M. *et al.*, 2008. Fruit and vegetable consumption and the risk of hypertension determined by self measurement of blood pressure at home: The Ohasama study. *Hypertension Research*, 31(7), 1435–1443.

Van de Vijver, S. *et al.*, 2013. Status report on hypertension in Africa – consultative review for the 6th Session of the African Union Conference of Ministers of Health on NCDs. *Pan African Medical Journal*, 16.

Vorster, H., Kruger, A., and Margetts, B., 2011. The nutrition transition in Africa: can it be steered into a more positive direction? *Nutrients*, 3(12), 429–441.

Wang, L., Manson, J.E., Gaziano, J.M. Buring, J.E., and Sesso, H.D., 2012. Fruit and vegetable intake and the risk of hypertension in middle-aged and older women. *American Journal of Hypertension*, 25(2), 180–189.

Wang, S. *et al.*, 2014. Obesity and its relationship with hypertension among adults 50 years and older in Jinan, China. *PLoS ONE*, 9(12), e114424.

World Health Organization, 2013. *A global brief on hypertension: silent killer, global public health crisis.* World Health Day 2013.

9 Racial and ethnic differences in non-communicable diseases in South Africa

Eric Y. Tenkorang

Introduction

Non-communicable diseases (NCDs) are the leading cause of mortality in both developed and developing countries (World Health Organization 2016). Described as one of the major public health and development challenges of our time, NCDs, including cancer, diabetes, and hypertension, are known to have caused 38 million deaths in 2012 (World Health Organization 2016). If not controlled, the death toll is projected to increase to 52 million by 2030 (World Health Organization 2010). While previous research and early theorising of these diseases indicate they are more prevalent in affluent societies, recent evidence shows low- and middle-income countries share the highest burden of NCDs (World Health Organization 2010).

In sub-Saharan Africa (SSA), several countries are experiencing epidemiologic shifts from infectious/communicable diseases to chronic diseases. Although infectious diseases such as tuberculosis and malaria remain relatively common, cardiovascular/circulatory diseases including diabetes, hypertension, stroke, and ischemic heart diseases have increased drastically (Holmes *et al.* 2010, Shona *et al.* 2011). For instance, a report by Marquez and Farrington (2013) indicates that the age-standardised mortality rate for cardiovascular diseases is highest (779 per 100,000 population) in SSA compared to other regions of the world. Similarly, it is projected that chronic diseases will be responsible for about 46% of all deaths in SSA in 2030 (Dalal *et al.* 2011).

The increasing prevalence of chronic diseases in SSA has largely been interpreted as part of its demographic transition, especially given the declining child mortality and fertility rates for several countries in the sub-region. Significant reductions in child mortality have meant that the majority of the populations in SSA can live into their adult years when chronic diseases are most prevalent (Penfold *et al.* 2013, Bongaarts and Casterline 2013, Shapiro 2015). However, the changing demography of countries in SSA and the shifting disease burden from communicable diseases to NCDs have important implications on affected countries. The shift in disease burden puts pressure on healthcare systems that are already under-resourced, under-funded, and have not been properly oriented towards dealing with the many challenges that chronic diseases pose for affected societies (Aikins *et al.* 2010). The changing dynamics also means governments in SSA will have to prioritise resources and make important decisions, especially in a context where NCDs coexist with communicable diseases.

Chronic and cardiovascular diseases in South Africa

South Africa, arguably the wealthiest country in sub-Saharan Africa, faces severe challenges with its public health systems given the growing prevalence of NCDs in the country. Currently, it is estimated that about 49% of deaths in South Africa are attributable to chronic diseases which cause premature deaths among young adults (World Health Organization 2015).

The causes of these chronic conditions in South Africa, especially cardiovascular diseases, are well documented. For instance, as an upper-middle income country, and given its fertility and mortality rates, South Africa has been described as undergoing an epidemiologic transition, where cardiovascular diseases coexist with or replace communicable and infectious diseases (Tollman *et al.* 2008, Mayosi *et al.* 2009, Dalal *et al.* 2011, Mayosi *et al.* 2012). For some, the sudden rise in NCDs partially reflects the commencement of an ageing process given South Africa's declining fertility rates (Tollman *et al.* 2008, Dalal *et al.* 2011, Mayosi *et al.* 2012). For others, the forces of globalisation, in tandem with structural transformation in the economy, have ushered drastic changes in both dietary habits and physical activity levels among South Africans leading to NCDs (Vorster *et al.* 2005, World Health Organization 2015).The latter argument is succinctly captured by Popkin's (2002) nutrition transition framework which explains that as countries develop, there are changes in food intake mostly from healthy organic foods to processed foods that are high in cholesterol, fat, and sugar. Consuming processed food leads to higher obesity rates, but this is made worse by physical inactivity at this stage of the development process. It is, thus, not surprising that over half of South Africans are overweight and approximately 25% of the population is obese (Tollman *et al.* 2008, Mayosi *et al.* 2009, World Health Organization 2015). These dynamics explain the high prevalence of NCDs such as hypertension (42.2%) and diabetes (7%) in the South African population (World Health Organization 2015).

Extant literature acknowledges, however, that South African society is very complex due to its politics, history, and ethnic and cultural diversity (Seekings 2008, World Election 2014). Demographically, four major racial and ethnic groups are identified: Africans (80.8%), Coloureds (8.8%), Whites (8.4%), and Indian/Asians (2.5%) (Statistics South Africa 2014). The political history of South Africa and its experiences with apartheid means members of some racial and ethnic groups, mainly Africans, have not developed equally compared to other racial/ethnic groups, especially in comparison to White South Africans. This lack of development has been documented as accounting for the differences in the demographic and epidemiologic transitions of these groups. For instance, while it is documented that White South Africans have completed their epidemiologic transition; Africans have yet to complete theirs (Myer *et al.* 2004). Differences in socioeconomic development existing between racial and ethnic groups in South Africa have implications for understanding chronic disease development, yet these complexities have rarely been brought to bear on analysing chronic/NCDs in sub-Saharan Africa, in general, and South Africa, specifically. This study seeks to fill this research gap by first examining if racial and ethnic differences exist among people living with NCDs in South Africa. Second, we shall explore if the racial and ethnic variations in NCDs can be explained by differences in physical activity levels, dietary habits, or the socioeconomic inequality among the various ethnic groups.

Race, ethnicity, and health

The literature on race, ethnicity, and health dates as far back as the nineteenth century when W.E.B. Du Bois, in his classic book *The Philadelphia Negro*, found that racial inequality in the United States reflected in the socioeconomic circumstances of Blacks and their health (Du Bois 1899, 1967). Also, focusing on gender and health, Du Bois (1899, 1967, p. 151) reported that the gender differences in health were larger for Blacks compared to Whites. Du Bois' explanation of the social patterning of disease and health has been embraced by sociologists as it challenged existing paradigms that reduced racial and gender differences in health to biological differences (Williams and Sternthal 2010). Even beyond the nineteenth century, Du Bois' theoretical insights continue to inform academic discourse on race, ethnicity, and health. For instance, it is widely documented in South Africa that – similar to the United States – race and ethnicity are intricately linked to health outcomes (Williams 1999, William *et al.* 2008). South Africa's history of apartheid left behind legacies of racial discrimination and social inequality (Williams *et al.* 2008). Several researchers have demonstrated that these legacies have had adverse health effects on groups that were racialised, in particular Coloureds and Africans (Dommisse 1986, Mohutsioa-Makhudu 1989, Turton and Chalmers 1990). It is, therefore, not too surprising that the Coloured and African populations perceived their health as poorer compared to Whites in South Africa (Statistics South Africa, 2004).

While previous literature examined the effects of race and ethnicity on several health outcomes, including mental health (Singh and Burns 2008, Stuber *et al.* 2008,William *et al.* 2008) in South Africa, very few focused on cardiovascular health outcomes. Furthermore, studies that explored chronic diseases in South Africa have often used race as a control variable and not as a focal independent variable of theoretical significance (Steyn *et al.* 2006, Cois and Ehrlich 2014). Thus, previous analysis of race may be problematic as it seems to underscore the position that risk factors for chronic diseases operate the same way in all South African racial groups (Steyn *et al.* 2006). For instance, in their analysis of self-reported chronic NCDs in older South Africans, Phaswana-Mafuya *et al.* (2013) found that Coloureds and Asians had a higher prevalence of NCDs compared to Africans. However, this study did not focus on the racial distinctions and the underlying reasons for these differences. Similarly, Cois and Ehrlich (2014) had employed structural equation modelling to examine hypertension in South Africa, but the focus was mainly on socioeconomic determinants, with race employed as a control variable. In addition, some studies show differences in physical activity and dietary habits among the various racial and ethnic groups in South Africa (Kruger *et al.* 2005, Steyn *et al.* 2006, McVeigh and Meiring 2014).

This study offers a different approach to examining NCDs in South Africa by focusing on racial and ethnic distinctions. The study also explores if racial variations in NCDs can be explained by differences in physical activity levels and nutrition intake or if these variations are an artefact of the socioeconomic differences among the various racial and ethnic groups.

Data and methods

Data used for this study came from wave 1 of the Study of Aging and Adult Health (SAGE) collected by the World Health Organization (WHO) between January 2007 and December 2008. The SAGE data are nationally representative and were conducted for six countries, including South Africa. The survey aims at monitoring the health and wellbeing of adults over 50 years of age, although the survey also included a sample of young adults aged 18–49 years. The SAGE data employed a two-stage probability sampling procedure. The first stage involved selecting 1,000 enumeration areas from the 2001 South African National Census. The second stage selected about 4,000 households from both rural and urban areas, out of which 4,227 respondents were interviewed. The analytical sample for this study was limited to the 4,026 respondents (Male = 1,923, Female = 2,103) who answered questions on hypertension, diabetes, and stroke which were used in creating the outcome variable for this study.

Measures

The dependent variable employed for analysis was based on biometric and self-report accounts of whether respondents had ever been diagnosed with hypertension, diabetes, and stroke. For instance, hypertension was measured through physical examinations using a Boso Medistar Wrist BP Monitor Model S (Minicuci *et al.* 2014). A respondent was considered hypertensive if the average of the three measurements was greater than or equal to 140 mmHg for (systolic BP) and/or greater than or equal to 90 mmHg for (diastolic BP). Diabetes and stroke were self-reported. The three variables were combined into a single outcome called 'living with NCDs' (coded no = 0 and yes = 1). Respondents with at least one of the health conditions identified were categorised as 'living with an NCD.' Otherwise, they were coded as 'not living with an NCD.' The reasons for pooling the outcomes into a single variable have been established elsewhere (see Tenkorang and Kuuire 2016).

The focal independent variable asked respondents to identify their ethnic/racial groups. This was dummy-coded into four main categories (coded Blacks/Africans = 0, Whites = 1, Coloureds = 2, Asians/Indians = 3). Nutrition was measured with two predictors that asked how many servings of fruits (banana, mango, apple, orange, papaya, tangerine, grapefruit, peach, pear, etc.) and vegetables (tomato, cauliflower, cucumber, peas, corn, lettuce, squash, bean, etc.) respondents ate on a typical day. Physical activity was also measured with two instruments that asked respondents if they walk or use a bicycle (pedal cycle) for at least ten minutes continuously to get to and from places, and if respondents do any vigorous /intensive sports, fitness, or recreational (leisure) activities that cause large increases in breathing or heart rate (like running or football) for at least ten minutes continuously. Socioeconomic inequality was measured with the income quintile of the household, educational background of respondents, and their employment status. Respondents' age, Body Mass Index, and religious denominations were used as statistical controls.

Analytical technique

Binary logit models were used given the dichotomous nature of our outcome variable. Also, gender-specific models were estimated given differences in prevalence and the fact that risk factors for NCDs may be gendered in South Africa (BeLue *et al.* 2009, Tenkorang and Kuuire 2016). In all, four multivariate models were estimated each for men and women. The first model examines the effects of ethnicity on NCDs with control variables, the second model added nutrition variables, the third model included variables on physical activity, and the final model, socioeconomic predictors. The pattern-matching imputation[1] technique obtained through PRELIS (the data processor for LISREL, version 8) was used to impute missing cases for about 21% of respondents who failed to identify their racial/ethnic group.

Results

Descriptive results in Table 9.1 indicate that about half of the South African population live with an NCD, albeit this is higher in men (54.8%) than women (47.9%). The racial/ethnic composition of the sample reflects what exists in the general South African population with Africans in the majority, followed by Coloureds, Whites, and Asians. Regarding nutrition and physical activity, it is clear that the majority of South African males and females do not eat the WHO's recommended daily servings of fruits and vegetables and are physically inactive. However, physical inactivity is higher among South African females than males. This probably explains the higher rates of obesity among the female population compared to males. Compared to 17.6% of males, 25.8% of females belonged to the richest wealth quintile.

Bivariate results are presented in Table 9.2. Results show racial/ethnic differences in NCDs. Compared to White males, Africans, Coloured, and Asian males are significantly more likely to report living with NCDS. For females, we observe that both Africans and Asians are more likely to live with NCDs compared to White females. However, Coloured females are significantly less likely to live

Table 9.1 Percentage distribution of selected dependent and independent variables

	Percentages/Mean	
	Male (N = 1923)	*Female (N = 2103)*
Variables		
Living with NCD		
No	45.2	52.1
Yes	54.8	47.9
Racial/Ethnic Background		
Whites	7.5	5.9
Africans	79.3	69.1
Coloreds	10.8	17.5
Asians/Indians	2.4	7.5
Nutrition		
Servings of fruits on a typical day?		
Less than 5 servings	82.1	80.7
5 or more servings	2.9	4.1
No stated	15.0	15.2
Servings of vegetables on a typical day?		
Less than 5 servings	89.5	88.0
5 or more servings	6.1	4.8
Not stated	4.4	7.2
Physical activity		
Walk or bike for at least 10 minutes?		
No	52.1	66.3
Yes	47.9	33.7
Vigorous intensity sports, fitness, recreation?		
No	90.8	94.9
Yes	9.2	5.1
Socioeconomic Controls		
Educational background		
No education	6.8	8.5
Primary education	31.3	25.8
Secondary education	54.7	58.2
Higher education	7.3	7.4
Wealth quintile		
Poorest	18.5	17.1
Poorer	22.5	16.4
Middle	18.3	23.8
Richer	23.2	16.8
Richest	17.6	25.8
Employment status		
No	31.6	37.7
Yes	44.0	38.0
Not stated	24.4	24.4

(*continued*)

Table 9.1 Continued

	Percentages/Mean	
	Male (N = 1923)	*Female (N = 2103)*
Control Variables		
Body Mass Index		
Normal weight	38.5	29.8
Underweight	5.8	2.3
Overweight	28.0	27.2
Obese	27.7	40.7
Average Age of respondent	41.4	42.1
Marital status		
Married	53.4	28.5
Never married	27.5	34.8
Divorced/Widowed/Separated	19.1	36.7
Residence		
Rural	30.4	29.2
Urban	69.6	70.8

with NCDs compared to Whites. The effects of nutrition and physical activity on the risks of living with NCDs are mixed. While having the recommended five servings of fruits daily reduces the risks of living with NCDs, eating the recommended servings of vegetables rather increases these risks. Socioeconomic variables are significantly associated with the risks of living with NCDS. Higher education and employment reduce the risks of living with NCDs for both males and females. However, living in wealthy households has differential effects for males and females. While males in wealthier households are more likely to live with NCDs, females in such households are rather less likely to live with such diseases.

As our main objective was to examine racial and ethnic differences, we estimated a series of multivariate models to determine if our bivariate findings still hold even after including theoretically relevant covariates. For instance, we estimated some multivariate models in Tables 9.3 and 9.4 to test if the racial and ethnic differences could be explained by differences in nutrition intake, physical activity levels or socioeconomic inequality among these groups. Multivariate results show that as we move from Model 1 to Model 2 where nutrition variables are added, not enough changes are observed in the odds ratios for the various racial/ethnic groups for both males and females. However, significant changes are observed, especially for males, when variables measuring physical activity are included in the model.

First, we observe that the health disadvantage between African men and their White counterparts vanish. Second, we see a significant decrease in the odds of living with NCDs for both Asians and Coloureds. For females, not enough differences occur when variables measuring physical activity are added. Significant

Table 9.2 Bivariate models of NCDS among South African men and women (SAGE 2008)

	Odds Ratios	
	Male (N = 1923)	*Female (N = 2103)*
Variables		
Racial/Ethnic Background		
Whites	1.00	1.00
Africans	2.03 (.189)***	1.71 (.196)***
Coloreds	7.80 (.258)***	.631 (.223)**
Asians/Indians	7.81 (.432)***	2.96 (.254)***
Nutrition		
Servings of fruits on a typical day?		
Less than 5 servings	1.00	1.00
5 or more servings	.433 (.286)***	2.11 (.228)***
No stated	1.40 (.132)***	2.06 (.125)***
Servings of vegetables on a typical day?		
Less than 5 servings	1.00	1.00
5 or more servings	8.68 (.319)***	3.84 (.237)***
Not stated	.928 (.223)	2.31 (.177)***
Physical activity		
Walk or bike for at least 10 minutes?		
No	1.00	1.00
Yes	1.43 (.092)***	1.44 (.093)***
Vigorous intensity sports, fitness, recreation?		
No	1.00	1.00
Yes	.725 (.158)**	.739 (.201)
Socioeconomic Controls		
Educational background		
No education	1.00	1.00
Primary education	.288 (.230)***	.829 (.192)
Secondary education	.284 (.224)***	.227 (.178)***
Higher education	.566 (.282)**	.245 (.236)***
Wealth quintile		
Poorest	1.00	1.00
Poorer	1.91 (.146)***	.574 (.153)***
Middle	5.45 (.168)***	.594 (.141)***
Richer	1.65 (.144)***	.525 (.152)***
Richest	1.65 (.154)***	.421 (.140)***
Employment status		
No	1.00	1.00
Yes	.668 (.109)***	.344 (.104)***
Not stated	.425 (.126)***	.553 (.115)***
Control Variables		
Body Mass Index		
Normal weight	1.00	1.00

(*continued*)

Table 9.2 Continued

	Odds Ratios	
	Male (N = 1923)	*Female (N = 2103)*
Underweight	1.79 (.208)***	1.51 (.318)
Overweight	2.07 (.117)***	1.73 (.125)***
Obese	3.36 (.122)***	5.37 (.117)***
Age of respondent	1.040 (.004)***	1.08 (.004)***
Marital status		
Married	1.00	1.00
Never married	.770 (.108)**	.318 (.116)***
Divorced/Widowed/Separated	.812 (.123)	.462 (.113)***
Residence		
Rural	1.00	1.00
Urban	.799 (.100)**	.538 (.097)***

Note: **$p < .05$; ***$p < .01$; Odds ratios reported and standard errors are in brackets.

changes are rather observed for females as socioeconomic variables are included. For instance, the health disadvantage between African women and their White counterparts vanish, and we also observe a significant decrease in the odds of living with NCDs for both male and female Asians (see Model 4 in Table 9.4). For males, a significant increase in the odds of living with NCDs is observed for Asians compared to their African counterparts. However, the odds of living with NCDs decrease for Coloureds (see Model 4 in Table 9.3).

Discussion

The literature on race, ethnicity, and health is fairly well established. However, this literature is daunted with two major limitations. First, and as Smaje (1996) argued, the documented work in this area draws largely on the efforts of epidemiologists and clinicians whose main focus has been to report patterns of diseases, and whose interpretations of these patterns have been bereft of the nuanced theoretical significance of how race and ethnicity affect the illness experience. Second, and equally important, is that the literature on race, ethnicity, and health mostly emanates from the West, specifically the United States, where racial inequality and discrimination are known to have affected racialised groups in all facets of their lives including health (Williams *et al.* 1994, Kawachi *et al.* 2005, Jackson *et al.* 2010). This study fills these two important research gaps as it offers one of the few sociological interpretations of the effects of race and ethnicity on a neglected dimension of health (cardiovascular) in sub-Saharan Africa. Also, the focus on South Africa is unique and relevant given its long history of racial segregation and discrimination.

Results from the analysis show significant differences in NCDs among the different racial groups in South Africa. Specifically, African, Coloured, and Asian men were more likely to live with NCDs compared to White men. For women, African and Asians were significantly more likely to live with NCDs compared to Whites. No significant differences were observed between Coloured and White

Table 9.3 Multivariate models of NCDS among South African men (SAGE 2008)

	Model 1	*Model 2*	*Model 3*	*Model 4*
Variables	*AOR*	*AOR*	*AOR*	*AOR*
Racial/Ethnic Background				
Whites	1.00	1.00	1.00	1.00
Africans	1.98 (.220)***	2.04 (.223)***	1.47 (.235)	1.16 (.286)
Coloureds	5.46 (.281)***	5.59 (.287)***	4.71 (.297)***	3.66 (.321)***
Asians/Indians	7.66 (.465)***	7.53 (.470)***	7.05 (.479)***	9.95 (.503)***
Nutrition				
Servings of fruits on a typical day?				
Less than 5 servings		1.00	1.00	1.00
5 or more servings		.233 (.352)***	.187 (.345)***	.316 (.409)***
No stated		1.05 (.169)	1.09 (.176)	1.42 (.201)
Servings of vegetables on a typical day?				
Less than 5 servings		1.00	1.00	1.00
5 or more servings		2.34 (.383)**	1.83 (.399)	1.53 (.419)
Not stated		.490 (.296)**	.537 (.301)**	.569 (.309)
Physical activity				
Walk or bike for at least 10 minutes?				
No			1.00	1.00
Yes			2.93 (.124)***	2.91 (.133)***
Vigorous intensity sports, fitness, recreation?				
No			1.00	1.00
Yes			.945 (.198)	1.04 (.220)

(continued)

Table 9.3 Continued

	Model 1	*Model 2*	*Model 3*	*Model 4*
Variables	*AOR*	*AOR*	*AOR*	*AOR*
Socioeconomic Controls				
Educational background				
No education				1.00
Primary education				.581 (.289)
Secondary education				.679 (.307)
Higher education				1.02 (.389)
Wealth quintile				
Poorest				1.00
Poorer				1.89 (.186)***
Middle				6.42 (.229)***
Richer				1.67 (.204)***
Richest				1.03 (.254)
Employment status				
No				1.00
Yes				1.12 (.169)
Not stated				.322 (.199)***
Control Variables				
Body Mass Index				
Normal weight	1.00	1.00	1.00	1.00
Underweight	1.30 (.231)	1.32 (.237)	1.43 (.245)	1.33 (.267)
Overweight	1.78 (.129)***	1.91 (.134)***	1.81 (.138)***	1.33 (.159)
Obese	3.49 (.149)***	3.65 (.156)***	4.45 (.164)***	4.81 (.181)***
Age of respondent	1.03 (,005)***	1.03 (.005)***	1.03 (.005)***	1.03 (.006)***

Marital status				
Married	1.00	1.00	1.00	1.00
Never married	1.02 (.151)	1.07 (.153)	.830 (.160)	1.23 (.184)
Divorced/Widowed/Separated	.997 (.139)	1.15 (.145)	.903 (.150)	1.18 (.167)
Residence				
Rural	1.00	1.00	1.00	1.00
Urban	.752 (.120)**	.752 (.123)**	.902 (.127)	.642 (.143)***
Nagelkerke R squared	.211	.231	.281	.368
–2 Log-likelihood ratio	2052.748	2022.971	1941.740	1794.168

Note: **p < .05; ***p < .01; Adjusted odds ratios reported and standard errors are in brackets.

Table 9.4 Multivariate models of NCDS among South African women (SAGE 2008)

Variables	*Model 1* AOR	*Model 2* AOR	*Model 3* AOR	*Model 4* AOR
Racial/Ethnic Background				
Whites	1.00	1.00	1.00	1.00
Africans	4.41 (.351)***	4.06 (.249)***	4.09 (.252)***	1.78 (.307)
Coloreds	1.43 (.271)	1.31 (.272)	1.38 (.278)	.799 (.315)
Asians/Indians	5.84 (.316)***	5.67 (.323)***	5.53 (.326)***	4.28 (.334)***
Nutrition				
Servings of fruits on a typical day?				
Less than 5 servings		1.00	1.00	1.00
5 or more servings		.660 (.385)	.639 (.380)	.609 (.389)
No stated		3.76 (.189)***	3.69 (.189)***	3.07 (.202)***
Servings of vegetables on a typical day?				
Less than 5 servings		1.00	1.00	1.00
5 or more servings		1.04 (.391)	1.07 (.388)	.840 (.408)
Not stated		.809 (.267)	.781 (.268)	.733 (.270)
Physical activity				
Walk or bike for at least 10 minutes?				
No			1.00	1.00
Yes			1.29 (.134)	1.24 (.142)
Vigorous intensity sports, fitness, recreation?				
No			1.00	1.00
Yes			.678 (.286)	.840 (.408)

Socioeconomic Controls				
Educational background				
No education				1.00
Primary education				1.06 (.264)
Secondary education				.518 (.272)**
Higher education				.906 (.365)
Wealth quintile				
Poorest				1.00
Poorer				.583 (.209)***
Middle				.748 (.202)
Richer				.484 (.222)***
Richest				.301 (.266)***
Employment status				
No				1.00
Yes				.877 (.162)
Not stated				1.02 (.171)
Control Variables				
Body Mass Index				
Normal weight	1.00	1.00	1.00	1.00
Underweight	.919 (.419)	.641 (.420)	.673 (.424)	.409 (.429)**
Overweight	1.65 (.153)***	1.57 (.157)***	1.55 (.157)***	1.53 (.164)***
Obese	4.35 (.143)***	4.72 (.149)***	4.70 (.151)***	5.22 (.160)***
Age of respondent	1.08 (.005)***	1.08 (.005)***	1.08 (.005)***	1.06 (.006)***
Marital status				
Married	1.00	1.00	1.00	1.00
Never married	.727 (.160)**	.743 (.162)	.696 (.166)**	.669 (.173)**

(continued)

Table 9.3 Continued

Variables	*Model 1* *AOR*	*Model 2* *AOR*	*Model 3* *AOR*	*Model 4* *AOR*
Divorced/Widowed/Separated	.312 (.152)***	.276 (.157)***	.270 (.157)***	.269 (.162)***
Residence				
Rural	1.00	1.00	1.00	1.00
Urban	.817 (.139)	.884 (.145)	.930 (.148)	1.14 (.161)
Nagelkerke R squared	.420	.449	.451	.478
–2 Log-likelihood ratio	1931.389	1868.206	1863.218	1785.810

Note: **$p < .05$; ***$p < .01$; Adjusted odds ratios reported and standard errors are in brackets.

women. Possible reasons for the observed racial and ethnic health differences are complex given that these differences are often rooted in biological, socioeconomic, and psychosocial factors. For instance, some research documents genetic and family history as important factors to consider in explaining chronic diseases, especially cardiovascular diseases (see Wilmot and Idris, 2014). In line with this, extant literature has shown that for some specific racial/ethnic groups, such as South Asians, including Indians, there is a single nucleotide polymorphism in their gene that affects fat mass and obesity placing them at higher risks of living with NCDs compared to other members of their racial/ethnic groups (Yajnik *et al.* 2009,Becerra and Becerra 2015, Tenkorang 2017).

It is the case, however, that genetic susceptibility combines with socioeconomic, lifestyle, cultural, and environmental factors in a complex manner to affect the risks of living with NCDs for the various racial/ethnic groups (Tenkorang 2017). In fact, our models demonstrate this complexity, as including variables on physical activity completely attenuated the health disadvantage between African and White men, and decreased the likelihood of living with NCDs for Asians and Coloureds. This finding could mean that the health disadvantage between African males and their White counterparts may be due to differences in levels of physical activity. Also, to the extent that the disease burden of the other racial groups is reduced means that physical activity is crucial to minimising the risks of living with chronic diseases for South African men. For women, we did not observe significant changes in the odds of living with NCDs as variables measuring physical activity were added to the models. However, as socioeconomic variables were added, the health disadvantage between White women and their African counterparts lost its statistical significance. This means that the chronic disease burden for African females may be largely due to poverty, especially against the backdrop that wealth is inversely related to the risks of living with NCDs for women Although we did observe loss of significance for Asian women, their odds of living with NCDs reduced as socioeconomic variables were included. This could mean that Asian women's vulnerability to NCDs may be partially related to poverty but goes beyond that to include other unobserved characteristics.

The strong effects of socioeconomic variables on the health outcomes of African and Asian women reflects how social inequality in South Africa, partially interpreted as one of the numerous legacies of apartheid, translates into health inequalities for specific racial groups that benefitted less from this regime (Dommisse 1986,Seekings 2008, Williams *et al.* 2008). Interestingly, the inclusion of socioeconomic variables had different effects for males, especially Asian males, as their risks of living with chronic diseases increased substantially. It should be noted that unlike females, males in wealthier households had higher risks of living with a chronic condition.

Taken together, the findings demonstrate complex theoretical pathways to explaining NCDs among the various racial groups in South Africa, especially as these are gendered and deeply rooted in socioeconomic differences. This means chronic disease management and interventions in South Africa should not only pay attention to gender differences but also the specific needs of the various racial and ethnic groups. Specific interventions that target wealthier men, in particular those of Asian descent, might help in reducing prevalence. Previous studies including ours showed limited physical activity among South Africans (see Kruger *et al.* 2002, Tenkorang 2017 *Physical Activity, Nutrition, and Hypertension in Sub-Saharan Africa* in this volume). Notwithstanding, physical activity is beneficial as it significantly reduced the risks of living with NCDs for specific racial groups including Africans. Our results demonstrated further that interventions for women that aim at tackling poverty and socioeconomic inequality might be extremely beneficial in reducing the risks of NCDs, especially for African women.

Limitations

In spite of these findings, some limitations are worth noting. The data used are cross-sectional and are limited in helping us disentangle the complex 'causal' connections between our predictors and the outcome variable. As a result, interpretations of these findings are limited to associations. Also, about 21% of the responses of participants who failed to identify their racial/ethnic backgrounds were imputed. While statistical imputations are often employed to avoid bias in samples, it can be an important source of bias itself, especially when not rigorously implemented. The study employed a sophisticated imputation technique in ensuring that the chances of bias through imputation are reduced. As a test of robustness, we performed sensitivity analysis of the sample with missing data on race and ethnicity and compared direction and significance of coefficients. There was no substantial difference between samples with missing and non-missing cases. However, it is important to note that this study provides the first comprehensive and documented account of racial and ethnic differences in NCDs using nationally representative data in South Africa.

Policy implications and recommendations

The findings from this study demonstrate once again that chronic diseases have multiple risk factors that operate at different levels, and that in most cases these risk factors are modifiable and preventable. Thus, there is evidence to show that in developing interventions towards chronic disease management in SSA and South Africa specifically, it is important to pay attention to both individual and contextual factors that serve as barriers to healthy lifestyles. Also, given that these preventable risk factors have different 'causal' paths for various socio-demographic and cultural groups; it is important for such interventions to be nuanced and acknowledge the complexities existing within African populations. Under its Integrated Chronic Disease Management Model (ICDM), the South African government has focused on chronic disease prevention at both individual and community levels, and by strengthening its health and surveillance systems (Mahomed and Asmall 2015). However, similar to the majority of population health interventions, the ICDM does not overtly acknowledge the gender and ethnic differences in the risks of living with chronic diseases. We draw the attention of policymakers in South Africa and SSA in general to these differences and their potential of enhancing chronic disease interventions and management on the continent.

Note

1 In the pattern-matching imputation approach, a case with missing data on a given question is matched to others with the same responses on all other questions. Under these circumstances, missing data are replaced with a score from another respondent who has a similar profile of scores across other variables (Carter 2006).

References

Aikins de-Graft, A., Unwin, N., Agyemang, C., Allotey, P., Campbell, C., and Arhinful, D. 2010. Tackling Africa's chronic disease burden: From the local to the global. *Globalization and Health*, 6(5).

Becerra, M.P. and Becerra, B.J., 2015. Disparities in age at diabetes diagnosis among Asian Americans: Implications for early preventive measures. *Preventing Chronic Disease*, 12 (E46).

BeLue, R., Okoror, T.A., Iwelunmor, J., Taylor, K.D., Degboe, A.N., Agyemang, C., Ogedegbe, G., 2009. An overview of cardiovascular risk factor burden in sub-Saharan African countries: A socio-cultural perspective. *Globalization and Health*, 5(10).

Bongaarts, J and Casterline, J., 2013. Fertility transition: Is sub-Saharan Africa different? *Population Development Review*, 38(Suppl 1), 153–168.

Carter, R.L., 2006. Solutions for missing data in structural equation modeling. *Research and Practice in Assessment*, 1, 4–7.

Cois, A. and Ehrlich, R., 2014. Analysing the socioeconomic determinants of hypertension in South Africa: A structural equation modelling approach. *BMC Public Health*, 14(1).

Dalal, S., Beunza, J.J., Volmink, J., Adebamowo, C., Bajunirwe, F., Njelekela, M., Mozaffarian, D., Fawzi, W., Willett, W., Adami, H.O., and Holmes, M.D., 2011. Non-communicable diseases in sub-Saharan Africa: What we know now. *International Journal of Epidemiology*, 40(4), 885–901.

Dommisse, J., 1986. The psychological effects of apartheid psychoanalysis social, moral, and political influences. *International Journal of Social Psychiatry*, 32(2), 51–63.

Du Bois, W.E.B., 1899. *The Philadelphia Negro: A Social Study*. 1967 ed. New York: Schocken Books.

Holmes, M.D., Dalal, S., Volmink, J., Adebamowo, C.A., Njelekela, M., Fawzi, W.W., Willett, W.C., and Adami, H.-O., 2010. Non-communicable diseases in sub-Saharan Africa: The case for cohort studies. *PLoS Med*, 7(5), e1000244.

Jackson, J.S., Knight, K.M., and Rafferty, J.A., 2010. Race and unhealthy behaviors: Chronic stress, the HPA axis, and physical and mental health disparities over the life course. *American Journal of Public Health*, 100(5), 933–939.

Kawachi, I., Daniels, N., and Robinson, D.E., 2005. Health disparities by race and class: Why both matter. *Health Affairs*, 24(2), 343–352.

Kruger, H.S., Puoane, T., Senekal, M., and van der Merwe, M.T., 2005. Obesity in South Africa: Challenges for government and health professionals. *Public Health Nutrition*, 8(05), 491–500.

Kruger, H.S., Venter, C.S., Vorster, H.H., and Margetts, B.M., 2002. Physical inactivity is the major determinant of obesity in black women in the North West Province, South Africa: The THUSA study. *Nutrition*, 18(5), 422–427.

Mahomed, O.H. and Asmall, S., 2015. Development and implementation of an integrated chronic disease model in South Africa: Lessons in the management of change through improving the quality of clinical practice. *International Journal of Integrated Care*, 1, e038.

Marquez, P.V. and Farrington, J.L., 2013. The challenge of non-communicable diseases and road traffic injuries in sub-Saharan Africa: An overview. Washington, D.C: The World Bank.

Mayosi, B.M., Flisher, A.J., Lalloo, U.G., Sitas, F., Tollman, S.M., and Bradshaw, D., 2009. The burden of non-communicable diseases in South Africa. *The Lancet*, 374(9693), 934–947.

Mayosi, B.M., Lawn, J.E., Van Niekerk, A., Bradshaw, D., Karim, S.S.A., Coovadia, H.M., and Lancet South Africa team, 2012. Health in South Africa: Changes and challenges since 2009. *The Lancet*, 380(9858), 2029–2043.

McVeigh, J. and Meiring, R., 2014. Physical activity and sedentary behavior in an ethnically diverse group of South African school children. *Journal of Sports Science Medicine*, 13(2), 371–378.

Minicuci, N., Biritwum, R.B., Mensah, G., Yawson, A.E., Naidoo, N., Chatterji, S., and Kowal, P., 2014. Sociodemographic and socioeconomic patterns of chronic non-communicable disease among the older adult population in Ghana. *Global Health Action*, 7.

Mohutsioa-Makhudu, Y.N.K., 1989. The psychological effects of apartheid on the mental health of Black South African women domestics. *Journal of Multicultural Counseling and Development*, 17(3), 134–142.

Myer, L., Ehrlich, R.I., and Susser, E.S., 2004. Social epidemiology in South Africa. *Epidemiologic Reviews*, 26(1), 112–123.

Penfold, S., Willey, B., and Schellenberg, J., 2013. Newborn care behaviours and neonatal survival: evidence from sub-Saharan Africa. *Tropical Medicine and International Health*, 18(11), 1294–1316.

Phaswana-Mafuya, N., Peltzer, K., Chirinda, W., Kose, Z., Hoosain, E., Ramlagan, S., Tabane, C., and Davids, A., 2013. Self-rated health and associated factors among older South Africans: Evidence from the study on global ageing and adult health. *Global Health Action*, 6.

Popkin, B.M., 2002. An overview on the nutrition transition and its health implications: the Bellagio meeting. *Public Health Nutrition*, 5(1A), 93–103.

Seekings, J., 2008. The continuing salience of race: Discrimination and diversity in South Africa. *Journal of Contemporary African Studies*, 26(1), 1–25.

Shapiro, D., 2015. Accelerating fertility decline in sub-Saharan Africa. *Population Horizons*, 12(1).

Singh, S.P. and Burns, T., 2008. Race and mental health: There is more to race than racism. *Mental Health Services Today and Tomorrow: Perspective on Policy and Practice*, 73, 648–651.

Smaje, C., 1996. The ethnic patterning of health: New directions for theory and research. *Sociology of Health and Illness*, 18(2), 139–171.

Statistics South Africa, 2004. Perceived health and other health indicators in South Africa. Pretoria. Available at http://www.statssa.gov.za/publications/HealthOHS/HealthOHS1999.pdf

Statistics South Africa, 2014. Mid-year population estimates 2014. Pretoria: Statistical release P0302. Available at https://www.statssa.gov.za/publications/P0302/P03022014.pdf

Steyn, N. P., Bradshaw, D., Norman, R., Joubert, J., Schneider, M., and Steyn, K., 2006. Dietary changes and the health. In *The double burden of malnutrition: Case studies from six developing countries.* FAO Food and Nutrition Paper, 84, 259–304.

Stuber, J., Meyer, I., and Link, B., 2008. Stigma, prejudice, discrimination and health. *Social Science and Medicine (1982)*, 67(3), 351–357.

Tenkorang, E.Y., 2017. Early onset of type 2 diabetes among visible minority and immigrant populations in Canada. *Ethnicity and Health*, 22(3), 266–284.

Tenkorang, E.Y. and Kuuire, V.Z., 2016. Noncommunicable diseases in Ghana: Does the theory of social gradient in health hold? *Health Education and Behavior*, 43(1 suppl), 25–36.

Tollman, S.M., Kahn, K., Sartorius, B., Collinson, M.A., Clark, S.J., and Garenne, M.L., 2008. Implications of mortality transition for primary health care in rural South Africa: A population-based surveillance study. *The Lancet*, 372(9642), 893–901.

Turton, R.W. and Chalmers, B.E., 1990. Apartheid, stress and illness: The demographic context of distress reported by South African Africans. *Social Science and Medicine*, 31(11), 1191–1200.

Vorster, H.H., Venter, C.S., Wissing, M.P., and Margetts, B.M., 2005. The nutrition and health transition in the North West Province of South Africa: A review of the THUSA (Transition and Health during Urbanisation of South Africans) study. *Public Health Nutrition*, 8(05), 480–490.

World Health Organization, 2010. Burden: mortality, morbidity and risk factors [Online]. *Global status report on noncommunicable diseases, 2011*. Geneva: World Health Organization.

World Health Organization, 2015. Non-communicable disease prevention and control (NCDs) [Online]. Available from http://www.afro.who.int/en/south-africa/country-programmes/4248-non-communicable-disease-prevention-and-control-ncds.html [Accessed 14 February 2017].

World Health Organization, 2016. NCD mortality and morbidity [Online]. *Global health observatory data.* Available from http://www.who.int/gho/ncd/mortality_morbidity/en/ [Accessed 12 February 2017].

Williams, D.R., 1999. Race, socioeconomic status, and health. The added effects of racism and discrimination. *Annals of the New York Academy of Sciences*,896(1), 173–188.

Williams, D.R., González, H.M., Williams, S., Mohammed, S.A., Moomal, H., and Stein, D.J., 2008. Perceived discrimination, race and health in South Africa. *Social Science and Medicine*, 67(3), 441–452.

Williams, D.R., Lavizzo-Mourey, R., and Warren, R.C., 1994. The concept of race and health status in America. *Public Health Reports*, 109(1), 26.

Williams, D.R. and Sternthal, M., 2010. Understanding racial-ethnic disparities in health sociological contributions. *Journal of Health and Social Behavior*, 51(1 suppl), 15–27.

Wilmot, E. and Idris, I., 2014. Early onset type 2 diabetes: Risk factors, clinical impact and management. *Therapeutic Advances in Chronic Disease*, 5(6), 234–244.

World Elections, 2014. Race, ethnicity and languages in South Africa [Online]. Available from https://welections.wordpress.com/guide-to-the-2014-south-african-election/race-ethnicity-and-language-in-south-africa/ [Accessed 12 February 2017].

Yajnik, C.S. *et al.*, 2009. FTO gene variants are strongly associated with type 2 diabetes in South Asian Indians. *Diabetologia*, 52(2), 247–252.

10 Weight status and uncontrolled urbanization in Cameroon

Current and future health challenges

Jude Saji, Felix K. Assah, Emmanuella N. Atanga, and Jean Claude Mbanya

Introduction

Once perceived as a problem exclusively affecting the affluent, excess weight – including overweight and obesity – has become a major cause for concern to populations worldwide, including those in low- and middle-income countries which were hitherto prone to malnutrition and underweight (Popkin 2002,BeLue *et al.* 2009, Popkin *et al.* 2012, Ellulu *et al.* 2014, World Health Organization 2015). Globally, more than half a billion people are currently affected, which, statistically, is almost twice the number in 1980 (World Health Organization 2015). Developing countries bear a significant burden of this problem as mortality from obesity-related pathologies such as diabetes and heart disease is now almost as high as from infectious diseases including HIV/AIDS and malaria (Frenk *et al.* 1989). Excess weight accounted for about 3.4 million deaths in 2010 (Lim *et al.* 2012), and by 2030 non-communicable diseases (NCDs) – a significant burden of which has been attributed to obesity – are expected to overtake infectious diseases as the number one health problem in developing countries should the present situation remain unchecked (BeLue 2009). This problem has been attributed, at least in part, to the rapid rate of urbanization which has brought about greater access to unhealthy foods (processed and high-fat content) and lower levels of physical activity (Popkin 2002, Popkin *et al.* 2012). It is arguably a huge and growing problem in Africa where urbanization is mostly spontaneous.

This article reviews patterns of anthropometry, related health risks, and urbanization in Cameroon. We explore the factors which have led to the current situation, as well as likely future trends. Mitigation strategies from a public health perspective are also presented. The issue of overweight, obesity, and NCDs is framed within the context of the Millennium Development Goals (MDGs) and the dialogue which led to the Sustainable Development Goals (SDG). We use Barry Popkin's revised demographic, epidemiologic, and nutrition transition model as a framework (Popkin 2002). The findings highlight the increased impact of NCDs on global health, particularly in developing countries like Cameroon, and justify the inclusion of NCDs with specific targets within the health objective of the new Sustainable Development Framework. NCDs are included as a specific SDG target (reducing premature mortality from NCDs by one-third by 2030) and are part of several other health targets (World Health Organization 2015). Without downplaying the benefits of urbanization, this paper focuses more on its negative influence on diet and exercise habits which contribute to weight gain.

Methodology

PubMed was searched for relevant publications available before January 2015. We used the search terms (singly or in combination) "weight", "overweight", "obesity", "obesity risk factors", "urbanization", "rural–urban", "health", and "Cameroon". We screened the titles and abstracts of the resulting literature for further review. Only papers published in English or French after 1989 and pertaining to humans were included. We also manually scanned reports and books in the relevant area. Content analysis was performed for all retrieved publications to identify relevant information that would paint a clear picture of the trend in urbanization and weight status in Cameroon over the last 25 years.

Urbanization and weight status trends in Cameroon

In 2013, Cameroon's population was estimated at 22.3 million with over half (53.3%) of this number residing in urban areas. This represents an increase from 40% in 1990 (The World Bank Group 2014) (see Figure 10.1). Like in other low- and middle-income countries, this trend is expected to persist. Rural exodus accounts for 2.8% of the urban population growth rate in Cameroon with the (mostly young aged 15–35) population leaving less rewarding rural farming activities to seek better education and job opportunities in urban settings such as Yaoundé, Douala, Bafoussam, Bamenda, and Garoua (Folefack 2015).

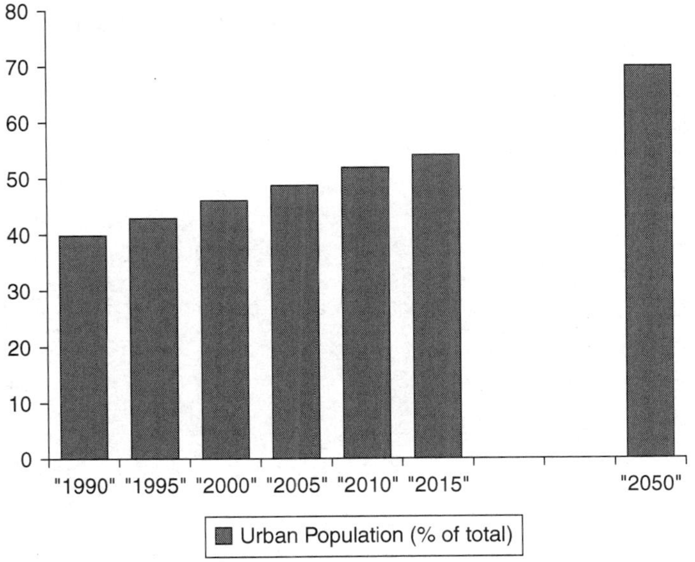

Figure 10.1 Urban population growth in Cameroon (1990–2050) (The World Bank 2014).

Urbanization has been defined as the process whereby large numbers of people congregate and settle in an area, eventually developing social institutions such as businesses and government to support themselves (Orum 2011). In many countries in sub-Saharan Africa, migrants are faced with enormous accommodation challenges upon arrival, mostly linked to housing shortage and/or unaffordable costs. This pushes them to migrate to surrounding rural areas and other built-up areas which develop into informal urban settlements from where they commute to the main city (Aluko and Amidu 2006). Such spontaneous urban settlements are often associated with numerous challenges with the potential to significantly interfere with health and wellbeing. Parts of residential settings in many Cameroonian cities are characterized by the absence of uniformity in housing infrastructure, low walkability, and the absence of basic sanitation facilities (Sikod 1995). This spontaneous expansion of urban settlements without commensurate economic and infrastructural development has led to increased pressure on available infrastructure and social services. Consequently, living conditions for many urban dwellers remain far from meeting acceptable standards for ensuring optimal health and wellbeing. The need for preemptive action has been acknowledged in Cameroon's Vision 2035 paper which draws attention to the fact that problems resulting from rapid and uncontrolled urban expansion already plaguing Douala and Yaoundé (Cameroon's two main cities) might worsen as 75% of Cameroon's population is expected, in the next 25–30 years, to live in urban areas (Ministry of Economy, Planning, and Regional Development 2009).

Through complex interactions between factors associated with uncontrolled urbanization together with increased income for urban residents (World Health Organization 2015), excess weight (overweight and obesity) is becoming a serious health concern in this milieu. The World Health Organization (WHO) has defined obesity as "a condition of abnormal or excessive fat accumulation in adipose tissue, to the extent that health may be impaired" (World Health Organization 2000).

Over the last few decades, with increasing urbanization, Cameroonian men and women, especially in urban areas, have simultaneously been increasing in girth (see Figure 10.2). Evidence suggests that with increasing socioeconomic status, the prevalence of overweight and obesity tends to increase or remain constant but hardly ever decreases (Hodge *et al.* 1996, Monteiro *et al.* 2000, Padez 2006, Villamor *et al.* 2006, Wang *et al.* 2007, Parikh *et al.* 2007, Reas *et al.* 2007, Schröder *et al.* 2007). In a study examining changes in central obesity and body mass index (BMI) in rural and urban Cameroon between 1994 and 2003, Fezeu *et al.* (2008) reported changes characterized by an increase in BMI in rural populations and of waist circumference in the urban populations studied. Specifically, they reported that the age-standardized prevalence of overweight and obesity (BMI $\geq$25 kg/m2) increased in the rural area by 54% for women ($P < 0.01$) and by 82% for men ($P < 0.01$), while the age-standardized prevalence of central obesity increased significantly in the urban population by 32% (see Figure 10.2). Their findings suggest that even without all the requirements for a place to be described as "urban", the advent of basic amenities such as electricity and roads

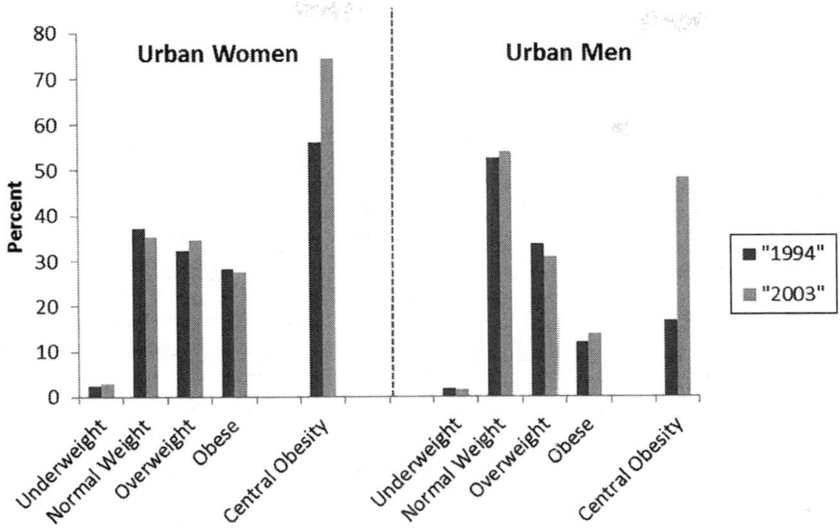

Figure 10.2 Changes in weight and central obesity: women and men in urban Cameroon (1994–2003) (Feuzeu 2008).

tends to have a negative impact on the weight status of its population, especially if accompanying awareness-raising and health promotion measures are not simultaneously introduced.

Factors related to excess weight in urban Cameroon

A person's genes can determine their risk of accumulating excess weight in adult life, but lifestyle – especially pertaining to habitual diet and physical activity levels – is a major determinant (Hoffman 2015). In Cameroon, like elsewhere in the developing world, cities tend to provide conditions that favourably skew an individuals' propensity towards weight gain (Popkin 1993, World Health Organization 1998, Cavalli-Sforza *et al.* 1996). Research findings from developing countries in general (Hoffman 2015) and from Cameroon in particular (Fezeu 2008) have highlighted a number of factors which could account for this. According to these findings, excess weight among urban dwellers is directly related to unhealthy diets and physical inactivity. Other indirect factors include the traditional perception of overweight/obesity as a desirable state, the transition from traditional eating patterns, increased socioeconomic status, inadequate awareness regarding the nutrient content of local food, improved transportation systems, increased ownership of motorized means of transportation, and excessive alcohol consumption.

Urbanization and diet

A positive imbalance between energy intake and expenditure for extended periods increases the likelihood of weight gain (Stunkard 1996). Improvements in individuals' socioeconomic status represent increased disposable income and access to processed foods. In Cameroon's urban settings, the relationship between dietary habits and weight gain can be observed in the following areas:

a *Poor adaptation of eating patterns to energy requirements of urban life:* Cameroon has a rich cultural diversity, characterized by over 240 ethnic groups, and almost as many culinary specialities. Most of these foods harvested from local farms are, to a large extent, healthy (various grains, fruits, and vegetables). Contrary to popular scientific knowledge, Mennen *et al.* (2000), in a study comparing energy, fat, and alcohol intake between rural and urban populations in Cameroon, reported higher food intake in the rural populations. They explained that despite this finding, the lower prevalence of the classic cardiovascular risk factors like BMI, blood pressure, and total cholesterol in rural areas could be attributed to high physical activity levels – mainly occupational – of the population. It can be deduced from this that as people migrate from rural to urban settings, their total physical activity energy expenditure tends to decrease due to the increased sedentary nature of urban life. As such, maintaining previous dietary habits would tilt the energy balance and increase the risk of weight gain. Also, "modern" or "urban" food culture is characterized by unhealthy foods (high in energy, sugar, refined grains, and fat). Eating in solitude and/or on-the-go are factors which favour the consumption of much larger quantities than required by the body for optimal health. Weight gain and related pathologies such as diabetes and heart disease have been attributed to such dietary habits (Popkin *et al.* 1995, Drewnowski and Popkin 1997).

b *Limited awareness of nutrient content of local foods:* This is particularly important in urban settings where, due to overall lower daily energy requirements, people generally require less energy compared to rural dwellers, who need higher levels to meet the demands of their mostly labour-intensive lifestyles. In an attempt to fill this gap, Sharma *et al.* (2007) conducted a study to determine the nutritional composition of 34 commonly consumed composite dishes from the Centre Region of Cameroon. It is, however, not clear to what extent this information has been disseminated to relevant stakeholders for use in the promotion of healthier food choices and meal planning.

c *Unhealthy dietary habits:* Urban populations have greater access to processed foods possibly due in part to the absence of, or inadequate implementation of regulatory policies, as well as the availability of more disposable income. Here, deep-fried foods sold by roadside vendors are a common phenomenon, accessible to all population segments (Mennen *et al.* 2000). In African cities in general, a lot of snacks and meals tend to be consumed outdoors, away from the home while this seldom occurs in rural settings (Ag Bendech *et al.* 1996). Furthermore, diets high in fruits and vegetables are known to favour healthy weight maintenance (Rodriguez *et al.* 2005, de Oliveira *et al.* 2008, Alinia *et al.* 2009). In Cameroon, there is evidence that these health-friendly foods are insufficiently consumed. Ntentie *et al.* (2014) found that more than 45%

of urban dwellers interviewed (n = 3,512) consumed green vegetable sauces at most once a week. Over 46% of this population also reported consuming fruits less than once per week. Participants in a qualitative study carried out in Yaoundé in 2010 (Awah 2008) exploring prevailing eating habits attributed their unhealthy food choices to poverty. They acknowledged awareness of the poor quality of their diets but were unable to change these habits given their economic conditions.

Urbanization and physical activity levels

Physical activity is a key determinant of energy expenditure and thus fundamental to energy balance and weight control (World Health Organization 2015b). Data on physical activity trends in Cameroon, like in other low- and middle-income countries, are still scarce (Halal *et al.* 2012). However, the evidence base is expanding. Existing data indicate a high prevalence of physical inactivity in Cameroon especially among urban populations (Sobngwi *et al.* 2002, Assah *et al.* 2011). In 2010, the World Health Organization reported physical inactivity prevalence rates of 30.5% for men and 47.6% for women in this setting (World Health Organization 2010). With increasing urbanization, Cameroonians are experiencing a shift from traditionally labour-intensive activities such as farm work to more sedentary urban activities such as store-keeping, factory, and office work. In a study examining the relationship between urbanization, physical activity, and metabolic health in sub-Saharan Africa, Assah *et al.* (2011) demonstrated that urban, compared with rural residence, is associated with lower physical activity energy expenditure (PAEE) and higher prevalence of metabolic syndrome. They reported significantly lower PAEE in urban compared to rural dwellers (44.2 ± 21.0 vs. 59.6 ± 23.7 kJ/kg/day, $P < 0.001$).

The following factors could potentially explain the observed relationship between urbanization and physical inactivity:

a *Improvements in the transportation system:* Among other recommendations, Cameroon's Vision 2035 strategy paper highlights the need to improve the nation's road infrastructure (Ministry of Economy, Planning, and Regional Development 2009). Ingram and Liu (2015) have reported that the length of paved road at the national level increases with income as does the number of vehicles. With the improvement in road infrastructure and the increase in the number of vehicles, the trend is for the population to opt for motorized transportation, even over distances previously considered walkable.

b *Increase in the number of motorcycles and other motorized means of transportation:* In a study conducted among Chinese adults to assess the impact of motorized vehicle ownership on the risk of becoming obese, Bell *et al.* (2002) found that, between 1989 and 1997, 14% of households acquired a motorized vehicle. They went on to show that compared with those whose vehicle ownership did not change, men who acquired a vehicle experienced a 1.8kg greater weight gain ($p < 0.05$) and had 2 to 1 odds of becoming obese.

Yet another study conducted in Colombia found evidence that household motor vehicle ownership is associated with overweight, obesity, and abdominal obesity among men but not women (Parra *et al.* 2009). While still not optimal, the improvement in road infrastructure in Cameroon has prompted an increase in the ownership of motorized means of transportation both for private as well as for commercial use. Between 2009 and 2012, according to Cameroon's Ministry of Transport, a total of 156,982 motorbikes came into circulation on Cameroon's roads (see Figure 10.3). The sharp decline in the number of new vehicles in 2011 as shown in the graph could be related to the fact that 2011 was an election year in Cameroon, a period usually characterized by timid economic activity (National Institute of Statistics 2013).

The proliferation of motorbikes has contributed to the reduction of transport-related physical activity in Cameroon, as city dwellers are able to easily access and cost-effectively reach most destinations, including those which were in the past inaccessible. In a recent review exploring the problem of obesity and type 2 diabetes in sub-Saharan Africa, Mbanya *et al.* (2014) highlighted the contribution of the urban proliferation of motorbikes in fuelling the obesity epidemic.

c *Sedentary occupations*: It is well known that low levels of physical activity are associated with an increased risk of weight gain and obesity (Jebb and Moore 1999). Various authors are in agreement regarding the role of sedentary occupations in promoting the obesity epidemic in Cameroon (Kamadjeu *et al.* 2006, Awah *et al.* 2008, Echouffo-Tcheugui and Kengne 2011).

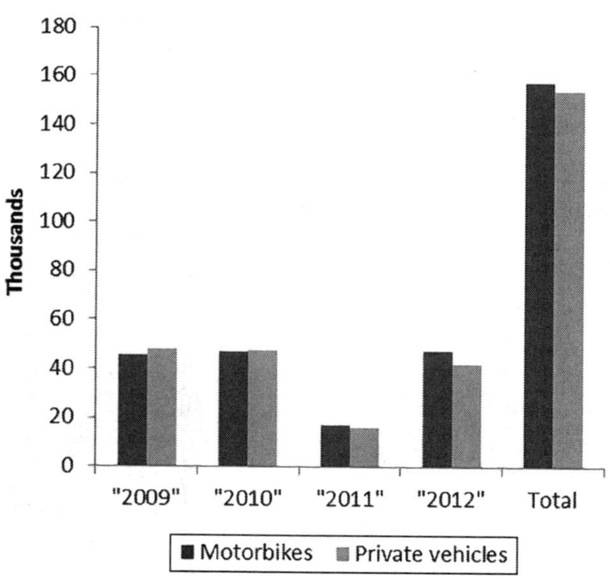

Figure 10.3 New motorized transportation means added to circulation in Cameroon: 2009–2012 (x1000) (National Institute of Statistics 2013).

Such sedentary occupations increase the likelihood of accumulating excess calories which, over time, increase the risk of weight gain.

d *Limited environment-related opportunities and motivation for outdoor physical activity*: Owing to the (mostly) spontaneous nature urbanization in Cameroon's cities, there is limited provision for outdoor parks, pedestrian pavements, and bicycle lanes. The near absence of pedestrian walkways, coupled with the rapid proliferation of commercial motorbike activities, greatly limits walkability in most neighbourhoods because of the increased risk of road traffic accidents involving pedestrians (Sobngwi-Tambekou 2010).

This has also pushed several parents to opt for motorized over active means (walking and cycling) of transportation for their children to travel to and from school (Nyagwui *et al.* 2016). For those interested in leisure time physical activity, the few dedicated fitness trails available in towns such as Yaoundé, Douala, and Bamenda are generally not fully accessible to the public as they are located a long way from residential areas (making transportation cost prohibitive).

e *Unfavourable school infrastructure and curricula*: Due to land scarcity in urban areas, schools are constructed with limited provision for playgrounds. Students are, therefore, compelled to spend their lunch breaks either sitting in their classrooms or standing in clusters in the corridors as they eat. Furthermore, most school curricula usually involve a single physical education session per week, and school health clubs tend to focus more on educating the students on issues related to HIV/AIDS and sexual health (Sobze 2016), omitting awareness raising related to overweight and obesity.

f *Improvement in socioeconomic status:* A number of studies have reported a positive relationship between socioeconomic status and adiposity, especially in urban settings (Bunker *et al.* 1992, Bovet *et al.* 2002). In a Cameroonian study (Fezeu *et al.* 2006), obesity in women was found to increase with household amenities and was positively associated with household amenities and occupation in men even after adjusting for age, physical activity, education, alcohol, and smoking. This is reiterated in Fezeu *et al.* (2008) which reports that increased electrification of different parts of Cameroon between 1994 and 2003 contributed to increasing the prevalence of overweight and obesity – likely due to the acquisition of such amenities as televisions sets, prompting a transition from active leisure time activities to sedentary television watching.

g *Misconceptions about obesity:* The coexistence of malnutrition and obesity in Cameroon, like elsewhere in the developing world, has given rise to misconceptions regarding excess weight. Based on a South African report, people view overweight as a sign of wealth (Kruger *et al.* 1994). Similar findings contained in a recent publication of the Cameroon Academy of Science indicates that Cameroonians attribute excess weight to "wealth", "peace of mind", "good living", "good health", " happiness ", and " authority" (Tanya *et al.* 2011). The negative perception of the lean body type as being associated with the HIV/AIDS pandemic in the 1990s played a major role in cementing this misguided perception of weight loss as linked to a seropositive status. Fezeu *et al.* (2008) also found that weight gain for a man after marriage represents an indicator of the degree of care the man is getting from his wife.

Weight Status, urbanization, and the Millennium Development Goals

The fact that health is wealth is critical to the economic, political, and social development of all countries. This was clearly reflected in the formulation of the MDGs at the Millennium Summit in 2000, with three of the eight goals focusing on health (United Nations General Assembly 2000, The NCD Alliance 2015): MDG 4 Reduce childhood mortality, MDG 5 Improve maternal health, and MDG 6 Combat HIV/AIDS, malaria, and other diseases. While having a strong focus on health, the global development agenda did not include any item directly targeting NCDs and related risks (including obesity), the greater burden of which is today borne by developing countries (World Health Organization 2002). It is likely that this has unfavourably affected the adoption of social, economic, environmental, and other relevant policies (as well as budget allocation) to address and mitigate the impact of NCDs in developing countries. As the prevalence of NCDs continues to increase in the developing world, it is clear that these chronic conditions are set to put a significant strain on already fragile economies and health systems. The increase in the rate of weight gain and unplanned urbanization has had multiple effects on all the MDG.

MDG 1: Eradicate extreme poverty and hunger

As opposed to living in a rural area, urban residence offers more opportunities to generate wealth (even if insufficient) that could have a favourable impact on individuals' economic empowerment and help eliminate extreme poverty. Cities and slums have therefore been described as the "first step" out of rural poverty as they offer more employment opportunities and better access to services such as healthcare and education (United Nations Habitat 2006). In Cameroon, such economic empowerment has been shown to precede weight gain and related health problems (Fezeu 2006). When this happens, the meagre resources that could otherwise have been used to procure food and other basic necessities are spent on healthcare, and the individual(s) concerned find themselves pushed further into poverty.

MDG 2: Achieve universal primary education

Excess weight is associated with increased risk of chronic NCDs, which often require expensive long-term care. The implication of chronic NCDs on families could mean a reduction in the capacity of affected parents to provide education for their offspring as medical expenses push families further into poverty or if children are forced to leave school to take care of sick parents despite the known negative consequences of using children as caregivers (Kangethe 2010). Also, achieving universal primary education implies more children spending more time in school settings. In Cameroon, where urban infrastructural development is poorly regulated, many school environments do not meet WHO standards for health-promoting schools (World Health Organization 2004). Also, school attendance could also imply lower physical activity energy expenditure and possible weight gain, hence the need for lifestyle-related health promotion, which is not popular in Cameroon's schools.

MDG 3: Promote gender equality and empower women

In Cameroon, studies show higher prevalence rates of overweight and obesity in women relative to men (17.1% vs. 5.8%) (World Health Organization 2014). This problem is more serious in urban settings, where there is greater access to unhealthy foods and fewer physical activity opportunities. This translates into a higher risk of developing conditions such as high blood pressure and diabetes (World Health Organization 2014). Furthermore, caregivers of most people with chronic diseases in many parts of the world are women – mothers, sisters, daughters, and aunts (Stone 1987, Akpinar *et al.* 2011, Sharma 2016). In Cameroon, this implies that active women have to stay at home to perform this function, a sacrifice which could hamper current efforts to curb gender inequality.

MDG 4 and 5: Reduce child mortality and improve maternal health

Overweight and obesity in women of reproductive age increase the risk of gestational diabetes, which could have life-threatening effects on both mother and baby (Vambergue 2011). The current high rates of obesity in Cameroon's urban women, as well as the predicted future trends, therefore constitute a major cause for concern.

MDG 6: Combat HIV/AIDS, malaria, and other diseases

Despite having greater access to health facilities compared to rural dwellers, residents of unplanned urban neighbourhoods in Cameroon still face challenges regarding health facility availability and accessibility. In Africa, people tend to view excess weight as an indicator of affluence and/or negative HIV status, and this poses a challenge to the fight against HIV and also to address the problem of overweight and obesity (Muhihi *et al.* 2012).

MDG 7: Ensure environmental sustainability

Well-designed towns and cities with good public transport and food systems are key in the fight against overweight and obesity in these settings (Rundle and Heymsfield 2016). At present, the low-walkability status of several urban neighbourhoods in Cameroon poses a major challenge to physical activity levels.

Effective sustainable urbanization policies that consider health as a cornerstone of urban planning still need to be fully incorporated by urban planners in most Cameroonian cities.

MDG 8: Develop a global partnership for development

Cardiovascular disease and other NCDs (related to obesity) account for 60% of all deaths in the developing world, but only 0.9% of US$22 billion international aid (Official Development Assistance) spent on health in developing countries is spent on NCDs (International Diabetes Federation 2015). With the current global consensus that these problems constitute a common threat, it has become clearer that all nations – developed and developing – need to work together in addressing and mitigating the impact of NCDs.

Next steps

By endorsing the *Global Action Plan for the Prevention and Control of Non-Communicable Diseases 2013–2020* (World Health Organization 2013), the World Health Assembly provided a road map and menu of policy options to guide member states in their effort to reduce the burden of NCDs and their risk factors. The plan operationalizes the commitments of the Political Declaration of the High-Level Meeting of the General Assembly on the Prevention and Control of non-communicable diseases. Of the nine targets adopted, two directly concern actions to lower the rising obesity epidemic by 2025: a 10% relative reduction in the prevalence of insufficient physical activity and halting the rise in diabetes and obesity. A number of calls have been made highlighting the necessity for the Sustainable Development Goals to include a specific target for the prevention and management of NCDs and related risks, including obesity (The NCD Alliance 2015, Fuster and Voûte 2005). The situation in Cameroon could be tackled from multiple angles:

a *Identifying and promoting healthy components of Cameroon's rich culture:* This includes improving the intake of vegetables and fruits and enhanced meal planning within families to ensure a balanced diet. Community cultural and development associations which exist in several Cameroonian cities could provide an effective platform for health promotion and awareness-raising on overweight and obesity.

b *Build the capacity of frontline healthcare personnel:* This knowledge would empower healthcare personnel to be able to educate patients at healthcare facilities, especially mothers attending antenatal clinics and infant welfare clinics, on the importance of healthy weight maintenance.

c *Increase overall health literacy:* Diverse communication channels – mass media, churches, and schools – could be used to improve the population's awareness of issues related to health and wellbeing. Such initiatives could be built into existing structures such as community exercise groups and school health clubs which are already working in the area of health promotion and only require targeted capacity-building and guidance.

d *Increased involvement from policymakers:* Policymakers would have to recognize the detrimental effects of obesity and related health problems on the nation's development and commit to adopting policies that would involve all sectors in curbing its rise. More work is needed to create conditions that would include appropriate policies and budgetary allocations to mitigate the impact of NCDs. Outputs from the work of researchers will provide the evidence to guide the development of policies and mitigation strategies. The ongoing decentralization effort of the government, currently underway in Cameroon, could also be an opportunity as local council mayors would be empowered and challenged to improve urban development in their individual municipalities. This may involve making provision for green spaces, sanctions against local schools with no playgrounds, and allowance for pavements and bicycle lanes – all measures to promote improved wellbeing and physical activity levels. The decentralization process would also offer public health workers a broader platform for dialogue with policymakers to ensure that health is adequately represented as a cross-cutting development enabler in every policy.

e *Capacity building and support for relevant organizations:* This includes providing support for community-based organizations and other groups working at grassroots level to promote healthy lifestyles.

f *Regulations on the sale of soft drinks:* In recent years, the soft drink market in Cameroon has witnessed the arrival of several companies selling diverse products (Silver 2015). There is an urgent need for regulatory mechanisms to be re-enforced so as to prevent negative health effects on the population.

Conclusion

This chapter has explored the interplay between weight changes and uncontrolled urbanization in the Cameroonian population. It has particularly highlighted the influence of lifestyle habits, notably diet and exercise, on anthropometry as more Cameroonians make the transition from rural to urban lifestyles without any accompanying sensitization on precautionary measures to ensure healthy weight maintenance. The importance of NCDs on global health has gained increased recognition over the last decade. As the evidence base of the impact of NCDs continues to grow especially in developing countries, so too do calls for a stronger policy response. The global burden and threat of NCDs pose a major challenge for development and undermines social and economic advancement around the world. As a result, we have shown that progress towards the attainment of the global development objectives outlined in the MDGs was hampered by the exclusion of NCDs in the original framework. The obesity pandemic is a threat to global health, and priority must be given to addressing this multi-sectoral problem through prevention, treatment, and care measures to consolidate efforts to make the world healthier.

References

Ag Bendech M., Gerbouin-Rerolle P., Chauliac M., and Malvy D., 1996. Approche de la consommation alimentaire en milieu urbain. Le cas de l'Afrique de l'Ouest. *Cahiers Santé*, 6, 173–179.

Akpınar B., Küçükgüçlü O., Yener G., 2011. Effects of gender on burden among caregivers of Alzheimer's patients. *Journal of Nursing Scholarship*, 43, 248–254.

Alinia, S., Hels, O., and Tetens, I., 2009. The potential association between fruit intake and body weight – a review. *Obesity Reviews: An Official Journal of the International Association for the Study of Obesity*, 10(6), 639–647.

Aluko, B.T. and Amidu, A., 2006. Urban low income settlements, land regulation, and sustainable development in Nigeria. *Promoting Land Administration and Good Governance.5th Regional Conference*. Accra, Ghana.

Assah, F.K., Ekelund, U., Brage, S., Mbanya, J.C., and Wareham, N.J., 2011. Urbanization, physical activity, and metabolic health in sub-Saharan Africa. *Diabetes Care*, 34(2), 491–496.

Assembly, U.G., 2000. United Nations millennium declaration. *United Nations General Assembly*.

Awah, P.K., Kengne, A.P., Fezeu, L.L.K., and Mbanya, J.C., 2008. Perceived risk factors of cardiovascular diseases and diabetes in Cameroon. *Health Education Research*, 23(4), 612–620.

Bell, A.C., Ge, K., and Popkin, B.M., 2002. The road to obesity or the path to prevention: motorized transportation and obesity in China. *Obesity Research*, 10(4), 277–283.

BeLue, R. *et al.*, 2009. An overview of cardiovascular risk factor burden in sub-Saharan African countries: a socio-cultural perspective. *Globalization and Health*, 5, 10.

Bovet, P. *et al.*, 2002. Distribution of blood pressure, body mass index and smoking habits in the urban population of Dar es Salaam, Tanzania, and associations with socioeconomic status. *International Journal of Epidemiology*, 31(1), 240–247.

Bunker, C.H. *et al.*, 1992. Factors associated with hypertension in Nigerian civil servants. *Preventive Medicine*, 21(6), 710–722.

Cavalli-Sforza, L.T., Rosman, A., de Boer, A.S., and Darnton-Hill, I., 1996. Nutritional aspects of changes in disease patterns in the Western Pacific region. *Bulletin of the World Health Organization*, 74(3), 307–318.

de Oliveira, M.C., Sichieri, R., and Venturim Mozzer, R., 2008. A low-energy-dense diet adding fruit reduces weight and energy intake in women. *Appetite*, 51(2), 291–295.

Drewnowski, A. and Popkin, B.M., 1997. The nutrition transition: new trends in the global diet. *Nutrition Reviews*, 55(2), 31–43.

Echouffo-Tcheugui, J.B. and Kengne, A.P., 2011. Chronic non-communicable diseases in Cameroon – burden, determinants and current policies. *Globalization and Health*, 7, 44.

Ellulu, M., Abed, Y., Rahmat, A., Ranneh, Y., and Ali, F., 2014. Epidemiology of obesity in developing countries: challenges and prevention. *Global Epidemic Obesity*, 2(1), 2.

Fezeu, L. *et al.*, 2006. Association between socioeconomic status and adiposity in urban Cameroon. *International Journal of Epidemiology*, 35(1), 105–111.

Fezeu, L. *et al.*, 2008. Ten-year changes in central obesity and BMI in rural and urban Cameroon. *Obesity*, 16(5), 1144–1147.

Folefack, A.J., 2015. The rural exodus of young farmers and its impact on the shortage of labor and food crop production in Cameroon: a computable general equilibrium model's analysis. *Journal of Human Ecology*, 49(30), 197–210

Frenk, J., Bobadilla, J.L., Sepuúlveda, J., and Cervantes, M.L., 1989. Health transition in middle-income countries: new challenges for health care. *Health Policy and Planning*, 4(1), 29–39.

Fuster, V. and Voûte, J., 2005. MDGs: chronic diseases are not on the agenda. *The Lancet*, 366(9496), 1512–1514.

Hodge, A.M. *et al.*, 1996. Incidence, increasing prevalence, and predictors of change in obesity and fat distribution over 5 years in the rapidly developing population of Mauritius. *International Journal of Obesity and Related Metabolic Disorders: Journal of the International Association for the Study of Obesity*, 20(2), 137–146.

Hoffman, D., 2015. Obesity in developing countries: causes and implications [Online]. *Food, Nutrition and Agriculture*. Available from: http://www.fao.org/docrep/003/Y0600M/y0600m05.htm. [Accessed 1 February 2016].

Ingram, G. and Liu, Z., 2015. Determinants of motorization and road provision. In World Resource Institute, ed. *World Resources 1996–97*. New York: Oxford University Press.

International Diabetes Federation, 2015. The millenium development goals and diabetes [Online]. Available from: https://www.un-ngls.org/IMG/pdf_MDGs_and_Diabetes_Factsheet.pdf [Accessed on 1 August 2017].

Jebb, S.A. and Moore, M.S., 1999. Contribution of a sedentary lifestyle and inactivity to the etiology of overweight and obesity: current evidence and research issues. *Medicine and Science in Sports and Exercise*, 31 (11 Suppl), S534–41.

Kamadjeu, R.M. *et al.*, 2006. Anthropometry measures and prevalence of obesity in the urban adult population of Cameroon: an update from the Cameroon Burden of Diabetes Baseline Survey. *BMC Public Health*, 6, 228.

Kangethe, S., 2010. The dangers of involving children as family caregivers of palliative home-based-care to advanced HIV/AIDS patients. *Indian Journal of Palliative Care*, 16(3), 117–122.

Kruger, H.S., Van Aardt, A.M., Walker, A.R.P., and Bosman, M.J.C., 1994. Obesity in African hypertensive women: problems in treatment. *South African Journal of Food Science and Nutrition*, 8, 106–112.

Lim, S.S. *et al.*, 2012. A comparative risk assessment of burden of disease and injury attributable to 67 risk factors and risk factor clusters in 21 regions, 1990–2010: a systematic analysis for the Global Burden of Disease Study 2010. *The Lancet*, 380(9859), 2224–2260.

Mbanya, J.C., Assah, F.K., Saji, J., and Atanga, E.N., 2014. Obesity and Type 2 Diabetes in Sub-Sahara Africa. *Current Diabetes Reports*, 14(7), 501.

Mennen, L. *et al.*, 2000. The habitual diet in rural and urban Cameroon. *European Journal of Clinical Nutrition*, 54(2), 150–154.

Ministry of Economy, Planning and Regional Development, 2009. *Cameroon Vision 2035*. Republic of Cameroon.

Monteiro, C.D., Benicio, M., Conde, W., and Popkin, B.M., 2000. Shifting obesity trends in Brazil. *European Journal of Clinical Nutrition*, 54(4), 342–346.

Muhihi, A.J. *et al.*, 2012. Obesity, overweight, and perceptions about body weight among middle-aged adults in Dares Salaam, Tanzania. *ISRN Obesity*, 2012, 368520.

National Institute of Statistics, 2013. *Cameroon Statistics Directory, 2013*. Yaounde, Cameroon: Republic of Cameroon.

The NCD Alliance, 2015. *NCD alliance report 2013–2014: Putting non-communicable diseases on the global agenda*. The NCD Alliance.

Ntentie, F.R. *et al.*, 2014. Urbanization and metabolic syndrome in Cameroon: alertness on less urbanised areas. *Endocrinology and Metabolic Syndrome*, 3, 137.

Nyagwui, A.E., Fredinah, N., Che, L.B., and Yulia, B., 2016. Motorcycle injury among secondary school students in the Tiko municipality, Cameroon. *The Pan African Medical Journal*, 24, 116.

Orum, A., 2011. Urbanization. In *Encyclopedia of Social Theory*. Sage publications.

Padez, C., 2006. Trends in overweight and obesity in Portuguese conscripts from 1986 to 2000 in relation to place of residence and educational level. *Public Health*, 120(10), 946–952.

Parikh, N.I. *et al.*, 2007. Increasing trends in incidence of overweight and obesity over 5 decades. *The American Journal of Medicine*, 120(3), 242–250.

Parra, D.C. *et al.*, 2009. Household motor vehicle use and weight status among Colombian adults: Are we driving our way towards obesity? *Preventive Medicine*, 49(2), 179–183.

Popkin, B.M., 2002. An overview on the nutrition transition and its health impacts. *Public Health Nutrition*, 5(1A), 93–103.

Popkin, B.M., 1993. Nutritional patterns and transitions. *Population and Development Review*, 19(1), 138–157.

Popkin, B.M., Adair, L.S., and Ng, S.W., 2012. Global nutrition transition and the pandemic of obesity in developing countries. *Nutrition Reviews*, 70(1), 3–21.

Popkin, B.M., Paeratakul, S., Zhai, F., and Ge, K., 1995. A review of dietary and environmental correlates of obesity with emphasis on developing countries. *Obesity Research*, 3 (S2), 145s–153s.

Reas, D.L. *et al.*, 2007. Changes in body mass index by age, gender, and socio-economic status among a cohort of Norwegian men and women (1990–2001). *BMC Public Health*, 7, 269.

Rodríguez, M.C. *et al.*, 2005. Effects of two energy-restricted diets containing different fruit amounts on body weight loss and macronutrient oxidation. *Plant Foods for Human Nutrition*, 60(4), 219–224.

Rundle, A.G. and Heymsfield, S.B., 2016. Can walkable urban design play a role in reducing the incidence of obesity-related conditions? *JAMA*, 315(20), 2175–2177.

Schröder, H. *et al.*, 2007. Secular trends of obesity and cardiovascular risk factors in a Mediterranean population. *Obesity*, 15(3), 557–562.

Sharma, N., Chakrabarti, S., and Grover, S., 2016. Gender differences in caregiving among family – caregivers of people with mental illnesses. *World Journal of Psychiatry*, 6(1), 7–17.

Sikod, F., 1995. Urbanization in sub-SaharanAfrica; A case study of the city of Yaounde, Cameroon. *Conference on Human Settlements in the Changing Global Political* and *Economic Processes*, Helsinki, Finland: 25–27 August.

Silver, M., 2015. Guess which country has the biggest increase in soda drinking [Online]. Available from: http://www.npr.org/sections/goatsandsoda/2015/06/19/415223346/guess-which-country-has-the-biggest-increase-in-soda-drinking [Accessed on 2 February 2016].

Sobngwi, E. *et al.*, 2002. Physical activity and its relationship with obesity, hypertension and diabetes in urban and rural Cameroon. *International Journal of Obesity*, 26(7), 1009–1016.

Sobngwi-Tambekou, J., Bhatti, J., Kounga, G., Salmi, L.R., and Lagarde, E., 2010. Road traffic crashes on the Yaoundé-Douala road section, Cameroon. *Accident Analysis and Prevention*, 42(2), 422–6.

Sobze, M.S., Tiotsia, A.T., Dongho, G.B.D. *et al.*, 2016. Youth awareness on sexually transmitted infections, HIV, and AIDS in secondary schools in the Dschang Municipality (Cameroon): The mobile caravan project. *Journal of Public Health in Africa*, 7(2), 614.

Spearing, K. *et al.*, 2013. Nutritional composition of commonly consumed composite dishes from rural villages in Empangeni, KwaZulu-Natal, South Africa. *Journal of Human Nutrition and Dietetics*, 26(3), 222–229.

Stone, R., Cafferata, G.L., and Sangl, J., 1987. Caregivers of the frail elderly: a national profile. *Gerontologist*, 27, 616–626.

Stunkard, A.J., 1996. Socioeconomic status and obesity. *Ciba Foundation Symposium*, 201, 174-182-187, 188–193.

Tanya, A.N.K., Lantum, D.N., and Tanya, V.N., 2011. Nutrition and public health in Cameroon: Combating the crisis, 54. *Yaoundé, Cameroon: Cameroon Academy of Sciences*.

UN Habitat, 2006. Proceedings of the Asia Pacific Conference on Housing and Human Settlements. Cities, Slums and the Millennium Development Goals, India.

Villamor, E. *et al.*, 2006. Trends in obesity, underweight, and wasting among women attending prenatal clinics in urban Tanzania, 1995–2004. *The American Journal of Clinical Nutrition*, 83(6), 1387–1394.

Wang, Y., Mi, J., Shan. X.Y., Wang, Q.J., and Ge, K.Y., 2007. Is China facing an obesity epidemic and the consequences? The trends in obesity and chronic disease in China. *International Journal of Obesity*, 31(1), 177–188.

The World Bank Group, 2014. *World Development Indicators 2013: Population dynamics*. The World Bank Group.

World Health Organization, 2000. *Obesity: Preventing and Managing the Global Epidemic*. World Health Organization.

World Health Organization, 2002. *The World Health Report 2002: Reducing Risks, Promoting Healthy Life*. World Health Organization.

World Health Organization, 2004. The Physical School Environment: An Essential Component of a Health-Promoting School. World Health Organization. Geneva, Switzerland.

World Health Organization, 2010. *Global Status Report on Noncommunicable Diseases*. Geneva: World Health Organization.

World Health Organization, 2013. *Global Action Plan for the Prevention and Control of NCDs 2013–2020*. Geneva: World Health Organization.

World Health Organization, 2014. *Global Status Report on Noncommunicable Diseases*. Geneva: World Health Organization.

World Health Organization, 2015a. Health in 2015: from MDGs, Millennium Development Goals to SDGs, Sustainable Development Goals. 2015. World Health Organization, Geneva, Switzerland.

World Health Organization, 2015b. Prevalence of insufficient physical activity: Situation and trends [Online]. Available from: http://www.who.int/gho/ncd/risk_factors/physical_activity_text/en/ [Accessed 22 December 2015].

11 Maternal perception about early childhood caries in Nigeria

Afolabi Oyapero

Introduction

Africa has 54 acknowledged sovereign states and countries with many ethnic groupings. The African region has the heaviest burden of disease in spite of important national and regional efforts (World Bank 1994). The World Bank states that 32 African countries are among the world's 48 least developed nations, and 80% of the people in the region are in the low socioeconomic category (World Bank 1994). Poverty is the most important determinant of ill health in Africa (World Health Organization 2002). Sub-Saharan Africa is the poorest region in the world, and it is marked by political and economic instability and great variations in natural resources with most of the poorest countries in the world residing in this region (Popkin 2002). The region faces an intense scourge of HIV/AIDS in addition to the rising prevalence of non-communicable diseases (NCDs) like hypertension, diabetes mellitus, and other medical conditions related to obesity (Popkin 2002).

Nigeria has a high burden of infectious diseases with related morbidity and mortality, while women and children bear the greatest impact of diseases in this region (Lambo 2015). About 70% of the population live in rural areas (World Bank 1987), and even though Nigeria has one of the highest numbers of health workers in Africa, there are disparities in their distribution between the rural and urban areas, geopolitical zones, states, and local government areas (Lambo 2015). These poorly motivated staff work in poor work environments with inadequate infrastructure, limited opportunity for continuing education, inadequate supervision, and low logistic support (Lambo 2015). This ultimately leads to poor retention and migration of health workers, further worsening the health indices of Nigeria. The presence of widespread poverty and underdevelopment in Nigeria, like other African countries, means that these communities are also exposed to the major environmental determinants of oral disease (Thorpe 2006).

The prevalence of dental diseases closely mimics the levels of social deprivation. Oral diseases are among the most common NCDs that affect people throughout their lifetime, causing pain, disfigurement, and even death. In the World Health Organization (WHO) African Region where above 80% of the population are of low socioeconomic status, these diseases affect the health and wellbeing of millions of people. Oral health disparities continue to widen, more so among disadvantaged and vulnerable groups where the vast majority experience the highest burden of oral diseases (Petersen *et al.* 2005, Watt 2005, Jin *et al.* 2011). Importantly, children in developing countries are known to suffer disproportionately from the burden of dental diseases. Children with poor oral health experience pain and tooth loss which affect their feeding and negatively impact their nutrition, self-esteem, speech, socialisation, quality of life, and school attendance (Edelstein *et al.* 2006). Worldwide, it is estimated that over 51 million school hours are lost annually from dental-related illness (Peterson *et al.* 2005, Pau *et al.* 2008, Rowan-Legg 2013). The majority of these children in Africa have no access to dental care (Petersen *et al.* 2005).

Barry Popkin's revised demographic, epidemiologic, and nutrition transition model (Popkin 2002) serves as the framework for this work, and it addresses how nutrition-related NCDs are now the main causes of disability and death globally. Dental caries is a unique diet-dependent disease which has a communicable component through the horizontal or vertical transmission of cariogenic bacteria such as streptococcus mutans (SM), especially from the mother. It can manifest due to a complex interplay between dietary and infectious processes in addition to other social and cultural influences. Mothers are key decision makers on children's oral healthcare, and maternal knowledge and attitudes to oral health have been reported to be an important predictor of children's oral health in Nigeria (Adeniyi *et al.* 2009). Since the critical focus of this model was on prevention and because many oral diseases have modifiable risk factors, it is desirable to determine maternal perception about early childhood caries in Nigeria and to explore preventive strategies that can be adopted.

Factors implicated for the high burden of oral diseases in Africa

In Africa, there is a low priority given to oral healthcare due to the prevalence of diarrhoeal diseases, acute respiratory infections, malaria, measles, HIV/AIDS, and other health conditions associated with high morbidity and mortality, in addition to other enormous developmental needs. There is also low oral health awareness, and many patients present in the hospital with advanced disease when expensive and rehabilitative procedures that require highly-skilled personnel are required. Socio-demographic characteristics have been known to influence oral health perception, behaviour, and oral care utilisation (Lambo 2015). Rural areas are often associated with lower education levels, which are often related to lower levels of health literacy and poor use of healthcare services (Booysen *et al.* 2007). Rural residents in East and West Africa were more likely to present with caries, have a high need for dental treatments, present poor oral hygiene, and have low utilisation of dental services compared with their urban counterparts (Varenne *et al.* 2006, Mashoto *et al.* 2010, Azodo and Amenaghawon 2013). In many African countries, hospitals offering dental services are few and are located mainly in urban centres; rural dwellers must travel several kilometres to access care. The few dental personnel are also less likely to work in rural areas which have the greatest burden of oral health needs due to lower remuneration, poorer standards of living, poor infrastructure, less professional support, and fewer choices and opportunities for specialisation (Price and Weiner 2005). In most African countries, there is the option to pay for healthcare out of pocket or with health insurance coverage, but health insurance rarely provides payment for oral health.

Priority oral diseases in Africa by the World Health Organization

WHO has given priority to seven oral diseases and conditions that represent the largest part of the oral disease burden in Africa. These are craniofacial malformations (cleft lip/palate), Noma, oral manifestations of HIV and AIDS, oral cancers, orofacial trauma from accidents/ violence, periodontal diseases, and dental caries. Most of these oral diseases are either preventable or treatable in their early stages. In addition, they share modifiable risk factors with many non-communicable diseases such as tobacco use, alcohol consumption, and poor diet. Orofacial clefts (OFCs) are the most common craniofacial birth defect (Peterka *et al.* 2000), occurring in approximately 1 in 700 live births (Mossey and Castillia 2003). Predisposing factors for OFCs' development include genetic factors, drug intake, smoking during pregnancy, radiation, poverty, trauma, and religious and cultural beliefs (Oginni *et al.* 2010, Sabbagh *et al.* 2012). OFC is associated with feeding difficulties, speech impairment, poor facial aesthetics, growth, psychological, and social problems. Children with OFCs require prolonged surgical, speech, and medical specialist care which can pose a lot of financial and emotional burden on their families (Nackashi *et al.* 2002, Jelliffe-Pawlowski *et al.* 2003).

Noma is a pathetic affliction that has been eradicated from the developed world but represents a scourge associated with poverty, and it remains a major public health problem in some African countries (United Nations 2012). The annual incidence of Noma is about 20 cases per 100,000 people, and it has a mortality rate of 70% to 90% where treatment is unavailable (United Nations 2012). It can be a pointer to HIV/AIDS infection, malaria, severe diarrhoea, measles, tuberculosis, and necrotizing ulcerative gingivitis (Ogbureke and Ogbureke 2010, World Health Organization 2017). Oral manifestations of HIV/AIDS are also important indicators of HIV infection and underlying immune suppression. Oral conditions associated with HIV disease include candidiasis, non-Hodgkin's lymphoma, linear gingival erythema, necrotizing ulcerative gingivitis, Hairy leukoplakia, and Kaposi's sarcoma. They are among the early clinical features of the infection and can predict progression of HIV to AIDS (Hodgson *et al.* 2006).

Oral cancer accounts for about 3.6% of all malignant tumours in Nairobi, Kenya (Onyango *et al.* 2004), while in Nigeria, it accounts for 36.8% of head and neck malignancies (Amusa *et al.* 2004). The high incidence rates relate directly to risk behaviours such as smoking, using smokeless tobacco, and consuming alcohol. In Africa, the prevalence of tobacco smoking ranges from 15% of adults in Nigeria to 67% in Kenya (Mackay and Eriksen 2002). Similarly, craniofacial trauma and traumatic dental injuries are a major public health problem in Africa that result from road traffic accidents, drowning, burns, interpersonal and self-inflicted violence, as well as casualties from warfare (Glendor 2008). The peak age of incidence of maxillofacial injuries is 21 to 30 years due to involvement in risky sporting activities, irresponsible driving of motor vehicles, and interpersonal violence (Adeyemo *et al.* 2005). The prevalence of gingival inflammation is also very high in several African countries, and it affects all age groups (Axelsson *et al.* 2002). The important risk factors for periodontal disease include poor oral hygiene, tobacco use, excessive alcohol consumption, stress, poor general health, diabetes mellitus, and HIV. Epidemiological studies in developing countries also show that the prevalence and severity of dental caries have increased with industrial development and exposure to Western diets (Mascarenhas 1999).

The Pattern 4 or the nutrition-related non-communicable diseases in Barry Popkin's (2002) revised demographic, epidemiologic, and nutrition transition model shows that a diet high in total fat, cholesterol, sugar and other refined carbohydrates, and low in polyunsaturated fatty acids and fibre and often accompanied by an increasingly sedentary life, is characteristic of most high-income societies and increasing portions of the population in low-income societies. Sugar and other refined carbohydrates are one of the three known primary risk factors in the development of dental caries. Sugar and refined carbohydrates is thus a common risk factor in the development or progression of many nutrition-related non-communicable diseases such as obesity, Type 2 diabetes mellitus, dental caries, certain types of cancer, and cardiovascular disease (Popkin 2002, Taubman and Nash 2006). Dental caries is one of the most common chronic infectious childhood diseases and individuals remain susceptible to the disease throughout their lifetime (Edelstein and Chinn 2009). Data from 39 African countries show that even though the majority of children have a low Decayed Missing Filled Teeth (DMFT), most of it represents untreated caries and reflects the inadequacy of the response to the problem by existing oral health services (Hobdell *et al.* 2003).

Early childhood caries

Early childhood caries (ECC) is the presence of one or more decayed (non-cavitated or cavitated lesions), missing (due to caries), or filled tooth surface on any primary tooth in children up to 71 months of age. Severe ECC, conversely, is the presence of smooth-surface caries in children younger than three years of age; the presence of one or more cavitated, missing (due to caries), or filled smooth surfaces in primary maxillary anterior teeth; or a decayed, missing, or filled score of ≥4 (age 3), ≥5 (age 4), or ≥6 (age 5) from ages three through five. ECC results from a prolonged discrepancy between risk factors and protective factors (American Academy of Pediatric Dentistry 2008). The socially disadvantaged segments of the society have the highest burden of disease, with the reported prevalence rate in developing countries as high as 70% (Milnes 1996). ECC is acknowledged as a serious public health problem due to its high prevalence, impact on quality of life, potential for increasing risk of caries in the permanent dentition, and role in oral health inequalities (Seow 2012).

Impact of ECC

In low-income African nations, 95% of the caries in the primary dentition and 89% in the permanent dentition remain untreated (World Health Organization 2017). Dental caries of the primary dentition can impair growth and cognitive development due to decreased nutritional intake associated with pain. Studies show an association between caries in the primary dentition with protein-energy malnutrition and stunted growth (Alvarez *et al.* 1993). Nigeria is facing a crisis of malnutrition and ranks second behind India among all countries with the highest number of stunted children (United Nations Children Fund 2013) with attendant susceptibility to illness and intellectual disability. Severe early childhood caries may be a possible aetiological or contributing factor to failure to thrive and poor weight gain in young children (Acs *et al.* 1992, Adeniyi *et al.* 2016). It was observed that dental rehabilitation in children led to weight gain (Acs *et al.* 1992) and improvement in eating preferences, quality of food eaten, social behaviour, and sleeping habits (Thomas and Primosch 2002).

Besides the morbidity and mortality from sepsis and general anaesthesia, the economic costs of ECC are also enormous (Cassamassimo *et al.* 2009). Many African nations have reduced their investment in the health sector and the proportion of the gross national product allocated to health has also been reduced (Van Der Gaag and Barham 1998). Traditional restorative treatment for oral disease is very costly, and it has been identified as the fourth most expensive category of disease to treat in most industrialised countries (Moynihan and Petersen 2004). The estimated cost for restorative dental treatment among children in the United States is over $2 billion annually, making it one of the single most expensive diseases of childhood (Berg and Stapleton 2012). Children with severe ECC are also at a higher risk for hospitalisation, emergency dental visits (Sheller *et al.* 1997), loss of school days, and a diminished ability to learn (Peterson *et al.* 1999). Treatment of ECC frequently necessitates extensive restorative treatment and extraction of teeth at an early age. In addition, general anaesthesia or deep sedation may be required, because such children may not tolerate surgical procedures under local anaesthesia. Due to the lack of public funding, many households in Africa are burdened with the costs of dental treatment. This is disheartening because a dental filling in Zimbabwe would cost US$2.50 in public funding which is equivalent to one to two kilogrammes of meat or two tubes of toothpaste (Bratthall and Barnes 1993). Low dental manpower in Africa also compounds this problem since the dentist to population ratio is 1:150,000 in Africa, while most industrialised countries have a ratio of 1:2,000 (Peterson 2003). ECC also affects the quality of life of children and their parents due to dental pain and subsequent tooth loss resulting in difficulty in eating, speaking, sleeping, and socialising (Edelstein *et al.* 2006). Dental caries can lead to decreased appetite and diminished ability to eat causing poor nutritional status, weight loss, impaired wound healing, and decreased resistance to infections, and poor general health (Palmer *et al.* 2010). These impacts may become apparent when changes are observed in sleeping patterns and eating behaviours (Low *et al.* 1999).

Risk factors for ECC

The risk factors associated with ECC can be generally classified into biological and social factors (Berg and Stapleton 2012). Biological risk factors include nutritional parameters, feeding habits, and early colonisation of cariogenic microorganisms, while social risk factors include low parental education, low socioeconomic status, and lack of awareness about dental disease (Hallett and O'Rourke 2003). A child's diet has a profound influence on cognition, behaviour, physical/emotional development, and physical growth and development. Diets rich in non-milk extrinsic sugars allow acidogenic bacteria to ferment carbohydrates to produce acids (Taubman and Nash 2006). Breastfeeding babies for up to 12 months of age is associated with significantly lower ECC experience because human breast milk contains protective elements against caries such as maternal immunoglobulins, enzymes, leucocytes, and specific antibacterial agents. Bottle feeding infants for prolonged periods with bovine milk and sweetened drinks, especially at night, has, however, been associated with a high ECC experience (Hallet and O'Rourke 2003).

Plaque, which acts synergistically with cariogenic diets, is a type of biofilm composed of a colonised bacteria and salivary protein (Fejerskov 2004) and traditional culture methods have estimated that around 700 bacterial types exist in the human oral microbiome. Streptococcus mutans (SM) and lactobacillus (Loesche *et al.* 1975) are, however, the predominant causative agent in the initiation and progression of dental caries (Nyvad and Kilian 1990). SM utilises sucrose to produce extracellular glucan, a water-insoluble polysaccharide, which enables bacterial adherence to the tooth surface and prevents salivary clearance and removal of the organism (Tinanoff *et al.* 2002). Research has shown that the child's acquisition of SM occurs either vertically when a child acquires the bacteria from the mother or caregiver or horizontally from members of the same group, most likely siblings or children of the same age at daycare. Initial colonisation by SM in the oral cavity of children occurs between the ages of seven months to three years (Law *et al.* 2007).

Social risk factors, however, play a major role in ECC prevalence in Nigeria and the rest of Africa. Parental, family, and cultural beliefs determine oral health behaviour and influence the likelihood of caries developing in children (Pine *et al.* 2004). The socioeconomic status of the family has also been shown to influence the caries risk in children. A systematic review by Reisine and Psoter (2001) observed an inverse relationship between socioeconomic status and the prevalence of caries among children less than 12 years of age. Studies show that children that experience ECC continue to be at high risk for new lesions as they grow older, both in the primary and permanent dentitions (O'Sullivan and Tinanoff 1996). This alarming observation is a cause for concern in Nigeria and other resource-constrained nations in Africa.

Maternal perception about early childhood caries in Nigeria

This research was conducted to determine maternal perception about ECC in Nigeria, because mothers are recognised as the key decision makers on children's oral healthcare. Moreover, maternal knowledge and attitudes to oral health have been reported to be an important predictor of children's oral health in Nigeria (Adeniyi *et al.* 2009). The concept of caries in children has evolved from that of ascribing the role of sole causation to inappropriate feeding methods to a focus on the multiple socioeconomic, behavioural, and psychosocial factors that contribute to it (Reisine and Douglass 1998). There are a number of significant events that result in caries in young children. The first is the acquisition of infection with SM, the bacteria highly implicated in caries initiation (Loesche 1975). It has been noted that children who have caries usually have mothers with poorer oral health compared to children without caries (Litt *et al.* 1995). The highest fidelity of SM transmission lies with the mother, given that DNA analysis shows the same sequence in maternal and infant SM (Li *et al.* 2000). The second event is the build-up of SM to pathogenic levels resulting from frequent and prolonged exposure to caries-promoting carbohydrates. Use of nursing bottles filled with sugar-containing fluids at sleep-time, frequent sugar-containing snacks, and lack of oral hygiene can lead to the build-up of SM. The third event is the rapid demineralisation of enamel, which if unchecked leads to cavitations. These events are strongly influenced by the child's social environment (Christensen *et al.* 2010).

Social risk factors implicated in ECC include low parental education, low socioeconomic status, lack of awareness about dental disease, childhood poverty, insufficient exposure to fluoride, and poor perinatal and prenatal health (Gussy *et al.* 2006). The rates of ECC are highest among socially disadvantaged groups (Harris 2004). Socioeconomic background and educational levels are known to correlate with maternal health beliefs and oral health practices (Adeniyi *et al.* 2009). Very little consideration has, however, been given to the effects of pre-existing parental attitudes, knowledge, and behaviours on dental outcomes in very young children. Oral health specific self-efficacy and knowledge measures are potentially modifiable cognitions and interventions that can lead to healthy dental habits (Finlayson 2007). If oral health promotion efforts are to be effective in improving the oral health of young children, it is essential to have an understanding of parental knowledge and attitudes (Schroth 2007). The aim of this study, therefore, is to determine the maternal knowledge, attitudes, and practices on early childhood caries and to assess its relationship with sociodemographic variables among mothers at the immunisation clinic of a Tertiary Hospital in Nigeria.

Methodology

The descriptive study was conducted at the immunisation clinic of the Lagos State University Teaching Hospital, Ikeja (LASUTH). The sample included all mothers who had their babies enrolled for care at the immunisation clinic at LASUTH during the period of study. The target sample size (n = 144) was selected based on the results from a similar study with a prevalence value of 91.2% (Schroth 2007). A simple random sampling technique using the balloting method was employed in selecting the study participants. Mothers with babies aged 3 to 12 months attending the clinic were included in the study after explaining the nature of the study to them and obtaining their informed consent. Mothers who were unwilling to give their informed consent and those who had recently had dental treatment or oral health education were excluded from the study. A total of 144 participants were recruited into the study during the period of January to May 2012. An interviewer-administered questionnaire was employed in obtaining information on the respondents' socioeconomic status, gender, age, marital status, level of education, and occupation. Responses on maternal perception of early childhood caries were also obtained. Data was analysed using Statistical Package for Social Sciences (SPSS) version 20. Frequency distribution tables were generated for all variables and measures of central tendency and dispersion were computed for numerical variables. The Chi-square test was used to determine the level of association between the variables. A 95% confidence interval and a 5% level of significance were adopted.

Result of survey

A total of 144 mothers took part in the study. Socio-demographic characteristics of the respondents are presented in Table 11.1. 73% of the population were in the 26–35 age category and 73.8% worked full time. The majority of the respondents (56.9%) had tertiary education, and most of them (43.1%) had two children.

Maternal knowledge and attitude about dental caries in children

The mothers had a low level of awareness (<30%) in one out of the twelve items included in the awareness and attitude section of the questionnaire. They had moderate awareness (30–60%) on seven of the items, while they had high perception (>60%) on four items (see Table 11.2). Most (87.5%) knew that decayed teeth can lead to illness and that the baby's mouth should be cleaned even if they had no teeth. Most of the respondents (86.1%) also believed that a child should not go to bed with a feeding bottle and that baby teeth are important (81.9%). Respondents had a similar attitude to breastfeeding and bottle-feeding babies at night with more than 50% of them believing it was proper to feed the baby all night. About 52.4% knew that problems with deciduous teeth could affect adult dentition, and 42.7% knew that fluoridated toothpaste could prevent dental caries. Only 29.2%, however, believed that they could infect their baby with cariogenic bacteria.

Association between mothers' perception of ECC and their socio-demographic variables

Tertiary educated respondents had significantly better knowledge about the importance of deciduous teeth ($p = 0.00$) with 92% of them providing the right response compared to 66.7% of those that had no schooling. There was also a significant difference between tertiary respondents (90.2%) who knew that babies should not sleep

Table 11.1 Socio-demographic characteristics of the respondents

Variable	*Frequency*	*Percentage*
Age category		
16–25	12	8.3%
26–35	106	73.6%
36–45	26	18.1%
Ethnicity		
Yoruba	84	53.3%
Ibo	28	19.4%
Hausa	10	6.9%
Others	22	15.3%
Nature of work		
Full time	106	73.6%
Part time	24	16.7%
Not at all	14	9.7%
Religion		
Christianity	92	63.8%
Islam	50	34.7%
Others	2	1.4%
Education		
None	6	4.2%
Primary	10	6.9%
Secondary	46	31.9%
Tertiary	82	56.9%
No of children		
1	60	41.7%
2	62	43.1%
3	10	6.9%
4	12	8.3%
Totals	144	100%

with feeding bottles in their mouths compared to uneducated mothers (p = 0.00). Similarly, 52.9% of tertiary respondents believed that fluoride prevents decay compared to 33.3% of uneducated mothers. Level of education had no significant impact on the knowledge of the respondents about the vertical transmission of cariogenic bacteria with only 33.3% of tertiary educated respondents giving the right response (p = 0.99). There was low association between the age of the respondents and their knowledge about the causation of dental caries in children. Similar responses were made to most questions except with the use of sweetened pacifiers (p = 0.00), mother's diet during pregnancy (p = 0.00), and dental caries causing illness (p = 0.02) where respondents in the 36–45 age group had significantly better responses (see Table 11.3).

Table 11.2 Maternal knowledge and attitude on early childhood caries

	Number of respondents who agreed (%)	*Number of respondents who disagreed (%)*	*Number of respondents who were unsure (%)*
I can pass on tooth decay to my baby	42 (29.2%)	56 (38.9%)	46 (31.9%)
Baby teeth are not important	14 (9.7%)	118 (81.9%)	12 (8.3%)
Problems with baby teeth can affect adult teeth later	78 (54.2%)	36 (25%)	30 (20.8%)
Decayed teeth can make a baby sick	126 (87.5%)	10 (6.9%)	8 (5.6%)
Babies' mouth should be cleaned even if they have no teeth.	126 (87.5%)	8 (5.6%)	10 (6.9%)
Fluoride toothpaste prevents tooth decay	68 (47.2%)	16 (11.1%)	60 (41.7%)
Mothers' diet during pregnancy can affect babies' teeth	62 (43.1%)	26 (18.1%)	56 (38.9%)
Sweetened pacifiers can be used for babies	18 (12.5%)	74 (51.4%)	52 (36.1%)
I can breastfeed or bottle-feed my baby all night	72 (50%)	46 (31.9%)	26 (18.1%)
A baby can use the feeding bottle anytime when he is old enough to hold it	74 (51.4%)	60 (41.7%)	10 (6.9%)
A baby can sleep with the feeding bottle in his mouth	16 (11.1%)	124 (86.1%)	4 (2.8%)
Babies should be seen by the dentist before his/her first birthday	48 (33.3%)	40 (27.8%)	56 (38.9%)

Maternal practices related to early childhood caries

The majority of the respondents (98.6%) maintained good oral hygiene for their babies by cleaning their mouths every day. They, however, had poor practices with regards to exposure to cariogenic bacteria and cariogenic diet. About 88% have regular oral contact with their babies' mouth, while 84.7% chew and soften meals for their babies. 86% allow their babies to sleep with feeding bottles, while 75% leave the bottle with babies for prolonged periods. None of the mothers had ever taken their babies for a dental check-up (see Table 11.4).

Discussion

Kay and Locker (1996) suggested that an understanding of maternal knowledge about oral health issues is crucial in order to modify their behaviour and

Table 11.3 Association between maternal perception and age/education

	% with right responses	*Age category*			*Educational qualification*			
		16–25	*26–35*	*36–45*	*none*	*Primary*	*Secondary*	*Tertiary*
Mothers can transmit tooth decay to the baby	Agree	33.3%	28.3%	33.3%	33.3%	20%	30.8%	29.4%
		P = 0.46			P=0.99			
Baby teeth are not important	Disagree	50 %	83%	91.7%	66.7%	60%	53.8%	92.2%
		P = 0.18			P = 0.00			
Problems with baby teeth can affect adult teeth later	Agree	66.7%	58.5%	25%	66.7%	60%	46.2%	54.9%
		P = 0.04			P = 0.79			
Decayed teeth can make a baby sick	Agree	66.7%	86.8%	100%	66.7%	84.6%	88.2%	100%
	P = 0.02				P = 0.07			
Babies' mouth should be cleaned even if they have no teeth	Agree	83.3%	86.8%	100%	66.7%	80%	84.6%	90.2%
	P = 0.54				P = 0.01			
Fluoride toothpaste prevents tooth decay	Agree	33.3%	47.2%	50%	33.3%	40%	30.8%	52.9%
	P = 0.16				P = 0.08			
Mothers' diet during pregnancy can affect babies' teeth	Agree	50%	43.4%	41.7%	33.3%	38.5%	40%	45.1%
	P = 0.00				P = 0.37			
Sweetened pacifiers can be used for babies	Disagree	16.7%	49.1%	75%	33.3%	40%	53.8%	52.9%
	P = 0.00				P = 0.09			
Babies can be breastfed or bottle-fed all night	Disagree	16.7%	30.2%	50%	33.3%	0%	30.8%	35.3%
	P = 0.227				P = 0.10			
A baby can sleep with the feeding bottle in his mouth	Disagree	66.7%	88.7%	91.7%	60%	66.7%	84.6%	90.2%
	P = 0.05				P = 0.00			
Babies should be seen by the dentist before his/her first birthday	Agree	16.7%	30.2%	50%	33.3%	40%	30.8%	33.3%
	P = 0.03				P = 0.57			

Table 11.4 Maternal practices on early childhood caries

	Number of respondents who do (%)	*Number of respondents who don't (%)*	*Number of respondents who were unsure (%)*
I kiss my baby on the mouth frequently	128 (88.9%)	12 (8.3%)	4 (2.8%)
I chew my baby's food before giving it to him/her to eat	122 (84.7%)	17 (11.8%)	5 (3.5%)
I would take my child for dental treatment immediately I see a hole in the teeth	30 (20.8%)	102 (70.8%)	12 (8.3%)
I clean my baby's mouth every day	142 (98.6%)	2 (1.4%)	0 (0%)
I use sweetened pacifiers for my baby	56 (38.9%)	88 (61.1%)	0 (0%)
I allow my baby to use the feeding bottle anytime when he is old enough to hold it	103 (71.5%)	33 (22.9%)	8 (5.6%)
I leave my baby asleep with a feeding bottle in the mouth	125 (86.6%)	15 (10.4%)	4 (2.8%)
I have taken my child for a dental check-up before	0 (33.3%)	141 (97.9%)	3 (2.1%)

encourage good health promotion. In the present study, the mothers had moderate to high levels of awareness on the effect of oral habits on the causation of dental caries. Most were aware that deciduous teeth were important, that carious teeth could lead to serious illness, and that it was unsafe to allow the baby to sleep with a feeding bottle in the mouth. About half of the respondents also knew that problems with deciduous teeth could affect permanent teeth and that fluoridated toothpaste can prevent dental caries. A low level of knowledge was observed, however, on the possibility of maternal transmission of cariogenic bacteria to their infants. Most mothers were unaware of the process whereby babies get infected with the cariogenic organisms through activities between an infant and its mother such as kissing and sharing utensils. It is concerning that most mothers did not know about the most important step in the initiation of ECC. A previous study in Nigeria had found aspects of maternal knowledge about infant oral health to be deficient (Orenuga and Sofola 2005). Poor health literacy is associated with poorer perceptions of health, decreased utilisation of services, and poorer understanding of verbal and written instructions of self-care (Jackson 2006). Maternal knowledge and attitude have also been found to be significantly correlated with the oral health of their children (Adeniyi *et al.* 2009).

Educational attainment was also significantly associated with maternal perception of the importance of deciduous teeth and late night bottle feeding even though there was no association between educational qualification and knowledge of vertical transmission of cariogenic bacteria. Overall, age was not significantly associated with maternal knowledge and attitude even though older respondents had better knowledge than the younger ones. Experience with other children could have modified the perception of those aged between 36 and 45 years. Epidemiological evidence suggests a significant link between maternal education and infant oral health. Kinirons and McCabe (1995) reported an ECC prevalence of 40% in preschool children whose mothers had low levels of education compared to only 10% in children whose mothers had high levels of education.

There was, however, a sharp contrast between the oral health practices of the respondents and their level of knowledge. The majority of the respondents practised good oral hygiene with their babies and cleaned their mouths regularly. However, their nutritional habits were much less favourable to their child's oral health. About 88% of them regularly kiss their children on the lips, while 84% soften food in their mouth before giving it to the baby. Close to 72% always put their child to bed for the night with a feeding bottle, and none of them had ever taken their child for a routine dental check-up. Bottle-feeding, especially nocturnal feeding when children are allowed to sleep with a bottle in their mouth has been considered a risk factor for caries. It has been shown that health education alone is inadequate when aiming to improve oral health. The most effective interventions for changing oral health behaviours are reinforcement of oral health education messages as well as individualised instruction accompanied by demonstration and skill building (Seow 1998). Aggressive health promotion activities should be targeted at mothers and would be mothers at ante- and post-natal clinics as well as in paediatric outpatients and the general community.

The limitation of this study was that it sampled mothers with young babies with few teeth, and dental examination was not done. Follow-up study including dental examination will be included in future studies to determine the effect of maternal perceptions and practices and possibly the impact of health promotion activities on the level of caries in older children.

Specific interventions to reduce the burden of ECC

Strategies to prevent oral diseases will be very effective when a multisectoral approach is utilised for population-based primary preventive interventions. Due to a shortage of dental manpower in Nigeria and the high cost of restorative treatment, refocusing of service provision towards preventive care is required through the utilisation of the existing primary health care system (PHC). From 1986 to 1992, Nigeria made tremendous progress in the implementation of the local government area-focused PHC, with extensive expansion in PHC infrastructure, establishment of community health workers training institutions, and production of a large number of community health workers (Lambo 2015). The PHC system has been relatively successful in making healthcare facilities accessible to most Nigerians, and many mothers that have no access to oral care attend antenatal and postnatal clinics in PHCs, especially in rural areas where there is no access to secondary and tertiary healthcare facilities (Lambo 2015). These PHC facilities can serve as a platform to provide preventive oral healthcare through the utilisation of trained auxiliary workers for health promotion.

Anticipatory guidance is an approach to childcare that has been utilised in ECC primary prevention efforts (Edelstein 2006). Primary prevention should involve efforts to limit bacterial transmission and proliferation to decrease the risk of demineralisation. This method of prevention should occur prior to oral bacterial colonisation of the mouth and should emphasise the importance of careful control of caries-promoting activities (Edelstein 2006). Early interventions to instil healthy behaviours and practices during childhood have been observed to be significantly more cost-effective than efforts applied after unhealthy behaviours are ingrained (Centers for Disease Control and Prevention 2009). A specific intervention would be oral hygiene motivation on tooth-brushing and flossing for mothers on a daily basis to dislodge food particles and reduce bacterial plaque levels. Dietary education for mothers should stress the cariogenicity of certain foods, role of frequency of consumption, and the demineralisation/remineralisation process. They should be advised to reduce on-demand bottle feeding and the use of sweetened drinks for their infants (EAPD 2008). Mothers should also be counselled to avoid salivary exchange activities like sharing spoons, cups, and other utensils, softening the babies' meal by pre-chewing, and cleaning a dropped pacifier or toy with their mouth.

The use of fluoridated toothpaste and a mouth rinse containing 0.05% sodium fluoride which have been recommended to reduce plaque levels and to promote enamel remineralisation should be encouraged in mothers. Brushing of the infants' primary teeth must also start as soon as the first tooth erupts. There is evidence from systematic reviews (Ammari *et al.* 2003) that fluoridated toothpaste gives the highest preventive effect against early childhood caries. It has also been documented that fluoride toothpaste is the most cost-effective home care preventive measure (Marinho *et al.* 2002). In addition to fluoride toothpaste, professional application of fluoride varnish at least twice yearly and depending on patient's caries risk, is recommended to prevent ECC. In addition to these primary preventive initiatives, effective ECC mitigation efforts should also include methods of secondary prevention. These should include early detection of non-cavitated lesions through active surveillance of disease and the application of preventive measures to arrest or reverse tooth decay (American Academy of Pediatric Dentistry 2012). Early intervention in the ECC disease process is preferred because the caries process can be arrested by timely and effective management, minimising the need for restorative care (Yoon *et al.* 2012). Routine professional dental care for the mothers should also be done for removal of active caries with subsequent restoration to suppress maternal SM reservoirs and to minimise the transfer of SM to the infant, thereby diminishing the infant's risk of developing ECC.

Conclusion

Mothers in our study had moderate to high levels of knowledge on the effect of oral habits on the causation of dental caries. This did not, however, translate into appropriate oral health behaviour in their care for their babies. Even though Nigeria recently released the first oral health policy document for the nation, a policy framework for combating early childhood caries is required. An attempt should be made to bridge the gap between maternal knowledge/attitude and their practices with regards to the oral health of their children. The physical, psychological, and economic consequences of ECC can be avoided through the education of prospective and new parents on good oral hygiene and dietary practices utilising the PHC system. Oral health assessments with counselling at regularly scheduled visits from the first year of life should be promoted as a strategy to prevent ECC. Maternal SM reservoirs can be suppressed by chewing xylitol-containing gums by the mother during the prenatal period and that of primary teeth eruption. Most importantly, mothers must be educated to avoid saliva-sharing behaviours such as kissing the baby on the mouth, sharing a spoon when tasting baby food, or wiping the baby's mouth with a cloth moistened with saliva.

References

Acs, G., Lodolini, G., Kaminsky, S., and Cisneros, G.J., 1992. Effect of nursing caries on body weight in a pediatric population. *Pediatric Dentistry*, 14(5), 303.

Adeniyi, A.A., Eyitope Ogunbodede, O., Sonny Jeboda, O., and Morenike Folayan, O., 2009. Do maternal factors influence the dental health status of Nigerian pre-school children? *International Journal of Paediatric Dentistry*, 19(6), 448–454.

Adeniyi, A.A., Oyapero, A., Ekekezie, O.O., and Braimoh, M.O., 2016. Dental caries and nutritional status of school children in Lagos, Nigeria – A preliminary survey. *Journal of the West African College of Surgeons*, 6(3), 15–39.

Adeyemo, W.L., Ladeinde, A.L., Ogunlewe, M.O., and James, O., 2005. Trends and characteristics of oral and maxillofacial injuries in Nigeria: A review of the literature. *Head Face Med*, 1, 7–12.

Alvarez, J. *et al.*, 1993. A longitudinal study of dental caries in the primary teeth of children who suffered from infant malnutrition. *Journal of Dental Research*, 72(12), 1573–1576.

American Academy of Pediatric Dentistry, 2008. Policy on early childhood caries (ECC): Classifications, consequences, and preventive strategies. *Pediatric Dentistry*, 30, 40–43.

American Academy of Paediatric Dentistry and the American Academy of Paediatrics, 2011. *Policy on early childhood caries (ECC): Classifications, consequences and preventive strategies*. Chicago: American Academy of Paediatric Dentistry.

Ammari, A.B., Bloch-Zupan, A., and Ashley, P.F., 2003. Systematic review of studies comparing the anti-caries efficacy of children's toothpaste containing 600 ppm of fluoride or less with high fluoride toothpastes of 1,000 ppm or above. *Caries Research*, 37(2), 85–92.

Amusa, Y.B. *et al.*, 2004. Pattern of head and neck malignant tumours in a Nigerian teaching hospital – A ten year review. *West African Journal of Medicine*, 23(4), 280–285.

Axelsson, P., Albandar, J.M., and Rams, T.E., 2002. Prevention and control of periodontal diseases in developing and industrialized nations. *Periodontol* 2000, 29, 235–246.

Azodo, C.C. and Amenaghawon, O.P.., 2013. Oral hygiene status and practices among rural dwellers. *European Journal of General Dentistry*, 2(1), 42.

Berg, J.H. and Stapleton, F.B., 2012. Physician and dentist: New initiatives to jointly mitigate early childhood oral disease. *Clinical Pediatrics*, 51(6), 531–537.

Booysen, F. *et al.*, 2007. Trends in poverty and inequality in seven African countries. *Cahiers de Recherche PMMA*, 6.

Bratthall, D. and Barnes, D., 1993. Oral health. In D. Jamison, W. Mosley, A. Mesham, and J.L. Bobadilla, eds. *Disease control priorities in developing countries*. New York: Oxford University Press, 647–660.

Casamassimo, P.S., Thikkurissy, S., Edelstein, B.L., and Maiorini, E., 2009. Beyond the dmft: The human and economic cost of early childhood caries. *The Journal of the American Dental Association*, 140(6), 650–657.

Centers for Disease Control and Prevention, 2009. *2009 National youth risk behavior survey overview*. Centers for Disease Control and Prevention.

Christensen, L.B., Twetman, S., and Sundby, A., 2010. Oral health in children and adolescents with different socio-cultural and socio-economic backgrounds. *Acta Odontologica Scandinavica*, 68(1), 34–42.

Edelstein, B., Vargas, C.M., Candelaria, D., and Vemuri, M., 2006. Experience and policy implications of children presenting with dental emergencies to US pediatric dentistry training programs. *Pediatric Dentistry*, 28(5), 431–437.

Edelstein, B.L. and Chinn, C.H., 2009. Update on disparities in oral health and access to dental care for America's children. *Academic Pediatrics*, 9(6), 415–419.

European Archives of Paediatric Dentistry, 2008. *EAPD guideline on prevention of early childhood caries: An EAPD policy document,* 2008. *European Archives of Paediatric Dentistry.*

Fejerskov, O., 2004. Changing paradigms in concept of dental caries: consequences for oral health care. *Caries Research*, 38(3), 182–191.

Finlayson, T.L., Siefert, K., Ismail, A.I., and Sohn, W., 2007. Maternal self-efficacy and 1–5-year-old children's brushing habits. *Community Dentistry and Oral Epidemiology*, 35(4), 272–281.

Glendor, U., 2008. Epidemiology of traumatic dental injuries – A 12 year review of the literature. *Dental Traumatology*, 24(6), 603–611.

Gussy, M.G., Waters, E.G., Walsh, O., and Kilpatrick, N.M., 2006. Early childhood caries: Current evidence for aetiology and prevention. *Journal of Paediatrics and Child Health*, 42(1–2), 37–43.

Hallett, K.B. and O'Rourke, P.K., 2003. Social and behavioural determinants of early childhood caries. *Australian Dental Journal*, 48(1), 27–33.

Harris, R., Nicoll, A.D., Adair, P.M., and Pine, C.M., 2004. Risk factors for dental caries in young children: A systematic review of the literature. *Community Dental Health*, 21(1), 71–85.

Hobdell, M.H. *et al.*, 2003. Oral diseases and socio-economic status (SES). *British Dental Journal*, 194(2), 91–96.

Hodgson, T.A., Greenspan, D., and Greenspan, J.S., 2006. Oral lesions of HIV disease and HAART in industrialized countries. *Advances in Dental Research*, 19(1), 57–62.

Jackson, R., 2006. Parental health literacy and children's dental health: Implications for the future. *Pediatric Dentistry*, 28(1), 72–75.

Jelliffe-Pawlowski, L.L., Shaw, G.M., Nelson, V., and Harris, J.A., 2003. Risk of mental retardation among children born with birth defects. *Archives of Pediatrics and Adolescent Medicine*, 157(6), 545–550.

Jin, L.J. *et al.*, 2011. Global oral health inequalities: Task group—periodontal disease. *Advances in Dental Research*, 23(2), 221–226

Kay, E.J. and Locker, D., 1996. Is dental health education effective? A systematic review of current evidence. *Community Dentistry and Oral Epidemiology*, 24(4), 231–235.

Kinirons, M. and McCabe, M., 1995. Familial and maternal factors affecting the dental health and dental attendance of preschool children. *Community Dental Health*, 12(4), 226–229.

Lambo, E., 2015. *Primary health care: Realities, challenges and the way forward*. Abuja, Nigeria: National Primary Health Care Development Agency.

Law, V., Seow, W.K., and Townsend, G., 2007. Factors influencing oral colonization of mutans streptococci in young children. *Australian Dental Journal*, 52(2), 93–100.

Li, Y., Wang, W., and Caufield, P.W., 2000. The fidelity of mutans streptococci transmission and caries status correlate with breast-feeding experience among Chinese families. *Caries Research*, 34, 123–132.

Litt, M.D., Reisine, S., and Tinanoff, N., 1995. Multidimensional causal model of dental caries development in low-income preschool children. *Public Health Reports*, 110(5), 607.

Loesche, W.J., Rowan, J., Straffon, L.H., and Loos, P.J., 1975. Association of Streptococcus mutans with human dental decay. *Infection and Immunity*, 11(6), 1252–1260.

Low, W., Tan, S., and Schwartz, S., 1999. The effect of severe caries on the quality of life in young children. *Pediatric Dentistry*, 21(6), 325–326.

Mackay, J. and Eriksen, M.P., 2002. *The tobacco atlas*. World Health Organization.

Marinho, V.C., Higgins, J.P., Logan, S., and Sheiham, A., 2002. Fluoride gels for preventing dental caries in children and adolescents. *Cochrane Database System Review*, 2.

Mascarenhas, A.K., 1999. Determinants of caries prevalence and severity in higher SES Indian children. *Community Dental Health*, 16(2), 107–113.

Mashoto, K.O., Astrom, A.N., Skeie, M.S., and Masalu, J.R., 2010. Socio-demographic disparity in oral health among the poor: A cross sectional study of early adolescents in Kilwa district, Tanzania. *BMC Oral Health*, 10(1), 7.

Milnes, A.R., 1996. Description and epidemiology of nursing caries. *Journal of Public Health Dentistry*, 56(1), 38–50.

Mossey, P. and Castillia, E., 2003. *Global registry and database on craniofacial anomalies*. Geneva: World Health Organization.

Moynihan, P. and Petersen, P.E., 2004. Diet, nutrition and the prevention of dental diseases. *Public Health Nutrition*, 7(1a), 201–226.

Nackashi, J., Rosellen, E., and Dixon-Wood, V., 2002. Health care for children with cleft lip and palate: Comprehensive services and infant feeding. In Wyszynski, D.F. ed. *Cleft Lip and Palate: From Origin to Treatment*. New York: Oxford University Press.

Nyvad, B. and Kilian, M., 1990. Microflora associated with experimental root surface caries in humans. *Infection and Immunity*, 58(6), 1628–1633.

Ogbureke, K.U. and Ogbureke, E.I., 2010. NOMA: a preventable 'scourge' of African children. *The Open Dentistry Journal*, 4, 201.

Oginni, F.O., Asuku, M.E., Oladele, A.O., Obuekwe, O.N., and Nnabuko, R.E., 2010. Knowledge and cultural beliefs about the etiology and management of orofacial clefts in Nigeria's major ethnic groups. *The Cleft Palate-Craniofacial Journal*, 47(4), 327–334.

Onyango, J.F., Omondi, B.I., Njiru, A., and Awange, O.O., 2004. Oral cancer at Kenyatta national hospital, Nairobi. *East African Medical Journal*, 81(6), 318–321.

Orenuga, O.O. and Sofola, O.O., 2005. A survey of the knowledge, attitude and practices of antenatal mothers in Lagos, Nigeria about the primary teeth. *African Journal of Medicine and Medical Sciences*, 34(3), 285–291.

O'Sullivan, D.M. and Tinanoff, N., 1996. The association of early dental caries patterns with caries incidence in preschool children. *Journal of Public Health Dentistry*, 56(2), 81–83.

Palmer, C.A., Burnett, D.J., and Dean, B., 2010. It's more than just candy: Important relationships between nutrition and oral health. *Nutrition Today*, 45(4), 154–164.

Pau, A., Khan, S.S., Babar, M.G., and Croucher, R., 2008. Dental pain and care-seeking in 11–14-yr-old adolescents in a low-income country. *European Journal of Oral Sciences*, 116(5), 451–457.

Peterka, M. *et al.*, 2000. Significant differences in the incidence of orofacial clefts in fifty-two Czech districts between 1983 and 1997. *Acta Chirurgiae Plasticae*, 42(4), 124–129.

Petersen, P.E., 2003. The World Oral Health Report 2003: continuous improvement of oral health in the 21st century – The approach of the WHO Global Oral Health Programme. *Community Dentistry and Oral Epidemiology*, 31(s1), 3–24.

Petersen, P.E., Bourgeois, D., Ogawa, H., Estupinan-Day, S., and Ndiaye, C., 2005. The global burden of oral disease and risk to oral health. *Bulletin of the World Health Organization*, 83(9), 661–9.

Peterson, J., Niessen, L., and Lopez, G.M., 1999. Texas public school nurses' assessment of children's oral health status. *Journal of School Health*, 69(2), 69–72.

Pine, C.M. *et al.*, 2004. Developing explanatory models of health inequalities in childhood dental caries. *Community Dental Health*, 21(Suppl.), 86–95.

Popkin, B.M., 2002. An overview on the nutrition transition and its health implications: The Bellagio meeting. *Public Health Nutrition*, 5(1A), 93–103.

Price, M. and Weiner, R., 2005. Where have all the doctors gone? Career choices of Wits medical graduates. *South African Medical Journal*, 95(6).

Reisine, S. and Douglass, J.M., 1998. Psychosocial and behavioral issues in early childhood caries. *Community Dentistry and Oral Epidemiology*, 26(S1), 32–44.

Reisine, S.T. and Psoter, W., 2001. Socioeconomic status and selected behavioral determinants as risk factors for dental caries. *Journal of Dental Education*, 65(10), 1009–1016.

Rowan-Legg, A., 2013. Oral health care for children – A call for action. *Paediatrics and Child Health*, 18(1), 37–43.

Sabbagh, H.J., Mossey, P.A., and Innes, N.P., 2012. Prevalence of orofacial clefts in Saudi Arabia and neighboring countries: A systematic review. *The Saudi Dental Journal*, 24(1), 3–10.

Schroth, R.J., Brothwell, D.J., and Moffatt, M.E.K., 2007. Caregiver knowledge and attitudes of preschool oral health and early childhod caries (ECC). *International Journal of Circumpolar Health*, 66(2), 153–167.

Seow, W.K., 1998. Biological mechanisms of early childhood caries. *Community Dentistry and Oral Epidemiology*, 26(S1), 8–27.

Seow, W.K., 2012. Environmental, maternal, and child factors which contribute to early childhood caries: A unifying conceptual model. *International Journal of Paediatric Dentistry*, 22(3), 157–168.

Sheller, B., Williams, B.J., and Lombardi, S.M., 1997. Diagnosis and treatment of dental caries-related emergencies in a children's hospital. *Pediatric Dentistry*, 19(8), 470–475.

Taubman, M.A. and Nash, D.A., 2006. The scientific and public-health imperative for a vaccine against dental caries. *Nature Reviews Immunology*, 6, 555–563.

Thomas, C.W. and Primosch, R.E., 2002. Changes in incremental weight and well-being of children with rampant caries following complete dental rehabilitation. *Pediatric Dentistry*, 24, 109–113.

Thorpe, S., 2006. Oral health issues in the African region: Current situation and future perspectives. *Journal of Dental Education*, 70(11), 8–15.

Tinanoff, N., Kanellis, M.J., and Vargas, C.M., 2002. Current understanding of the epidemiology, mechanisms, and prevention of dental caries in preschool children. *Pediatric Dentistry*, 24, 543–550.

United Nations, 2012. *Study of the Human Rights Council Advisory Committee on severe malnutrition and childhood diseases with children affected by Noma as an example*. Geneva: United Nations Human Rights Council.

United Nations Children Fund, 2013. *Improving child nutrition: The achievable imperative for global progress*. New York: UNICEF.

Varenne, B., Petersen, P.E., and Ouattara, S., 2006. Oral health behaviour of children and adults in urban and rural areas of Burkina Faso, Africa. *International Dental Journal*, 56(2), 61–70.

Watt, R.G., 2005. Strategies and approaches in oral disease prevention and health promotion. *Bulletin of the World Health Organization*, 83(9), 711–718.

World Bank, 1987. *World development report*. New York: Oxford University Press.

World Bank, 1994. *World development report 1994: Infrastructure for development*. New York: Oxford University Press.

World Health Organization, 2002. *Poverty and health: A strategy for the African region*. Regional Office for Africa. Brazzaville, Republic of Congo: World Health Organization, Regional Office for Africa.

World Health Organization, 2017. Oral health databases [Online]. Available from www.who.int/oral_health/databases [Accessed 27 July 2017].

Yoon, R.K., Smaldone, A.M., and Edelstein, B.L., 2012. Early childhood caries screening tools: A comparison of four approaches. *Journal of the American Dental Association (1939)*, 143(7), 756–763.

12 Health data collection efforts and non-communicable diseases

A case study in Uganda

Mary J. Christoph, Diana S. Grigsby-Toussaint, and James M. Ntambi

Introduction

Sub-Saharan Africa (SSA) is currently undergoing sweeping demographic, epidemiologic, and economic transitions contributing to a change in the prevalence of infectious to non-communicable diseases (NCDs) (Byass *et al.* 2014). While substantial inroads have been made to combat infectious diseases and increase life expectancy, in the wake of these improvements, the burden of morbidity and mortality is shifting to NCDs (Marquez and Farrington 2012). Contributing to this increase in NCD prevalence is the worldwide upsurge in overweight and obesity prevalence in the past several decades. In low- and middle-income countries, overweight is especially related to urbanization and changes in diet and physical activity (Popkin, Adair, and Ng 2012). Changes in the global food system including food processing, availability, and relative cost have led to many areas experiencing a "nutrition transition", a shift characterized by changing from diets rich in legumes, other vegetables, and coarse grains to those rich in refined carbohydrates, added sugars, fats, and animal-source foods (Popkin, Adair, and Ng 2012; Popkin 2017). Concurrently, technological advances have led to decreases in transportation and occupational physical activity (Hallal *et al.* 2012).

These changes in morbidity, mortality, and NCD risk factors have led to increased pressure on SSA health systems, which have been inundated with increased numbers of patients with complex NCDs including Type 2 diabetes and cancers. Some SSA countries experienced a 35% increase in adult overweight prevalence between 1992 and 2005 (Ziraba *et al.* 2009), while hypertension increased four-fold from 2005 to a region-wide prevalence of 16.2% in 2008 (Twagirumukiza *et al.* 2011); however, the paucity of reliable data limits knowledge (Byass 2009). Health information systems vary widely between and even within countries (Byass *et al.* 2014) and are often plagued by poor implementation and lack of coordination, standardization, continuous funding, adequate training, and infrastructure (Gold and John 2013).

Many countries in SSA rely on surveys focused on infectious diseases and maternal-child health. Demographic and Health Surveys (DHS) are nationally representative, population-based surveys on indicators of demographics, health, nutrition, and risk factors for chronic conditions (Demographic and Health

Surveys 2014). However, DHS has historically focused on women of childbearing age (Popkin *et al.* 2012), and the infrequency (every five years) and lack of data collection on critical NCDs and risk factors including diabetes and high cholesterol (Demographic and Health Surveys 2014) limit data utility. Community-based Health and Demographic Surveillance Sites (HDSS) have also shown emerging risk of obesity and hypertension (InDepth Network 2014, Tollman *et al.* 2008); but as these sites are not necessarily nationally representative, findings may not be generalizable (In Depth Network 2014, Maher *et al.* 2011). Influenced by international partners, priorities for health data collection systems remain in flux and change over time (St. Louis *et al.* 2010); further, funding may be inconsistent, resulting in discontinuation of surveillance efforts. Finally, the focus of most surveillance systems in SSA is communicable diseases (St. Louis *et al.* 2010, Magadi 2011); surveillance targeting NCDs and risk factors is limited and non-continuous, hindering an effective public health response.

Health information systems in Uganda

Uganda is a low-income developing country in eastern SSA with a population of 37.6 million people (World Bank 2012). English and Swahili are the official languages, and regional languages are also spoken (Gordon 2014). The economy is primarily agricultural with an estimated GDP of $21.5 billion (World Bank 2012), of which 9% is spent on health and healthcare (World Health Organization 2014), which is free for residents. As of the 2011 Uganda DHS, Uganda was divided into 112 districts (Uganda Bureau of Statistics and Macro International 2011) with a District Health Officer (DHO) (Gladwin *et al.* 2003) to whom each clinic, health center, and hospital reports information on maternal-child health, infectious diseases, and pharmaceutical usage. The DHO summarizes information and reports to the District Council (local government) and the national Ministry of Health (MoH). In 2000, Uganda implemented the Integrated Disease Surveillance and Response (IDSR) strategy, which has successfully improved disease reporting, response times, and analysis; however, a ten-fold decrease in government funding, and limited reach and analytical capacity has hindered its performance (Lukwago *et al.* 2013).

Prevalence of obesity and related NCDs, including Type 2 diabetes (Lasky *et al.* 2002) and hypertension, which affect 22% of the population (Demographic and Health Surveys 2014), have substantially increased over the past several decades in Uganda. In adult women, overweight and obesity are now more common than underweight. In the capital of Kampala, 27% of women are overweight, and an additional 13% are at a weight classified as obese; among men, the combined overweight/obesity prevalence is 12%. Countrywide, the combined prevalence of overweight and obesity is 19% among adult women and 4.4% among adult men (Uganda Bureau of Statistics and Macro International 2011).

A national data reporting system currently collects NCD morbidity and mortality data, but does not assess specific risk factors such as obesity and diet, nor is there a national targeted policy or program to address NCD risk (World Health Organization 2014). This qualitative study was undertaken to assess health data collection initiatives and determine how current systems could be improved and leveraged to survey NCDs and associated risk factors.

Methods

Informants with experience in the healthcare, academic, and government settings were recruited in Kampala, Kabale, and Lyontonde, Uganda (see Figure 12.1). Each was interviewed in English using a semi-structured guide including questions about the state of health and NCD information systems, current projects related to NCDs, perceptions about weight, NCDs as a public health issue, and the role of the MoH and non-governmental organizations (NGOs) in improving data collection efforts related to overweight and nutrition-related NCDs in Uganda. Interviews also explored successes and challenges associated with data collection efforts from the collaboration between government, healthcare sector, and NGOs. Interviews were recorded, professionally transcribed, and coded with qualitative analysis software to identify themes.

Figure 12.1 Markers are shown in the Ugandan cities where interviews were conducted.

Results: Challenges and opportunities

Participants identified challenges and opportunities associated with improving surveillance systems in Uganda. Major challenges included: 1) lack of coordination between organizations and the government and different departments within the state and local governments, 2) inadequate funding, and 3) a shortage of human resources, particularly professionals trained in data management software systems. Informants suggested improvements by implementing technical training programs (e.g., database management) by governmental agencies and NGOs to build capacity, building public health education programs, reallocating resources, and involving community leaders in health promotion. Table 12.1 lists quotes regarding core public health surveillance, while Table 12.2 lists representative quotes concerning supporting activities.

Core public health surveillance activities

Reporting

Reporting of basic health statistics (e.g., births, mortality, malaria, HIV, etc.) is performed at all levels of health clinics and hospitals. Patients and conditions are recorded daily, and monthly reports are sent to the DHO and to the MoH. Informants noted maternal-child health factors (e.g., marasmus, Kwashiorkor) and infectious diseases (e.g., malaria, tuberculosis, and HIV) were the primary conditions reported. One government official explained: "The information system … is scattered, not targeted, irregular, and most of it is actually project-bound … But there is a lot of information".

While Type 2 diabetes is not yet reported in many regions, informants indicated that NCDs, diabetes in particular, are a burgeoning public health problem in Uganda. Because of the prevalence of malnutrition, overweight and obesity are often regarded as signs of good health; one physician noted that in a recent experience teaching health professionals about the negative health outcomes associated with overweight, several health professionals found it surprising that overweight was not a sign of good health. Another noted that: "due to a long-standing focus on preventing malnutrition and stunting, overweight and obesity are generally thought of as either a non-issue or as a sign of prosperity". Another physician explained that about 60% of his diabetic patients were obese, and many also hypertensive. One physician mentioned that patients who had been severely malnourished as children were at particular risk developing Type 2 diabetes. Informants agreed that reporting of infectious diseases and maternal-child health factors was generally done well, but several noted that health centers were ill-equipped to test for and treat chronic conditions such as cancer and diabetes. They also expressed that while health centers routinely collected data, analysis and feedback were not always present. Local officials did not always have the capacity and training to analyze surveillance data, but analyses by district or governmental agencies were not always shared with local offices. Informants in the Southwest noted significant contributions from NGOs, who often performed their own data collection

Table 12.1 Representative quotes concerning the core public health surveillance activities: reporting, analysis, and feedback

Core public health surveillance activities	
Reporting	[Data is reported to] the MOH [Ministry of Health]; but also we have other branches as you scale down. Here [a regional hospital] we have the record persons; they compile all the information … and reports; we have sub-departments, then for general hospital, inpatient, we have the general records person. So at each level we have someone, and then it gets send to the Ministry. Even the Ministry is responsible for regional referrals; these hospitals report to the district, which is the local government.
	The government is doing that [data collection] and it goes on reporting, you have master reports. They are collected from health centers [monthly] indicating the prevalence of every disease … They've got so many patients a day, such a number has malaria, a number for that and so.
	When they discharge they take all of the work to the record person … Then the records person also makes a report . . . After making that report they send it to the Ministry of Health, even the facility itself, I think they use it for making requisitions for drugs; then also the Ministry of Health also reprise that report; they do reprise it for policy formulation and also to see how the facility is performing. The institutions, those in universities, they also use it for research, literature. Even publications, publications like the media; they also use that information.
	That [Uganda Demographic Health Survey] is done regularly as well, expertly, very expertly done, and very well coordinated … But having said that … if you're trying to follow epidemiological trends, you do not wait for five years; lots of things happen within five years.
Analyses	We used to send [data reports] to both the district and the ministry and even UNICEF, but now the district is not utilizing those reports you would send, you wouldn't receive any feedback; they had no interest, so I just left it. So now we just send to the Ministry and our partners who are funding us, Ministry of Health, UNICEF, and LUSU.
	Here, the one who coordinates – the DHO, the District Health Officer, is the officer of the Minister of Health … The data is already sent to us [by the DHO]. Whether it is level 2, 3, 4, or hospitals, they all contribute. Health Centers 2, 3, and 4 … are in charge of their sub-district, and they analyze data for their own use; then we submit the data to the district. But also they look through at the growth charts to see how we are moving, the trend of the problems affecting us, whether it is malaria, whether it is diabetes, we track it, all of those we know where we stand. Even with the drugs; also we look at the consumption rates of the medicine to see which is being consumed, which is not, so that I know what to request. Unfortunately I do have about 60 patients who have it [diabetes], so those patients will need so much for a certain period; so that's how we analyze our data.

(*continued*)

Table 12.1 Continued

Core public health surveillance activities	
	The Bureau of Statistics is collecting data using partners and whatever to get quotas; the Ministry of planning should have this data, because they're working with the various ministries, to plan effectively using the requested data, so then you find the ministry of spending, the ministry of health and the ministry of agriculture [are all involved].
	NGOs usually have the information systems in place, really are in the most demand. NGOs, the local ones, NGOs in general; that information is targeted toward whatever they are doing and whatever they get the money for. Once that has come through, usually that's the end. Some local NGOs, which deal with specific issues like especially HIV-AIDS – those have a better-managed system.
Feedback	I don't know where it [malnutrition data] goes in particular; I don't know what happens to it, where it's reported, because if you look at any news, any document, any report on malnutrition in Uganda between 2006 and 2011, they report on the 2006 DHS throughout the five years, then they start reporting that 2011 one, but you don't get any reporting in between, apart from the Karamoja one [a northern region].
	[After collection and analysis] then make conclusions of what the data means, and get something to feedback into the implementation they have, because it is useless to send it to a national level; it takes six months to add up the data and by that time the epidemic is gone … If you are talking measles, and the latest problems like measles, would have passed. Use it at the source, and then the rest you can send for long-term plans for policy formulation. So the feedback really, you should have the feedback initially at the source.
	Well I think basically for them [NGOs], they can focus; they can write out a project and focus on collecting data for a, b, c, d but then they're supposed to feed this into the government, but if the government is not forthcoming then what's the point of collecting data and giving it to someone who is not going to use it? But I think they simply get discouraged along the way, when they don't see any light at the end of the tunnel. What they have done in terms of data collection is not going to be utilized. On the contrary, they would prefer to use the data on their own and do their small-scale implementation; that is to collect it for themselves and say okay, we've found so many children that need this help; what can we do? So they ask maybe another NGO or themselves actually, could their plan implement what they think; the data tells them.
	Now even the Ministry of Health has gotten more statisticians working with them. So the result is data. And then you also find that in different projects we have statisticians who do that, write reports and do what; though the dissemination of them is not that good.

Table 12.2 Quotes regarding the public health surveillance supporting activities: communication, training, supervision, resource provision

Supporting public health activities	
Communication	We've been having interventions for mothers and children, like vitamin A supplementation, immunization, then also vaccination for mothers ... Also we do health and nutrition education sessions, then we also counsel exclusive breastfeeding before even the mother gives birth we prepare them with the exclusive breastfeeding, then we do growth monitoring and promotion, that is maternal child health, all those are MCH, then we also do HIV counseling and testing for all mothers who are attending [sessions].
	The Ministry of Education ought to be very important in that it is directly involved with the children; and these children need be taught about nutrition for the schools. Before these activities good nutrition was lacking in the schools, especially the non-residents, in that they come to school without hardly any food. We find it very challenging for a child to perform well in education without food ... We even encourage that they pack food in their homes.
	They are using soccer players and celebrities in Uganda to talk about family planning ... so they can also do that [to talk about chronic disease risk factors].
	Of course there was a lot of awareness on the radio to make sure that people were aware of what causes malnutrition and what are the preventive measures, what are the corrective measures you can take; there were a lot of radio programs on awareness of this.
Training	[NGO] has employed nutritionists and they have even gone ahead to network with other NGOs to give training, even staff because most of the training the MOH is conducting, they are being supported by these partner organizations.
	One of the strongest partners we had was New Life, who did nutrition interventions for Uganda but that one scaled down there and no longer the project ended, it was a five-year project and it was – it was giving us support in terms of training in nutrition care for people living with HIV; then it also assisted the Ministry of Health, produced locally-made [nutritional] feeds, gathered from local industries, so it supported us also in that and also in coaching and mentoring but that one, the project ended.
Supervision	I think one of the things is usually it is tough; who is the head? Who is the leader? Both the Ministry of Health and Agriculture, they want to lead; I guess that's why that food and nutrition center was put on the Prime Minister's office so that it's [not disputed].
	In addition to the government, many projects and health information systems were also overseen by NGOs and funding partners. So being one, USAID funded being under one umbrella has made them become successful because they are avoiding replication of services.

(*continued*)

Table 12.2 Continued

Supporting public health activities	
	UNICEF works well, but it works well in the sense that they take a project in almost its own totality, that is to collect the information and then find another NGO or themselves to actually do the implementation, rather than saying we collect this information; we'll pass it on to the government for implementation; no, they don't … Of course, it's realizing that government cannot implement a number of these things, yeah. But of course government now and then uses the data because it is an open source to use the data to do their own some implementation, but purposely UNICEF would collect its' own data for their own use.
Resource Provision	We feel that if they could decentralize the system, rather than like using a worker to come from Kampala to do the work which actually the nutritionist would have done; like coaching … these trainings, they could be organized at a regional level, the trainers are there, rather than having it at the center … why not ask to train, to train from within our localities?
	We haven't had very much challenge in collecting and in testing our patients on nutrition. We even have nutritional materials which are everywhere in our clinics … but in other health centers, like in Health Center 2 [smaller clinic], the challenge is that they are not well staffed; they don't have enough personnel.
	If we had Village Health Teams in place that would be the key people to help us. Then of course NGOs, you cannot really depend on them because NGOs come with their own interests, agenda; they have their programs so we can't really run them, but I would prefer government put much emphasis on Village Health Teams. They are all trained and then we coordinate with them, their units coordinate with the group teams; that's when we shall have a strong health system in this country.
	Suppose USAID wanted to sponsor something. They may say they want to recruit their own staff. They may hire Ugandans but then you have to run yourself; why do you want to hire these other people? Because the village health worker could be able to do that, if you provide the resources. So if you provide a bicycle for this person to move, let the others not provide a bicycle, provide something else instead. But they may not want to do that.
	They know that most people will converge toward a referral hospital, and I think because of human resources, an inability to track every place, they actually do it [tracking] in places where there are referral hospitals. They have a big shortage of human resources, especially in terms of professionals who are able to do [Microsoft] Access and that kind of proper – of course payment by government is so low so many people are trained and then they disappear; they go out to do other things.

(*continued*)

Table 12.2 Continued

Supporting public health activities	
	Basically I hate this kind of business, sitting here within the facility that you are waiting for the next child to come and to give treatment, rather than preventing it from ever seeing one. Sustain [an NGO] are those supporting us; it is also supporting some volunteers, though the support is on and off; and that's why you see this HIV screening, pediatric HIV screening is not being done this month because those volunteers have been a long time without getting some more allowances [pay] and so they tended to run out and to survive.

and analysis, while those in Kampala noted more governmental oversight. One informant in the government noted that:

> Because of strikes and particularly lack of money, this information is never brought to the center; and therefore never used. If it is, sometimes it is not properly analyzed … So they do collect data; they collect a lot of data but its management is what the problem is, especially the analysis. Then feedback to where things happen that was in the communities, in the districts.

Feedback

Although all districts have a system for health data collection, efforts are sporadic and uncoordinated. Consequently, it can be difficult to follow-up and provide feedback to effectively implement programs based on data. One hospital nutrition officer explained that he no longer sent data to his health district because it was not utilized. Informants noted that the absence of a centralized data collection system with clear goals posed a significant handicap to the improvement of health information systems, with one referring to the information systems in Uganda as a "patchwork", showing intermittent data with little coordination that addressed specific projects rather than planned national surveillance.

Coordination was challenging when NGOs had different agendas from the government. Nearly all informants commenting on collaborating with NGOs explained that most funds were conditional, limiting discretionary spending and ameliorating less-publicized issues. For example, NGOs may have programs focused on HIV/AIDS with a nutrition component, but upon funding cuts, the ancillary nutrition program might be eliminated even if the program is crucial.

Public health surveillance and action support activities

Communication

Several informants proposed improving data coordination efforts between the government, NGOs, and district health centers to improve health information systems assessing NCDs. One informant also suggested partnerships with groups to improve the health of Ugandans by speaking at gatherings (celebrations for weddings, anniversaries, birthday parties, etc.), or by having celebrities and athletes act as public health promotion agents.

They felt that while messages regarding infectious diseases and preventing malnutrition were well-communicated, chronic disease etiology, treatment, and prevention was not well understood. At the national level, the MoH Nutrition Action Plan, which is currently geared toward maternal-child health and preventing malnutrition, could be leveraged to incorporate classes focused on cooking methods and encouraging healthy eating to prevent obesity. Informants also believed primary education and schools were an important setting for interventions and communicating public health messages. One suggested that national immunization days, which generally occur twice a year in schools, could be an opportunity for educating about obesity prevention. Due to the current prominence of malnutrition, one university researcher felt that it was important not to give mothers mixed messages about feeding by introducing messages about obesity. For this same reason, another informant suggested targeting physicians rather than mothers with obesity prevention campaigns.

Training

All informants noted that NGOs contribute significantly to monitoring and treating health conditions in Uganda, and many said they generally work well with the government-run healthcare system. Several informants noted recent capacity-building efforts by NGOs and the MoH, particularly in hiring and training data analysts to compile and analyze health data. The U.S. Agency for International Development (USAID), the United Nations Children's Fund (UNICEF), the World Health Organization (WHO), and smaller Ugandan-based NGOs such as Community Based Integrated Nutrition (CoBIN) are particularly important for nutrition programs. One informant noted that working with a large funding organization such as USAID was helpful because having one umbrella organization with many different projects allowed for the program planners to create a more coordinated system and avoid redundancy in training and programmatic offerings.

Supervision

Supervision emerged as one of the weaker supporting activities; while some informants noted that NGOs, the MoH, local health centers, and district officers provided supervision for programs and surveillance activities, several explained that supervision was not necessarily constant or consistent, particularly when it was unclear which agency had jurisdiction or when NGOs pulled out. One major concern regarding supervision was that services were occasionally replicated or not performed at all because of the lack of coordination between different NGOs, universities, or governmental agencies; USAID was mentioned as an agency that avoided this pitfall by being an umbrella organization for many others. Another informant noted that while supervision within an NGO was often good, it was of limited use because of the inability to coordinate:

> The NGOs seem to work on their own, of course they're directly assisting the government, but what they do is to collect their data, use it and make implementations of what they just have collected, rather than saying we collected the data and helped the government to use it.

Resource provision

Providing resources, particularly funding and skilled staff, was a major concern for all informants. One of the most common suggestions, especially in the Southwest, was funding village health teams composed of two community leaders who serve as the first health contact for each village. Although these teams are highly effective, it can be difficult to recruit team members and provide for the medical needs of the community without funding. A hospital nutritionist highlighted the necessity of engaging village health workers in prevention, which has the potential to conserve resources by reducing costs associated with treating long-term conditions including Type 2 diabetes.

Both understaffing of health facilities and under-training of staff on database management were indicated as critical impediments to health data collection efforts. In rural clinics, informants indicated there were simply too few workers, but countrywide, there is a lack of adequately trained healthcare professionals and data analysts. One participant expressed his view that many professionals obtain adequate training for analyzing and managing health data, but move out of the government sector to accept jobs with more lucrative compensation. One informant noted that university researchers were underutilized and could greatly improve human resource capacity, particularly as researchers are highly invested in their work: "[for] these public servants it's routine work, why should I bother? But for the researcher, it's a must-do."

Discussion

Ugandan health professionals with an average of 20 years of experience in rural and urban healthcare settings highlighted challenges with initiating and sustaining health data surveillance. In low-resource settings, limited human and material resources, as well as the dependence on external funding from NGOs were particular challenges. Informants in multiple settings explained that while reporting of infectious diseases and maternal-child health indicators was generally quite good, providing timely analysis of data and feedback based on the results was an ongoing struggle. Informants further noted the difficulty of effectively educating individuals about NCD risk factors when malnutrition and infectious diseases are still highly prevalent. Despite these challenges, informants were hopeful that capacity building efforts by the government and NGOs would result in the improved training of individuals with technical expertise in database management and analysis.

Inadequate funding was identified by all participants as an ongoing and systemic challenge. While well-publicized diseases including HIV have received robust funding, community interventions and surveillance related to NCDs have not received comparable support. This may be partially due to the dramatic effects and often rapid transmission of communicable diseases, but also likely stems partially from the fact that NCDs have only recently risen to prominence (Schwartz *et al.* 2014). One informant mentioned the inconsistencies created by receiving funds from different agencies and NGOs; if an NGO lost funding or ended operations in the region, hospitals were forced to go without the nutritional formula it had been providing, for example. Participants also cited the lack of funding for the recently developed Nutrition Action Plan by the MoH.

A major untapped resource for strengthening health information systems was the university system, listed by three different informants. One noted that universities used to be more involved in research and healthcare, but collaboration with the government has since dwindled. University involvement has the potential to both enhance university research as well as health information systems. Informants also suggested primary schools and the Ministry of Education as potential partners in the assessment of chronic disease risk factors.

Limitations

This study had several limitations and constraints; the sample size was small, largely due to challenges with identifying of individuals familiar with the various stages of surveillance and other health data collection (e.g., a survey) efforts in Uganda. Interviews were conducted in English, primarily by a non-Ugandan researcher, although a translator was available. All interviewees spoke English, and since the participants were generally highly educated, language did not seem to inhibit responses. Third, although we seemed to reach saturation with themes in our sample, it is possible there might have been other themes that could have been expressed with an increased sample size. Despite these limitations, our study provides a basis for guiding systematic data collection efforts to address the increasing prevalence of NCDs in Uganda.

Conclusion

The DHS is currently the main population-based data source concerning chronic disease prevalence, which typically occurs every five years (Demographic and Health Surveys 2014), while infectious diseases and maternal-child health indicators are regularly reported and analyzed as a part of the IDSR strategy (Institute for Health Metrics and Evaluation 2014). With the establishment of the Access, Bottlenecks, Costs, and Equity (ABCE) project in 2011 (Institute for Health Metrics and Evaluation 2014), greater focus is being placed on health systems delivery. However, due to the increasing prevalence of NCDs, it is crucial to establish a more comprehensive and timely surveillance system, for which the CDC Behavioral Risk Factor Surveillance System (BRFSS) (Centers for Disease Control and Prevention 2014) or the WHO STEPS instrument (World Health Organization 2012) could be models. Byass *et al.* recommend establishing a minimum dataset of continuously tracked cause-specific mortality, exposure risk, and health interventions focused on both prevention and treatment (Byass *et al.* 2014). All of our informants indicated that challenges existed with successfully implementing ongoing surveys and collecting NCD data, corroborating other investigations of health systems in Uganda (Gladwin *et al.* 2003, Lukwago *et al.* 2013); however, they also expressed great optimism about capacity-building efforts and the significant improvements in data collection, reporting, and response times occurring in the past decade.

Given limited resources for health data collection in Uganda, an integrated data collection approach which builds on the success of the IDSR and other systems and cohorts currently in place for HIV and other communicable diseases may be the most feasible. Recent studies examining chronic disease risk in Uganda utilized an HIV testing campaign and cohort with serum samples to estimate risk for hypertension and diabetes (Maher *et al.* 2011, Chamie *et al.* 2012). This approach provides an opportunity to effectively utilize resources to track co-existing chronic and infectious diseases, and may be a better model for sustaining long-term data collection efforts. Given the existing support through the Paris Declaration on Aid Effectiveness (Organization for Economic Co-operation and Development 2005) and the International Health Regulations (IHR) (World Health Organization 2008), which were both initiated in 2005 to help developing countries build capacity and coordinate surveillance efforts with international aid organizations, this is an opportune time to initiate these plans. By addressing the challenges associated with maintaining a comprehensive surveillance system and building capacity by employing universities, gatekeepers, and extension workers, Uganda can take a multi-sector approach in order to strengthen health information systems used to track NCD risk. As NCDs and risk factors including overweight and obesity become more prevalent in Uganda and other countries in SSA, it will be increasingly important to implement health information systems that adequately measure risk factors to account for the shifting burden of disease.

Acknowledgements

We would like to thank all of the Ugandan participants for taking the time to grant interviews, and our guide and translator, Ronald Nsimbe. This research was supported by the Agriculture and Food Research Initiative of the USDA National Institute of Food and Agriculture as part of the AFRI Childhood Obesity Prevention Challenge (2011-67001-30101) to the Division of Nutritional Sciences at the University of Illinois at Urbana–Champaign.

References

Byass, P., 2009. The unequal world of health data. *PLoS Medicine*, 6(11), e1000155.

Byass, P., de Savigny, D., and Lopez, A.D., 2014. Essential evidence for guiding health system priorities and policies: Anticipating epidemiological transition in Africa. *Global Health Action*, 7, 23359.

Centers for Disease Control and Prevention, 2014. *Behavioral Risk Factor Surveillance System (BRFSS)* [Online]. Available from: http://www.cdc.gov/brfss/ [Accessed 14 October 2014].

Chamie, G. *et al.*, 2012. The SEARCH collaboration. Leveraging rapid community-based HIV testing campaigns for non-communicable diseases in rural Uganda. *PLoS ONE*, 7(8), e43400.

Demographic and Health Surveys, 2014. *DHS Overview* [Online]. Calverton (MD): ICF International. Available from: http://dhsprogram.com/What-We-Do/Survey-Types/DHS.cfm. [Accessed 14 October 2014].

Gladwin, J., Dixon, R.A., and Wilson, T.D., 2003. Implementing a new health management information system in Uganda. *Health Policy and Planning*, 18(2), 214–224.

Gold, O.P. and John, E.U., 2013. Accelerating empowerment for sustainable development: The need for health systems strengthening in sub-Saharan Africa. *American Journal of Public Health Research*, 1.7, 152–158.

Gordon, R.G (ed.), 2014. *Ethnologue: Languages of the World*. Fifteenth edition [Online]. Dallas, Texas: SIL International. Available from: http://www.ethnologue.com [Accessed 14 October 2014].

Hallal, P.C. *et al.*, 2012. Global physical activity levels: Surveillance progress, pitfalls, and prospects. *The Lancet*, 380(9838), 247–257.

Health Metrics Network, 2007. *Assessment of the Health Information System in Uganda* [Online]. Kampala: Health Metrics Network. Available from: http://www.who.int/healthmetrics/library/countries/hmn_uga_his_2007_en.pdf [Accessed 14 October 2014].

InDepth Network, 2014. *Accra, Ghana* [Online]. Available from: http://www.indepth-network.org [Accessed 14 October 2014].

Institute for Health Metrics and Evaluation (IHME), 2014. *Health Service Provision in Uganda: Assessing Facility Capacity, Costs of Care, and Patient Perspectives*. Seattle, WA: IHME.

Lasky, D., Becerra, E., Boto, W., Otim, M., and Ntambi, J., 2002. Obesity and gender differences in the risk of type 2 diabetes mellitus in Uganda. *Nutrition*, 18(5), 417–21.

Lukwago, L. *et al.*, 2013. The implementation of integrated disease surveillance and response in Uganda: A review of progress and challenges between 2001 and 2007. *Health Policy Plan*, 28, 30–40.

Magadi, M.A., 2011. Household and community HIV/AIDS status and child malnutrition in sub-Saharan Africa: Evidence from the demographic and health surveys. *Social Science Medicine*, 73(3), 436–46.

Maher, D., Waswa, L., Baisley, K., Karabarinde, A., and Unwin, N., 2011. Epidemiology of hypertension in low-income countries: A cross-sectional population-based survey in rural Uganda. *Journal of Hypertension*, 29(6), 1061–8.

Marquez, P. and Farrington, J., 2012. Africa's next burden: Non-infectious disease. *BMJ*, *345*(7873), 24–27.

Organisation for Economic Co-operation and Development (OECD), 2005. *The Paris Declaration on Aid Effectiveness* [Online]. Paris: OECD. Available from: http://www.oecd.org/development/aideffectiveness/34428351.pdf [Accessed 14 October 2014].

Popkin, B.M., 2017. Relationship between shifts in food system dynamics and acceleration of the global nutrition transition. *Nutrition Reviews*, 75(2), 1, 73–82.

Popkin, B.M., Adair, L.S., and Ng, S.W., 2012. Global nutrition transition and the pandemic of obesity in developing countries. *Nutrition Reviews*, 70(1), 3–21.

Schwartz, J.I., Guwatudde, D., Nugent, R, and Kiiza, C.M., 2014. Looking at non-communicable diseases in Uganda through a local lens: An analysis using locally derived data. *Globalization and Health*, 10, 77.

St. Louis, M.E. *et al.*, 2010. Surveillance in low-resource settings: Challenges and opportunities in the current context of global health. In Lee, L.M., Teutsch, S.M., Thacker, S.B., and St. Louis, M.E. (eds.). *Principles and Practice of Public Health Surveillance*. New York: Oxford University Press, 357–380.

Tollman, S.M. *et al.*, 2008. Implications of mortality transition for primary health care in rural South Africa: A population-based surveillance study. *Lancet*, 372(9642), 893–901.

Twagirumukiza, M. *et al.*, 2011. Current and projected prevalence of arterial hypertension in sub-Saharan Africa by sex, age and habitat: An estimate from population studies. *Journal of Hypertension*, 29(7), 1243–1252.

Uganda Bureau of Statistics (UBOS) and Macro International Inc., 2011. *Uganda Demographic and Health Survey*. Calverton, MD, USA: UBOS and Macro International Inc., 143–162.

Uganda Bureau of Statistics (UBOS), 2010. *TP5: 2010 Mid-year Projected Population for Town Councils District*. Kampala, Uganda: UBOS. Available from: http://www.ubos.org/onlinefiles/uploads/ubos/pdf%20documents/TP52010.pdf [Accessed 14 October 2014].

World Bank, 2012. *Countries: Uganda* [Online]. Washington, DC. Available from: http://www.worldbank.org/en/country/uganda [Accessed 14 October 2014].

World Health Organization (WHO), 2014. *WHO Global InfoBase* [Online]. Geneva: WHO. Available from https://apps.who.int/infobase/ [Accessed 14 October 2014].

World Health Organization, 2008. *World Health Assembly. International Health Regulations*. 2nd ed. Geneva: World Health Organization.

World Health Organization, 2012. *Chronic Diseases and Health Promotion: STEPwise Approach to Chronic Disease Risk Factor Surveillance (STEPS)* [Online]. Geneva: WHO; 2012. Available from: http://www.who.int/chp/steps/riskfactor/en/index.html [Accessed 14 October 2014].

World Health Organization, 2012. *Global Health Observatory Data Repository. Country Statistics: Uganda 2012* [Online]. Geneva: WHO. Available from: http://apps.who.int/ghodata/?vid=20300&theme=country [Accessed 3 February 2016].

Ziraba, A.K., Fotso, J.C., and Ochoko, R., 2009. Overweight and obesity in urban Africa: A problem of the rich or poor? *BMC Public Health*, 9, 465.

Part IV

Food security, nutrition, and health

13 Nutritional and nutraceutical properties of traditional African foods

John H. Muyonga, Ogugua C. Aworh, John Kinyuru, Marena Manley, Sophie Nansereko, and Dorothy N. Nyangena

Introduction

Significant dietary changes have taken place in Africa in the past decades with many elements of African diets being replaced by more convenient and 'Western' alternatives. Traditional African diets mainly comprising small grain cereals, dark green leafy vegetables, tropical fruits, legumes, and starchy stem and root tubers, have largely been replaced with conventional cereals, mainly maize, wheat, and rice; introduced vegetables (e.g., cabbage); and energy-dense foods containing higher levels of saturated fats, added sugars, and salt (Uusitalo *et al.* 2002, Stamoulis *et al.* 2004). As a result of these dietary changes and sedentary lifestyles, diet-related non-communicable diseases such as obesity, heart disease, hypertension, and diabetes are on the rise (Naicker *et al.* 2015, Stamoulis *et al.* 2004). The prevalence of obesity and over-nutrition on the continent rose from 18% in 1980 to 30% in 2008 (Finucane *et al.* 2011). At the same time, prevalence of under-nutrition (FAO, IFAD, and WFP 2014) and micronutrient deficiencies (World Health Organization 2008, 2009) remains high. This places a double burden on the African healthcare system. The current situation in Africa seems reflective of the coexistence of stage three and four of the nutrition transition described by Popkin (2002), with under-nutrition mainly among the poor and over-nutrition among the not-so-poor. While the demographic and epidemiological transition is in early stages, the rise in non-communicable diseases requires urgent attention in light of the poor healthcare system. Nutrition interventions addressing both under- and over-nutrition offer a sustainable and relatively inexpensive option for the management of the prevalent health challenges.

Fresh fruits and vegetables, legumes, roots, and tubers are important components of traditional African diets while high intakes of fat, meat, especially fatty meat, sugar, and salt with low intake of fibre are characteristics of urban diets. As people move from rural areas in search of job opportunities in urban centres, their food consumption pattern changes from traditional African diets that involve time-consuming food preparation towards precooked, convenient foods consumed at home and outside meals in the form of snacks, fast foods, and street foods with adverse consequences for nutrition and health (Stamoulis *et al.* 2004). There is close association between traditional diet and traditional medicine in

Africa where many indigenous African fruits and vegetables and wild food plants are used for the treatment of a wide variety of ailments, including abdominal pains, diarrhoea, dysentery, haemorrhoids, wounds, malaria, intestinal worms and other parasitic diseases, anaemia, diabetes, asthma, leprosy, gonorrhoea, syphilis, and other sexually transmitted diseases, cough, eye and throat infections, and many other diseases (Muyonga *et al.* 2017). In this chapter, we present a description of the nutritional and nutraceutical properties of different groups of traditional African foods, providing a basis for understanding their potential to contribute to the sustenance of nutrition and health in the period of demographic, epidemiological, and nutrition transition.

African traditional fruits and vegetables

Importance

Fruits and vegetables are important components of African diets, adding a variety of colours, unique flavours, and a range of textures. However, consumption of fruits and vegetables in many African countries is below the 400g World Health Organization (WHO) recommendation (Ruel *et al.* 2004). In addition to common, well-known vegetables such as tomatoes, carrots, onions, and cabbage and fruits such as banana, mango, papaya, pineapple, watermelon, and oranges, there are hundreds of lesser-known indigenous fruits and vegetables as well as other food plants gathered from the wild that contribute to food and nutrition security in Africa, especially among the rural populace. These crops thrive with little care, are available even during the hungry seasons, and are the cheapest sources of ascorbic acid, pro-vitamin A carotenoids, folic acid, and many minerals, including calcium, iron, and zinc in many localities. They are also rich in fibre, antioxidant pigments, and health-promoting components. Their dry nuts and seeds may be rich in protein, carbohydrates, and fat and may contribute significantly to protein and energy intakes (Aworh 2014).

Diversity

Whilst there are some traditional vegetables whose cultivation and use cut across most of the continent such as okra (*Abelmoschus esculentus*), *Amaranthus* spp., *Celosia* spp., *Corchorus* spp., *Solanum* spp., *Brassica* spp., *Hibiscus* spp., *Curcubita* spp., *Cucumis* spp., and *Basella* spp., others are restricted to certain sub-regions (Guarino 1995). *Amaranthus cruentus* is the most commonly grown Amaranth in Africa. African eggplant or garden egg (*Solanum aethiopicum, Solanum macrocarpon*) is widely distributed in the areas between the Guinea belt and the Savanna (Schippers and Budd 1997). Baobab (*Adansonia digitata*) leaves are popular in West Africa (National Reseach Council 2006) and the savanna zone of other African countries, including the Sudan (Wiehle *et al.* 2014). Roselle (*Hibiscus sabdariffa*) is better known for its food value than for fibre in Africa and as a vegetable is mainly associated with the savanna and semi-arid areas of Africa

(National Research Council 2006). Fluted pumpkin (*Telfairia occidentalis*) most probably originated in the high-rainfall forest belt of West Africa, and together with bitter leaf (*Vernonia amygdalina*), *Gnetum africanum, Gnetum buchholzianum*, *Celosia argentea*, *Corchorus olitorius*, and *Talinum triangulare* are important leafy vegetables in Nigeria and other West African countries (Aworh 2014). *Gynandropsis gynandra*, *Crotalaria brevidens*, *Asystasia shimperi*, and *Vigna unguiculata* (cowpea leaves) are popular in Kenya (Guarino, 1995). Leaves of *Phaseolus vulgaris*, *Phaseolus lunatus*, *Vigna unguiculata*, *Sesamum indicum*, and *Manihot esculenta* (cassava leaves) are important vegetables in Uganda. *Eruca sativa*, *Portulaca oleracea*, and *Vigna unguiculata* are among vegetables that are popular in the Sudan, while *Ensete ventricosum* is among the cultivated indigenous vegetables widely used in Ethiopia (Guarino 1995). These traditional vegetables are used as part of the main dishes or as condiments in traditional African diets (Aworh 2014).

A number of indigenous fruits of Africa are commonly consumed in the fresh form (Chothani and Vaghasiya 2011, Keay 1989). The tree fruit African star apple (*Chrysophyllum albidum*) is mainly found in the lowland rain forest extending from Sierra Leone to East Africa. The African or wild mango (*Irvingia gabonensis* var. *gabonensis*) grows naturally in the forest habitat of parts of Africa extending from Senegal to Sudan and south to Angola. *Balanites aegyptiaca* (desert date), on the other hand, is widely distributed in the dry Sahel-savanna regions of Africa, while *Vitex doniana* (black plum) is widely distributed in Africa and is found naturally in a variety of habitats from forest to savanna. *Spondias mombin* (hog plum), which is believed to have originated from tropical central America, is found in several parts of the humid tropics, including west, east, and central Africa (Keay 1989). The young leaves of some of these tree fruits such as *Vitex doniana* and *Balanites aegyptiaca* are commonly boiled and used as a vegetable in many African countries. The fruit of African pear (*Dacryodes edulis*) that grows naturally in the forest habitat extending from south-western Nigeria to Zambia and Angola, as well as in Sao Tome and Principe (Keay 1989), is first softened by dipping in hot water or hot ash for a few minutes before consumption. The dry fruit pulps of velvet tamarind (*Dialium guineense*) that grows naturally in forests and forest outliers of the woodland savanna extending from Senegal to Nigeria and also in Sao Tome and Principe (Keay 1989) and of baobab (*Adansonia digitata*) are consumed in that form. The nuts of African walnut (*Coula edulis*) are consumed in the fresh form while conophor nut (*Tetracarpidium conophorum*) is first boiled before it is eaten.

The seeds of the African breadfruit (*Treculia africana*) that grows naturally in the forest habitat extending from Senegal to the Sudan and south to Angola and Mozambique (Keay 1989) are cooked and eaten as the main meal. They may also be dried and milled into flour for various applications or they may be roasted or fried and eaten as nuts. The dried, milled, oil-rich cotyledons of *Irvingia gabonensis* var. *excelsa* are used in soups for their stringy, gum-like properties and to impart flavour. They may also be used for the preparation of dika fat. The seeds of the African locust bean (*Parkia biglobosa*) are fermented to produce one of the most important condiments in West Africa known as 'iru' or 'dawadawa' and the sweet-tasting pulp is eaten or used for preparing a fruit drink (Campbell-Platt 1980, Odunfa 1985). The seeds of the African oil bean (*Pentaclethra macrophylla*) are fermented to produce 'ugba' or 'ukpaka', a delicacy popular in the Igbo-speaking areas of south-eastern Nigeria (Okafor 1987).

The seeds, leaves, fruits, and other parts of several lesser-known species of the family Cucurbitaceae that include many species such as watermelon (*Citrullus lanatus*), 'egusi' melon (*Colocynthis citrullus*), cucumber (*Cucumis sativus*), various gourds, and pumpkins are consumed in various forms in many countries of sub-Saharan Africa. 'Egusi' melon is used for the preparation of 'egusi' soup which is very popular throughout Nigeria and the seeds of bottle gourd (*Lagenaria sicceraria*) are used for the preparation of 'oseani', a very popular soup among the Igbo-speaking people of Delta State, Nigeria (Aworh 2014). 'Egusi' melon and fluted pumpkin seeds are fermented to produce a popular soup flavouring condiment known as 'ogiri' or 'ogili' in Nigeria (Okafor 1987). Vegetable oil may also be extracted from 'egusi' melon, *Vitellaria paradoxa* (shea butter), *B. aegyptiaca* kernel and other oil-bearing seeds and nuts of several lesser-known indigenous fruits and vegetables. The red calyxes of roselle (*H. sabdariffa*) are used for the production of a popular drink known as 'soborodo' or 'zobo' in Nigeria and as tea in the Sudan, Egypt, and Central Africa, where the species originated (Aworh 2014). The dry pulp of tamarind (*Tamarindus indica)* which is indigenous to the drier savanna zone of Africa, from Sudan, Ethiopia, Kenya, and Tanzania, westward through sub-Sahelian Africa to Senegal is used for the production of traditional drinks known as 'tsimi' or 'tsamia' in northern Nigeria (Adeola and Aworh 2010). Similarly, traditional drinks are also produced from the fresh fruits of *V. doniana*, *B. aegyptiaca,* and many other lesser-known indigenous fruit trees in many parts of Africa. Natukunda, Muyonga, and Mukisa (2015) proposed use of tamarind seed, a rich source of phytochemicals, as a source of natural antioxidants in other food products such as cookies and juices.

Nutritional and nutraceutical properties

Some components of the nutrient composition of some lesser-known African fruits and vegetables have been documented (Guarino 1995, Stadlmayr *et al.* 2012, 2013, Charrondiere *et al.* 2013, Aworh 2014, 2015). Data on the proximate composition, minerals, and vitamin C content of several lesser-known indigenous African fruits have been recently reviewed (Stadlmayr *et al.* 2013, Aworh 2015). The fresh fruits are generally rich in carbohydrates, are good sources of fibre, but are low in protein and fat with a few exceptions, such as *D. edulis* which is high in fat (21%). Baobab (*A. digitata*) fruit pulp is one of the best sources of ascorbic acid (300 mg/100 g). Traditional African fruits and vegetables that are high in vitamin C include *T. occidentalis* (129 mg/100 g), *C. olitorius* (78 mg/100 g), *G. africanum* (56 mg/100 g), *A. cruentus* (56 mg/100 g), *I. gabonensis* var. *gabonensis* (54 mg/100 g), *C. albidum* (48 mg/100 g), and *A. digitata* leaf (47 mg/100 g). Vitamin A deficiency and nutritional anemias due to iron and folate deficiencies are among the important micronutrient deficiencies (hidden hunger) in Africa (World Health Organization 2008, 2009). Even though red palm oil, an important component of traditional diets in West Africa, is the world's best source of pro-vitamin A carotenoids, leafy vegetables such as *C. olitorius* (3130 μg/100 g), *A. cruentus* (2890 μg/100 g), and *H. sabdariffa* leaf (2580 μg/100 g) are good sources of β-carotene and contribute significantly to vitamin A intake in Africa (Stadlmayr *et al.* 2012). Some of the best sources of folic acid in traditional African diets include *C. olitorius* (118 μg/100 g), *A. digitata* leaf (118 μg/100 g), and *A. cruentus* (79 μg/100 g). Apart from vitamins, traditional African vegetables also contribute significantly to the intake of essential minerals. In a preliminary assessment of the nutritional value of 20 traditional leafy vegetables in KwaZulu-Natal, South Africa, 12, namely *Amaranthus dubius*, *Asystasia gangetica*, *Amaranthus hybridus*, *Amaranthus spinosus*, *Cucumis metuliferus*, *Cleome monophylla*, *Ceratotheca tribola*, *Galinsoga parviflora*, *Justica flava*, *Momordica balsamina*, *Physalis viscosa*, and *Wahlenbergia undulata*, had mineral concentrations exceeding 1% of plant dry weight and these were much higher than typical mineral concentrations in conventional leafy vegetables (Odhav *et al.* 2007). Deficiency of iron in most traditional African diets is due to low intake of animal protein. Some of the good sources of iron include *S. aethiopicum* (18 mg/100 g), *C. argentea* (13 mg/100 g), *T. occidentalis* (10 mg/100 g), *A. digitata* fruit pulp (9 mg/ 100 g), *A. cruentus* (6 mg/100 g), and *C. olitorius* (6 mg/100 g).

Bitter leaf (*V. amygdalina*), which is commonly used for stomach pains and for treating wounds, is also used as a laxative herb and an antimalarial and is reported to be used by wild chimpanzees for the treatment of parasitic diseases in Tanzania (Schippers and Budd 1997). *Ocimum gratissimum*, *Crassocephalum rubens*, *G. africanum*, *G. buchholzianum,* and *C. argentea* are used for sore eyes, stomach pains, diarrhoea, haemorrhoids, and cough. Similarly, fruits, leaves, barks, and other plant parts of numerous trees that are sources of fruits and vegetables, including *S. mombin*, *I. gabonensis*, *C. albidum*, *T. africana*, *P. macrophylla*, *D. edulis*, *V. doniana,* and *B. aegyptiaca* are used for medicinal purposes in Africa. The chemotherapeutic properties of some of these food plants have been scientifically investigated and documented (Farombi 2003, Chothani and Vaghasiya 2011).

Roots and tubers

The now less grown traditional African tuber crops, including wild yam species (*Dioscorea spp*), African potato (*Plectranthus esculentus*), *Disa spp*, *Habenaria walleri*, and *Satyrium spp*, are nutritionally superior to the commonly grown roots and tuber crops like cassava, potato, and sweet potato (Maliro 2001). The African potato, for example, contains twice the protein of common potatoes and is a rich source of calcium, vitamin A, and iron (Stone *et al.* 2011). Yam tubers have various bioactive components, namely, mucin, dioscin, dioscorin, allantoin, choline, polyphenols, diosgenin, carotenoids, and tocopherols. Mucilage of yam tuber contains soluble glycoprotein and dietary fibre. Yam extracts have been shown to exhibit nutraceutical properties, including hypoglycemic, antimicrobial, and antioxidant activities (Chandrasekara and Kumar 2016). Anchote (*Coccinia abyssinica*) which is indigenous to the western parts of Ethiopia, contains high calcium and proteins. According to local farmers, it helps in the fast mending of fractures and makes lactating mothers healthier and stronger (Fekadu *et al.* 2013). Dawit and Estifanos (1991) reported that the juice prepared from tubers of Anchote has saponin as an active substance and is used to treat gonorrhoea, tuberculosis, and tumours.

Mushrooms

Although little data is available on the composition of traditional African mushrooms, some, including *Termitomyces mammiformis*, *Lactarius triviralis*, and *Russula vesca*, are reported to be good sources of carbohydrates, protein, dietary fibre, calcium, iron, copper, magnesium, and manganese (Adejumo and Awosanya 2005, Mshandete and Cuff 2007). Edible mushrooms also provide a nutritionally significant content of vitamins (B1, B2, B12, C, D, and E) (Heleno *et al.* 2010). Many traditional African mushroom species have been shown to contain bioactive compounds, particularly polysaccharides and phenolics, and to exhibit activity against diarrhoea, renal infections, epilepsy, headaches, colds, fever, cardiovascular diseases, high blood pressure, and arthritis, as well as antioxidant, anti-mutagenic, anti-tumour, anti-cholesterol, anti-allergic, anti-aging, antiviral, antifungal, anti-inflammatory, hypoglycemic, and hypolipidemic activities (Nabubuya *et al.* 2010, Oyetayo 2011, Hamzah *et al.* 2013, Obodai *et al.* 2014, Hussein *et al.* 2015).

Legumes

The main African legumes include marama beans (*Tylosema esulantum*), cowpeas (*Vigna unguiculata* (L.) Walp.), Bambara groundnut (*Vigna subterranea*), and chickpeas (*Cicer arietinum*). Marama beans, a perennial, tuberous, drought-resistant legume which grows in Southern Africa, have a protein content of 29–38% (which is comparable to soybean), while also containing high levels of fibre (18.5–25.8%), vitamin E (21.4–67.1 α tocopherol equivalents), zinc (12–40 mg/100 g), and iron (31–39 mg/100 g) (Normann *et al.* 1996, Holse *et al.* 2010). Unlike most other legumes, the marama bean is also rich in the essential amino acid lysine and moderately high in methionine (Normann *et al.* 1996). The African yam bean (AYB) (*Sphenostylis stenocarpa*) is another nutritionally important native legume, which flourishes all through tropical Africa. Both seed and tuber are rich in protein. According to Oshodi, Ipinmoroti, and Adeyeye (1997), the whole seeds on average contain 20.50% protein, 8.25% fat, 59.72% total carbohydrate, 649.49 mg/100 g potassium, and 241.21 mg/100 g phosphorus. Eneh, Orjonwe, and Adindu (2015) reported use of the AYB in treatment and management of gout and arthritis. According to Stone *et al.* (2011), approximately

200 million people in Africa subsist on a diet containing cowpea. Cowpea has also been found to improve the body's absorption and breakdown of other staple foods (Stone *et al.* 2011). Cowpea is a rich protein source and also contains phenolics which are antioxidants (Siddhuraju and Becker 2007). Bambara groundnuts have higher levels of methionine than other legumes. It also has very high concentration of soluble fibre, which has been linked to the reduction of heart disease and cancer (Stone *et al.* 2011). In addition to their nutritional advantage, African legumes are well adapted to drought and low nutrients (Sprent *et al.* 2009). They can, therefore, thrive where conventional crops cannot.

Oil seed crops

Sesame is an oil seed which is traditionally consumed in parts of Africa in the form of spreads, as part of traditional sauces, or as snacks. Sesame seeds contain 50–60% oil and 19–25% protein with antioxidants and lignans, such as sesamolin and sesamin, which prevent rancidity and exhibit antioxidant and anti-cancer properties. The lignin contents have useful physiological effects in human health (Ashakumary *et al.* 1999). The principal unsaturated fatty acids are oleic and linoleic with about 40% of each and about 14% saturated acids (Were *et al.* 2006, Kafiriti and Mponda 2016). Sesame also contains high levels of iron (7.75 mg/100 g), zinc (7.29 mg/100 g), phosphorous (629 mg/100 g), magnesium (351 mg/100 g), and substantial levels of vitamin B (thiamine) and vitamin E (tocopherol) (Nagendra *et al.* 2012). Sesame seeds also contain phytosterols which are associated with reduced levels of blood cholesterol (Bedigian 2004).

The African oil bean (*Pentaclethra macrophylla* Benth) is a tropical tree crop found mostly in the Southern and Middle Belt regions of Nigeria and in other coastal parts of West and Central Africa. The seed is a source of edible oil as well as protein (Fungo *et al.* 2016) and can be used for the preparation of many African delicacies, including African salad, soups, and sausages for eating with different staples (Enujiugha 2003). According to Ogueke *et al.* (2010), the oil bean seeds contain 4–17% carbohydrate, 36.2–23.89% protein, and 44–47% oil, which has been found to be rich in oleic, linoleic, and lignoceric acids. The protein in oil seeds is relatively low in sulphur amino acid but high in other essential amino acids (Ogueke *et al.* 2010).

Cereals

Africa is the centre of origin and also a major producer of several cereals like sorghum (*Sorghum bicolor*), pearl millet (*Pennisetum glaucum*), finger millet (*Eleucine corocana*), teff (*Eragrostis tef*), fonio (*Digitaria spp*), and African rice (*Oryza glaberrima*) (Macauley 2015). These cereals formed the basis of African diets for centuries. Africa's production of fonio, sorghum, and millet represents 100%, 42%, and 51% respectively of the global production, whereas the contribution towards maize, rice, and wheat production lies at 7%, 4%, and 4% respectively (Food and Agriculture Organization 2016). Traditional cereals have a high agronomic resilience, which increases their significance in terms of food security in Africa (Food and Agriculture Organization 1995, Dlaminiand Siwela 2015). Traditional African cereals (TACs) are also less prone to rot and insect damage during storage. Some, like millet, are less prone to mycotoxin contamination compared to maize (Bandyopadhyay *et al.* 2007).

Compared to maize (3.2–4.1 mg/100g), rice (0.7–2.5 mg/100g), and wheat (2 mg/100g), TACs, finger millet (10–20.7 mg/100g), pearl millet (9.76–20.9 mg/100g), fonio (9.4–10 mg/100g), and sorghum (3.4–8.7 mg/100g), contain higher levels of iron (Food and Agriculture Organization, 2010). Fonio and some sorghum and millet varieties are also excellent sources of calcium, whereas sorghum and millet are rich sources of B vitamins (Solange *et al.* 2014), while teff is high in calcium and iron (Gebremariam *et al.* 2014). The protein content of TACs depends on the type of cereal but tends to be higher than that of the major cereals (Food and Agriculture Organization 2010). The proteins of TACs generally have higher proportions of lysine, which is the most limiting essential amino acid among cereals compared to rice, maize, and refined wheat (Lukmanji *et al.* 2008). Fonio, which has been named one of the most nutritious of all grains, is rich in important amino acids (e.g., methionine and cysteine) which are limiting in wheat, rice, or maize (Stone *et al.* 2011). Finger millet has also been considered to be superior to wheat in terms of its protein digestibility (Stone *et al.* 2011). Consumption of sorghum and millet and their products has been shown to contribute to improved health and wellness due to their health-promoting bioactive compounds, such as phenolic acids, flavonoids, condensed tannins, anthocyanins, phytosterols, and policosanols (Awika and Rooney 2004, Awika *et al.* 2005, Dykes and Rooney 2006, Mathanghi and Sudha 2012, Shahidi and Chandrasekara 2013, Taylor *et al.* 2014). Sorghum and millets also contain resistant starches, hence their digestion results in slow release of glucose into the bloodstream, a desired attribute for the management of diabetes (Dlaminiand Siwela 2015). Many TACs are consumed whole, and the phenolics, tocotrienols, and tocopherols found in the whole cereals are associated with prevention of cardiovascular diseases, stroke, hypertension, metabolic syndrome, type 2 diabetes mellitus, obesity, as well as different forms of cancer (Borneo and Leon 2012).

Edible insects

About 1,900 species of insects are consumed as part of diet, and a projected two billion people consume insects on a daily basis (Food and Agriculture Organization 2013, van Huis 2013). The insects widely consumed belong to the orders *Lepidoptera*, *Coleoptera, Hymenoptera*, *Orthoptera*, *Hemiptera*, *Isoptera*, and *Diptera* (Kelemu *et al.* 2015). In East Africa, termites, grasshoppers, crickets, ants, and mayflies are traded and consumed (Food and Agriculture Organization 2013, Kinyuru *et al.* 2015). These insects are generally served roasted or pan-fried, accompanied with a staple or as a snack. Current trends are showing growing advocacy and interest in commercial farming of edible insects. Cricket farming is being piloted in Kenya and Uganda (Ayieko *et al.* 2016). In other parts of the world, successful ventures have farmed black soldier flies, mealworms, and crickets among others (Charlton *et al.* 2015).

The nutritional composition of edible insects is highly dependent on insect species, their metamorphic stage, habitat, diet, and processing methods (Kinyuru *et al.* 2010, van Huis 2013). In the recent past, the nutritional profiles of various edible insects have been included in a food composition database (Kinyuru *et al.* 2015). Insects are rich sources of proteins, fat, fibre, and ash. A protein content of up to 81% has been reported (van Huis 2013). Adult insects have been found to contain higher protein values as compared to their respective instars (Belluco *et al.* 2013). Insect proteins are highly digestible, with *in vitro* digestibility of up to 98% (Belluco *et al.* 2013), but this is reduced by some processing and preservation techniques (Kinyuru *et al.* 2009, van Huis *et al.* 2013). Protein from insects is characterised by an essential amino acid score higher than that for most other protein sources (Smith and Pryor 2013). A number of authors have concluded that insects can be comparable to, if not superior to, traditional protein sources such as beef, chicken, and soybean (Kinyuru *et al.* 2010, Yi *et al.* 2013).

Edible insects are also good sources of fat (9–77%) (Ramos-Elorduy *et al.* 1997, Kinyuru *et al.* 2015). The essential unsaturated fatty acids linoleic and linolenic acids are also high. For maximum health benefits, the lipid quality should maintain a balance between the saturated and the unsaturated fatty acid groups. Toasted chrysalis (*Bombyx mori*) was found to contain a polyunsaturated/saturated fatty acid ratio of 0.99, which is greater than the recommended minimum ratio of 0.45 (Pereira *et al.*2003). Most insects have a higher ratio than 0.45 and therefore consumption of insects is associated with low serum cholesterol levels and low risks of coronary heart diseases (Kinyuru *et al.* 2013). Some edible insects, like caterpillars and grasshoppers, also contain substantial levels of polyunsaturated fatty acids and health-promoting omega-3 fatty acids (Womeni *et al.* 2009). Due to their high fat content, edible insects are highly susceptible to lipid oxidation, which renders the product rancid quickly.

Edible insects also contain a significant amount of fibre, mainly constituted of chitin, an insoluble exoskeleton derivative (Finke 2002, Brownawell *et al.* 2012). Although chitin has been equated to indigestible cellulose in plants, human gastric juice has been found to contain chitinase enzyme which has the ability to digest chitin (Brownawell *et al.* 2012). Chitin promotes good digestion as it acts as a dietetic fibre. It is also closely associated with high immunity as it offers protection against both parasitic and pathogenic bacteria, hence maintaining the health of probiotic bacteria within the gut (Lertsutthiwong *et al.* 2002). Initial studies have indicated that chitosan, a component of chitin, has anti-tumour, cholesterol lowering, antifungal, and antibacterial effect (Lee *et al.* 2002). With the rise in non-communicable diseases that are highly linked to red and processed meat consumption, edible insects can be used in an effort to reduce the prevalence of these diseases.

The B group of vitamins are generally higher in insects than in meat (Paoletti and Dufour 2002). Significant levels of vitamins such as riboflavin, pantothenic acid, biotin, and in some cases folic acid have been reported in various insects (Rumpold and Schlüter, 2013). Among seven vitamins studied, riboflavin (4.18mg/100g) was reported highest in *Macrotermes subhylanus* while niacin (3.61mg/100g) was the dominant vitamin in *Ruspolia differens* (Kinyuru *et al.* 2010).

Insects are rich in minerals including copper, manganese, selenium, iron, calcium, zinc, and phosphorus, with a particularly high content of iron and zinc (Barker *et al.* 1998, Christensen *et al.* 2006, Rumpold and Schlüter 2013). Kinyuru *et al.* (2013) reported a calcium content of 42.89–63.60mg/100g, iron (53.33–115.97mg/100g), and zinc (7.10–12.86mg/100g) in edible winged termites. Incorporation of edible insects within the human diet can, therefore, help solve micronutrient deficiencies (Christensen *et al.* 2006). Some insects have however been reported to contain anti-nutrients such as phenols, hydro-cyanide, oxalates, thiaminase, phytates, and tannins (Nishimune *et al.* 2000, Shantibala *et al.* 2014). These anti-nutrients are nonetheless below toxic levels and are eliminated during processing.

Challenges and safety issues associated with production and consumption of traditional African foods

Despite their numerous advantages, many traditional African foods have not been fully exploited. This is partly because of the underdeveloped production and marketing systems of these foods, inadequate awareness of the nutritional and health benefits they impart, limited processing, and geographically confined cultural acceptance (Ayieko *et al.* 2010, Kinyuru *et al.* 2010, Abukutsa-Onyango 2011). Some traditional African foods have also been reported to have high levels of anti-nutrients and should be appropriately processed to increase nutrient bioavailability (Taylor *et al.* 2014).

Conclusion

Because of their unique nutritional and nutraceutical properties, traditional African foods have the potential to contribute to the alleviation of the under- and overnutrition that Africa faces. Despite this potential, consumption of these foods is low and seems to be declining further. Interventions aimed at enhancing production and improving understanding of the nutritional and nutraceutical value of traditional African foods are required to reverse this trend and gain from these foods.

References

Abukutsa-Onyango, M., 2011. Researching African indigenous fruits and vegetables – Why? [Online]. Available from http://knowledge.cta.int/Dossiers/Commodities/Fruits/Feature-articles/Researching-African-Indigenous-Fruits-and-Vegetables – Why [Accessed 6 February 2017].

Adejumo, T.O. and Awosanya, O.B., 2005. Proximate and mineral composition of four edible mushroom species from south-western Nigeria. *African Journal of Biotechnology*, 4(10), 1084–1088.

Adeola, A.A. and Aworh, O.C., 2010. Development and sensory evaluation of an improved beverage from Nigeria's tamarind (*Tamarindus indica* L.) fruit. *African Journal of Food, Agriculture, Nutrition and Development*, 10, 4079–4092.

Ashakumary, L. *et al.*, 1999. Sesamin, a sesame lignan, is a potent inducer of hepatic fatty acid oxidation in the rat. *Metabolism: Clinical and Experimental*, 48(10), 1303–1313.

Awika, J.M. and Rooney, L.W., 2004. Sorghum phytochemicals and their potential impact on human health. *Phytochemistry*, 65(9), 1199–1221.

Awika, J.M., McDonough, C.M., and Rooney, L.W , 2005. Decorticating sorghum to concentrate healthy phytochemicals. *Journal of Agricultural and Food Chemistry*, 53(16), 6230–6234.

Aworh, O.C., 2014. *Lesser-known Nigerian Fruits and Vegetables: Post-harvest Handling, Utilization and Nutritional Value.* Ibadan, Nigeria: Ibadan University Press.

Aworh, O.C., 2015. Promoting food security and enhancing Nigeria's small farmers' income through value-added processing of lesser-known and under-utilized indigenous fruits and vegetables. *Food Research International*, 76, 986–991.

Ayieko, M.A., Ogola, H.J., and Ayieko, I.A., 2016. Introducing rearing crickets (gryllids) at household levels: Adoption, processing and nutritional values. *Journal of Insects as Food and Feed*, 2(3), 203–211.

Ayieko. M.A., Oriaro. V., and Nyambuga, I., 2010. Processed products of termites and lake flies: Improving entomophagy for food security within the Lake Victoria region. *African Journal of Food Agriculture, Nutrition and Development*, 10(2), 2085–2098.

Bandyopadhyay, R., Kumar, M., and Leslie, J.F., 2007. Relative severity of aflatoxin contamination of cereal crops in West Africa. *Food Additives and Contaminants*, 24(10), 1109–1114.

Barker, D.H., Logan, B.A., Adams III, W.W., and Demmig-Adams, B., 1998. Photochemistry and xanthophyll cycle-dependent energy dissipation in differently oriented cladodes of Opuntia stricta during the winter. *Functional Plant Biology*, 25(1), 95–104.

Bedigian, D., 2004. History and lore of sesame in southwest Asia. *Economic Botany*, 58(3), 329–353.

Belluco, S. *et al.*, 2013. Edible insects in a food safety and nutritional perspective: A critical review. *Comprehensive Reviews in Food Science and Food Safety*, 12(3), 296–313.

Borneo, R. and Leon, A.E., 2012. Whole grain cereals: Functional components and health benefits. *Food and Function*, 3, 110–119.

Brownawell, A.M. *et al.*, 2012. Prebiotics and the health benefits of fiber: Current regulatory status, future research, and goals. *The Journal of Nutrition*, 142(5), 962–974.

Campbell-Platt, G., 1980. African locust bean (*Parkia* species) and its west African fermented food product, dawadawa. *Ecology of Food and Nutrition*, 9, 123–132.

Chandrasekara, A. and Kumar, T.J., 2016. Roots and tuber crops as functional foods: A review on phytochemical constituents and their potential health benefits. *International Journal of Food Science*, 2016, doi.org/10.1155/2016/3631647.

Charlton, A.J. *et al.*, 2015. Exploring the chemical safety of fly larvae as a source of protein for animal feed. *Journal of Insects as Food and Feed*, 1(1), 7–16.

Charrondiere, U.R. *et al.*, 2013. FAO/INFOODS food composition database for diversity. *Food Chemistry*, 140, 408–412.

Chothani, D.L. and Vaghasiya, H.U., 2011. A review on *Balanites aegyptiaca* Del (desert date): Phytochemical constituents, traditional uses, and pharmacological activity. *Pharmacognosy Review*, 5, 55–62.

Christensen, M.R. *et al.*, 2006. Multiple anthropogenic stressors cause ecological surprises in boreal lakes. *Global Change Biology*, 12(12), 2316–2322.

Dawit, A. and Estifanos, H., 1991. Plants as a primary source of drugs in the traditional health practices of Ethiopia. In: Engels, J.M.M., Hawkes, J.G, and Worede, M., eds. *Plant Genetic Resources of Ethiopia*. Cambridge University Press.

Dlamini, N.R. and Siwela, M., 2015. The future of grain science: The contribution of indigenous small grains to food security, nutrition, and health in South Africa. *Cereal Foods World*, 60(4), 177–180.

Dykes, L. and Rooney, L.W., 2006. Sorghum and millet phenols and antioxidants. *Journal of Cereal Science*, 44(3), 236–251.

Eneh, F.U, Orjionwe, R.N., and Adindu, C.S., 2015. Effect of African Yam Bean (Sphenostylis stenocarpa) on serum calcium, inorganic phosphate, uric acid, and alkaline phosphatase concentration of male albino rats. *Journal of Agricultural Science*, 8(1), 148–153.

Enujiugha, V.N., 2003. Nutrient changes during the fermentation of African oil bean (*Pentaclethra macophylla* Benth) seeds. *Pakistan Journal of Nutrition*, 2(5), 320–323.

Farombi, E.O., 2003. African indigenous plants with chemotherapeutic potentials and biotechnological approach to the production of bioactive prophylactic agents. *African Journal of Biotechnology*, 2, 662–671.

Fekadu, H., Beyene, F., and Desse, G., 2013. Effect of traditional processing methods on nutritional composition and anti-nutritional factors of anchote (*Coccinia Abyssinica* (lam.) Cogn) tubers grown in western Ethiopia. *Journal of Food Processing Technolology*, 4, 249.

Finke, M.D., 2002. Complete nutrient composition of commercially raised invertebrates used as food for insectivores. *Zoo Biology*, 21(3), 269–285.

Finucane, M.M. *et al.*, 2011. Global burden of metabolic risk factors of chronic diseases collaborating group (body mass index). National, regional, and global trends in body-mass index since 1980: Systematic analysis of health examination surveys and epidemiological studies with 960 country-years and 9.1 million participants. *Lancet*, 377(9765), 557–67.

Food and Agricultural Organization, 1995. Sorghum and millets in human nutrition [Online]. Available from http://www.fao.org/docrep/t0818e/t0818e00.htm [Accessed 20 April 2015].

Food and Agricultural Organization, 2010. *Composition of Selected Foods from West Africa*. Rome: FAO.

Food and Agricultural Organization, 2013. *Edible Insects: Future Prospects for Food and Feed Security*. Rome: FAO.

Food and Agricultural Organization, 2016. FAOSTAT [Online]. Available from http://www.fao.org/faostat/en/#compare [Accessed 27 January 2017].

Food and Agricultural Organization (FAO), International Fund for Agricultural Development (IFAD), and the World Food Programme (WFP), 2014. *The State of Food Insecurity in the World 2014. Strengthening the Enabling Environment for Food Security and Nutrition*. Rome: FAO.

Fungo, R. *et al.*, 2016. Nutrients and bioactive compounds of *Baillonella toxisperma, Trichoschypha abut* and *Pentaclethra macrophylla* from Cameroon. *Food Science and Nutrition*, 3(4), 292–301.

Gebremariam, M.M., Zarnkow, M., and Becker, T., 2014. Teff (*Eragrostis tef*) as a raw material for malting, brewing and manufacturing of gluten-free foods and beverages: A review. *Journal of Food Science and Technology*, 51(11), 2881–95.

Guarino, L., ed., 1995. *Traditional African Vegetables. Proceedings of the International Plant Genetic Resources Institute (IPGRI) International Workshop on Genetic Resources of Traditional Vegetables in Africa: Conservation and Use*, 29–31 August, 1995, Nairobi, Kenya. International Plant Genetic Resources Institute.

Hamzah, R.U., Jagam, A.A., Makun, H.A., and Engwin, E.C., 2013. Antioxidant properties of selected African vegetables, fruits and mushrooms: A review. In Makun, H., ed. *Mycotoxin and Food Safety in Developing Countries*. InTech. Available from http://www.intechopen.com/books/mycotoxin-and-food-safety-in-developing-countries/antioxidant-properties-of-selected-african-vegetables-fruits-and-mushrooms-a-review [Accessed 16 January 2017].

Heleno, S.A., Barros, L., Sousa, M.J., Martins, A., and Ferreira, I.C.F.R., 2010. Tocopherols composition of Portuguese wild mushrooms with antioxidant capacity. *Food Chemistry*, 119(4), 1443–1450.

Holse, M., Husted, S., and Hansen, A., 2010. Chemical composition of Marama beans (*Tylosema esculentum*): A wild African bean with unexploited potential. *Journal of Food Composition and Analysis*, 23(6), 648–657.

Hussein, J.M., Tibuhwa, D.D., Mshandete, A.M., and Kivaisi, A.K., 2015. Antioxidant properties of seven wild edible mushrooms from Tanzania. *African Journal of Food Science*, 9(9), 471–479.

Kafiriti, E. and Mponda, O., 2016. Growth and production of sesame, encyclopedia of life support systems (EOLSS), soils, plant growth and crop production [Online]. Available from http://www.eolss.net/sample-chapters/c10/e1-05a-46.pdf [Accessed 10 October 2016].

Keay, R.W.J., 1989. *Trees of Nigeria*. Oxford, UK: Clarendon Press.

Kelemu, S. *et al*., 2015. African edible insects for food and feed: Inventory, diversity, commonalities and contribution to food security. *Journal of Insects as Food and Feed*, 1(2), 103–119.

Kinyuru, J. N. *et al*., 2013. Nutrient composition of four species of winged termites consumed in western Kenya. *Journal of Food Composition and Analysis*, 30(2), 120–124.

Kinyuru, J.N., Kenji, G.M., Muhoho, S.N., and Ayieko, M., 2010. Nutritional potential of longhorn grasshopper (*Ruspolia differens*) consumed in Siaya district, Kenya. *Journal of Agriculture Science and Technology*, 12(1), 32–46.

Kinyuru, J.N., Kenji, G.M., Njoroge, S.M., and Ayieko, M., 2009. Effect of processing methods on the in vitro protein digestibility and vitamin content of edible winged termite (*Macrotermes subhylanus*) and grasshopper (*Ruspolia differens*). *Food and Bioprocess Technology*, 3(5), 778–782.

Kinyuru, J.N., Mogendi, J.B., Riwa, C.A., and Ndung'u, N.W., 2015. Edible insects – A novel source of essential nutrients for human diet: Learning from traditional knowledge. *Animal Frontiers*, 5(2), 14–19.

Lee, C.Y., Cooksey, B.A., and Baehrecke, E.H., 2002. Steroid regulation of midgut cell death during Drosophila development. *Developmental biology*, 250(1), 101–111.

Lertsutthiwong, P., How, N.C., Chandrkrachang, S., and Stevens, W.F., 2002. Effect of chemical treatment on the characteristics of shrimp Chitosan. *Journal of Metals, Materials and Minerals*, 12(1), 11–18.

Lukmanji, Z. *et al*., 2008. *Tanzania Food Composition Tables*. Dar es Salaam, Tanzania: Muhimbili University of Health and Allied Sciences, Tanzania Food and Nutrition Centre, and Harvard School of Public Health.

Macauley, H., 2015. Background paper: Cereal crops: rice, maize, millet, sorghum, wheat. *Feeding Africa*, 21–23 October 2015, Dakar, Senegal: United Nations Economic Commission for Africa.

Maliro, M.F.A., 2001. The role and potential of traditional tuber crops in Malawi. In Pasternak, D. and Schlissel, A., eds. *Combating Desertification with Plants*. Dordrecht, the Netherlands: Kluwer Academic Press, 55–64.

Mathanghi, S.K. and Sudha, K., 2012. Functional properties of finger millet (Eleusine coracana L.) for health. *International Journal of Pharmaceutical Chemical and Biological Sciences*, 2(4), 431–438.

Mshandete, A.M. and Cuff, J., 2007. Proximate and nutrient composition of three types of indigenous edible wild mushrooms grown in Tanzania and their utilization prospects. *African Journal of Food Agriculture Nutrition and Development*, 7(6), 1–16.

Muyonga, J.H. *et al.*, 2017. Traditional African foods and their potential to contribute to health and nutrition. In Shekhar, H.U, Howlader, Z.H., and Kabir, Y., eds. *Exploring the Nutrition and Health Benefits of Functional Foods*. Hershey, P.A.: IGI Global, 332–358.

Nabubuya, A., Muyonga, J.H., and Kabasa, J.D., 2010. Nutritional and hypocholesterolemic properties of Termitomyces microcarpus mushrooms. *African Journal of Food, Agriculture, Nutrition and Development*, 10(3), 2235–2257.

Nagendra, P.M.N. *et al.*, 2012. A review on nutritional and nutraceutical properties of sesame. *Journal of Nutrition and Food Sciences*, 2(127).

Naicker, A., Venter, C.S., MacIntyre, U.E., and Ellis, S., 2015. Dietary quality and patterns and non-communicable disease risk of an Indian community in KwaZulu-Natal, South Africa. *Journal of Health, Population and Nutrition*, 33, 12.

National Research Council, 2006. *Lost Crops of Africa. Volume II. Vegetables*. Washington, DC: National Academy Press.

Natukunda, S., Muyonga, J.H., and Mukisa, I.M., 2015. Effect of tamarind (Tamarindus indica L.) seed on antioxidant activity, phytocompounds, physicochemical characteristics, and sensory acceptability of enriched cookies and mango juice. *Food Science and Nutrition*, 4(4), 494–507.

Nishimune, T., Watanabe, Y., Okazaki, H., and Akai, H., 2000. Thiamin is decomposed due to *Anaphe spp*. entomophagy in seasonal ataxia patients in Nigeria. *The Journal of Nutrition*, 130(6), 1625–1628.

Normann, H., Snyman, I., and Cohen, M., eds., 1996. *Indigenous Knowledge and its Uses in Southern Africa*. Vol. 61. Pretoria, South Africa: HSRC Press.

Obodai, M. *et al.*, 2014. Evaluation of the chemical and antioxidant properties of wild and cultivated mushrooms in Ghana. *Molecules (Basel, Switzerland)*, 19(12), 19532–19548.

Odhav, B., Beekrum, S., Akula, U.S., and Baijnath, H., 2007. Preliminary assessment of nutritional value of traditional leafy vegetables in KwaZulu-Natal, South Africa. *Journal of Food Composition and Analysis*, 20, 430–435.

Odunfa, S.A., 1985. African fermented foods. In Wood, B.J.B., ed. *Microbiology of Fermented Foods Vol. 2*. Amsterdam, the Netherlands: Elsevier Applied Science Publishers, 155–191.

Ogueke, C.C, Nwosu, J.N., Owuamanam, C.I., and Iwouno, J.N., 2010. Ugba, the fermented african oil bean seeds; its production, chemical composition, preservation, safety and health benefits. *Pakistan Journal of Biological Sciences*, 13(10), 489–496.

Okafor, N., 1987. Upgrading local technologies of food processing with emphasis on fermented foods: The case of oils and fats. *Proceedings 11th Annual Conference of the Nigerian Institute of Food Science and Technology, Port Harcourt, Nigeria*, 61–74.

Oshodi, A.A., Ipinmoroti, K.O., and Adeyeye, E.I., 1997. Functional properties of some varieties of African yam bean (*Sphenostylis stenocarpa*) flour – III. *International Journal of Food Science and Nutrition*, 48(4), 243–50.

Oyetayo, O.V., 2011. Medicinal uses of mushrooms in Nigeria: Towards full and sustainable exploitation. *African Journal of Traditional, Complementary, and Alternative Medicines*, 8(3), 267–274.

Paoletti, M.G. and Dufour, D.L., 2002. Minilivestock. In: Pimental, D., ed. *Encyclopedia of Pest Management*. New York: Marcel Dekker, 487–492.

Pereira, N.R., Ferrarese-Filho, O., Matsushita, M., and de Souza, N.E., 2003. Proximate composition and fatty acid profile of Bombyx mori *L. chrysalis* toast. *Journal of Food Composition and Analysis*, 16(4), 451–457.

Popkin, B.M., 2002. An overview on the nutrition transition and its health implications: The Bellagio meeting. *Public Health Nutrition*, 5(1A), 93–103.

Ramos-Elorduy, J. *et al.*, 1997. Nutritional value of edible insects from the state of Oaxaca, Mexico. *Journal of Food Composition and Analysis*, 10(2), 142–157.

Ruel, M.T., Minot, N., and Smith, L., 2004. Background paper for the joint FAO/WHO workshop on fruit and vegetables for health [Online]. Available from www.who.int/dietphysicalactivity/publications/f%26v_africa_economics.pdf [Accessed 20 January 2017].

Rumpold, B.A. and Schlüter, O.K., 2013. Potential and challenges of insects as an innovative source for food and feed production. *Innovative Food Science and Emerging Technologies*, 17, 1–11.

Schippers, R. and Budd, L.A., eds., 1997. *Proceedings Workshop on African Indigenous Vegetables, 13–18 January 1997*, Limbe, Cameroon: National Resources Institute, International Plant Genetic Resources Institute Sub-Saharan Office, and the Centre for Plant Breeding and Reproduction Research.

Shahidi, F. and Chandrasekara, A., 2013. Millet grain phenolics and their role in disease risk reduction and promotion: A review. *Journal of Functional Foods*, 5(2), 570–581.

Shantibala, T., Lokeshwari, R.K., and Debaraj, H , 2014. Nutritional and antinutritional composition of the five species of aquatic edible insects consumed in Manipur, India. *Journal of Insect Science*, 14(1), 14.

Siddhuraju, P. and Becker, K., 2007. The antioxidant and free radical scavenging activities of processed cowpea (*Vigna unguiculata* (L.) Walp.) seed extracts. *Food Chemistry*, 101(1), 10–19.

Smith, R. and Pryor, R., 2013. Work Package 5: *Pro-Insect Platform in Europe*. PROteINSECT.

Solange, A. *et al.*, 2014. Review on African traditional cereal beverages. *American Journal of Research Communication*, 2(5), 103–153.

Sprent, J.I., Odee, D.W., and Dakora, F.D., 2009. African legumes: A vital but underutilized resource. *Journal of Experimental Botany*, 61(5): 1257–1265.

Stadlmayr, B. *et al.*, 2012. *West African Food Composition Table*. Rome: Food and Agriculture Organization of the United Nations (FAO).

Stadlmayr, B. *et al.*, 2013. Nutrient composition of selected indigenous fruits from sub-Saharan Africa. *Journal of the Science of Food and Agriculture*, 93, 2627–2636.

Stamoulis, K.G., Pingali, P., and Shetty, P., 2004. Emerging challenges for food and nutrition policy in developing countries. *e-Journal of Agricultural and Development Economics*, 1, 154–167.

Stone, A. *et al.*, 2011. *Africa's Indigenous Crops*. Washington DC: Worldwatch Institute.

Taylor, J.R.N., Belton, P.S., Beta, T., and Duodu, G.K., 2014. Increasing the utilization of sorghum, millets and pseudocereals: Developments in science of their phenolic phytochemicals, biofortification and protein functionality. *Cereal Science*, 59(3), 257–275.

Uusitalo, U., Pietinen, P., and Puska, P., 2002. Dietary transition in developing countries: Challenges for chronic disease prevention. In *Globalization, Diets and Noncommunicable Diseases.* Geneva: World Health Organization.

van Huis, A., 2013. Potential of insects as food and feed in assuring food security. *Annual Review of Entomology*, 58, 563–83.

van Huis, A. *et al.*, 2013. *Edible Insects: Future Prospects for Food and Feed Security.* Rome: FAO.

Were, B.A., Onkware, A.O., Gudu, S., Welander, M., and Carlsson, A.S., 2006. Seed oil content and fatty acid composition in East African sesame (*Sesamum indicum* L.) accessions evaluated over 3 years. *Field Crops Research*, 97(2–3), 254–260.

Wiehle, M., Prinz, K., Kehlenbeck, K., Goenster, S, AliMohammed, S., Finkeldey, R., Buerker, A., and Gebauer, J., 2014. The African Baobao (*Adansonia digitata*, Malvaceae): Genetic resources in neglected populations of the Nuba Mountains, Sudan. *American Journal of Botany*, 101, 1498–1507.

Womeni, H.M. *et al.*, 2009. Oils of insects and larvae consumed in Africa: Potential sources of polyunsaturated fatty acids. *Oilseeds and Fats Crops Lipids*, 16, 230–235.

World Health Organization, 2008. Worldwide prevalence of anaemia 1993–2005: WHO global database on anaemia [Online]. Available from http://www.who.int/vmnis/anaemia/prevalence/en/ [Accessed 15 July 2015].

World Health Organization, 2009. Global prevalence of vitamin A deficiency in populations at risk 1995–2005 [Online].WHO Global Database on Vitamin A Deficiency. Available from http://www.who.int/vmnis/database/vitamina/x/en/ [Accessed 15 July 2015].

Yi, L. *et al.*, 2013. Extraction and characterisation of protein fractions from five insect species. *Food Chemistry*, 141(4), 3341–8.

14 Multi-level participatory approaches to mobilize dietary diversity for improved infant and young child feeding in banana-based agri-food systems of rural East Africa

Beatrice Ekesa, Deborah Nabuuma, Samuel Mpiira, Vincent Johnson, Domina Nkuba, Gina Kennedy, and Charles Staver

Introduction

Nutrition in rural households

Although the share of global population affected by hunger significantly reduced from 60% to 15% during the past five decades (Godfray *et al.* 2010), about 795 million people are still chronically undernourished (FAO, IFAD, and WFP 2014). The majority of those affected are mostly from developing countries – notably in sub-Saharan Africa (Sibhatu *et al.* 2015, Global Panel 2016). Globally, three billion people have low-quality diets and 45% of deaths among children under five years of age are linked to undernutrition (Global Panel 2016). Two billion people are deficient in one or more micronutrients especially iron and vitamin A (FAO, IFAD, and WFP 2014). To address this issue, WHO recommends that infants and young children (IYC) meet their nutrient needs through a varied diet, including animal-source foods and vitamin A-rich fruits and vegetables daily or as often as possible.

In many developing countries, undernutrition is significantly higher in rural communities than in urban and peri-urban communities. These are countries that depend on agriculture for both food and income, as well as bear the larger burden of child malnutrition (Sibhatu *et al.* 2015, Gewa and Leslie 2015). In Uganda, rural stunting rates exceed 36% compared to 19 % in urban areas (USAID 2016). Although regionally variable, undernutrition is mainly caused by lack of availability and access to food, poor dietary diversity, and cultural and social traditions (USAID 2016). In addition, IYC in rural settings are the most vulnerable to undernutrition, and studies show that the special nutrient requirements of IYC are not considered, with children often left to depend on the meals consumed by the general household. Meals are predominantly starchy staples (maize-based porridge or boiled banana) with little or no animal-source foods, vegetables, or fruit (Kizza 2014). A potential pathway to eliminate IYC dietary shortfalls is the enhancement of farm production diversity. Although rural households may have lower levels of schooling and often limited food availability (Kizza 2014), they have land resources and a daily routine of crop production activities which could be explored to increase the supply of diverse food and meet recommendations for IYCF (Signorelli *et al.* 2016).

Identifying viable local agrobiodiversity-based food production alternatives to improve IYC dietary diversity calls for addressing malnutrition both as a health problem, focusing on food consumption guidelines (Black and Dewey 2014), and as a nutrition-sensitive rural development opportunity, addressing key underlying determinants of nutrition and enhancing coverage and effectiveness of nutrition-specific interventions (Ruel and Alderman 2013). The malnutrition conceptual framework (UNICEF 2014) proposes that nutritional status is affected by factors that extend from societal conditions, such as poverty alleviation and women's empowerment, to household considerations, such as mealtime organization and family feeding interactions, as well as access to and utilization of healthcare. Ruel and Alderman (2013) highlight that, "Nutrition-sensitive programmes draw on complementary sectors such as agriculture, health, social protection, early child development, education, and water and sanitation to affect the underlying determinants of nutrition, including poverty; food insecurity; scarcity of access to adequate care resources; and to health, water, and sanitation services". Food and nutrition strategies not only help ensure food security for all, but also support achieving consumption of adequate quantities of safe and good quality foods that together make up a healthy diet.

Finding local agrobiodiversity-based food production alternatives to improve IYC dietary diversity also calls for fully engaging end users and the multiple sectoral stakeholders in a participatory learning and action approach (Pretty *et al.* 1995). The shift from nutrition as a *health* problem, to improved diets as a component in rural development, has been accompanied by an overall shift from a "pipeline", or technology-transfer model for technology generation (Biggs 1989). In this approach scientists developed technologies which were then made available to end users through extension and training programs to participatory technology development or co-design processes (Meynard *et al.* 2012).

Participatory methods were initially developed in the 1970s starting with participatory and rapid rural appraisal (PRA and RRA), and have continued to expand from that time to address a broad range of development challenges, including nutrition (Gonsalves *et al.* 2005, Food and Agriculture Organization 1993). Methods for end-user participation expanded to include not just diagnosis and prioritization, but also technology development such as plant and animal breeding, plant nutrition, and pest management. Farmer field schools, used for both training based on a pre-developed curriculum and for research, highlighted the role of the learning cycle in achieving improved outcomes from technology adoption (Food and Agriculture Organization 2016). The successive steps in the learning cycle are *observation* of relevant variables, *analysis* of the observations in relation to different options for improving the situation and finally the *decision of an action*, often in the form of an experiment. These in turn generate the basis for a new learning cycle following the sequence of observation, analysis, and action. Farmer skills in observation and decision-making are particularly relevant in such areas as pest management and improved animal feeding based on on-farm resources. These complex production and rural development challenges are well-suited for the participatory learning and action methods. This approach brings together scientists and end users in all stages of the learning cycle. The time frame for such learning may be the crop cycle with specific observation points and decision-making for action at each relevant crop phase. It could also be an annual cycle focusing on animal feeding or IYC nutrition involving multiple observation and decision-making moments in both wet and dry seasons. Over time, the participatory learning method evolved into a multi-sectoral approach that provides a thorough and multi-faceted pathway to improving IYC dietary outcomes (Garrett 2014).

Banana-based food systems in East Africa as a test case

This section provides a detailed description of a research for development project that worked to improve the quality of diets given to children from rural households within banana-based farming and food systems in East Africa. Banana is an important livelihood and food security crop for millions in Uganda. It is a perennial crop and intercropping with other food crops such as leafy vegetables, yam, and legumes within banana systems in East Africa is a traditional practice (Ouma 2009). Certain tree crops such as jackfruit, avocado, and papaya have also been reported to remain in the banana plantation to provide windbreaks, fruits, and fodder and feed for small ruminants (Ouma 2009). Research also confirms that Uganda is the highest banana consumer in the world with an average per capita consumption of 0.70kg/day/person (Charlie 2011). Despite this, only 13% of the children 6–23 months in Uganda meet the minimum dietary diversity of four food groups and 45% are fed the minimum number of times (2–3 for breastfed infants and 4 for non-breastfed infants) (UBOS and ICF 2012). These two indicators provide a glimpse into the fact that children not meeting the minimum requirements are unlikely to consume quality diets that meet their nutrient requirements for optimum growth. Indeed, micronutrient deficiencies are high with 53% of children under five years anaemic (UBOS and ICF 2017) and 38% having vitamin A deficiencies (UBOS and ICF 2012); and national levels of chronic malnutrition are at 29% (UBOS and ICF 2017). Continued exploration of ways in which IYCF practices and nutrition status, especially in rural communities, is needed.

In the Austria Development Agency-financed Bioversity International project, *Developing agro biodiversity-based strategies for alleviation of micronutrient and protein deficiencies among smallholder households from banana growing regions of East Africa*, Bioversity International identified that project stakeholders deployed diverse formal and participatory research methods to prototype rural community-based alternatives for using dietary diversity to remedy dietary shortfalls among IYC in East Africa. This experience provides useful input into the search for more effective research-for-development strategies to address rural undernutrition and micronutrient deficiencies through the increased on-farm production of more diverse nutrition-rich foods. The approach, laid out along a project timeline, began with initial diagnostic studies as illustrated later in Figure 14.1. This was followed by participatory design and testing of experimentation and learning activities, adaptive monitoring and integration of emerging results in IYC health, food preparation and food production, and finally by the scaling out and up of useful lessons in these three areas. These two levels of participatory processes 1) Multi-sector stakeholder engagement for learning, feedback, and scaling; and 2) Household experimentation and learning will be described and illustrated in this chapter. First, the multi-stakeholder component begins with mapping local and national stakeholders' interest and engagement with IYC nutrition. It then identifies local partners and dietary and production diversity entry points, and culminates in organizing a community learning alliance to scale out and further develop dietary diversity approaches. Second, at the community level, four sections describe the processes used in engaging the community members and specific households throughout the project. To begin with, perspectives and interests of men and women in different types of households are mobilized through household learning and action groups. This is a core process in which the interaction among IYC nutrition, dietary, and production diversity and other household priorities are addressed. The participants in the household group follow multiple learning cycles to facilitate identifying

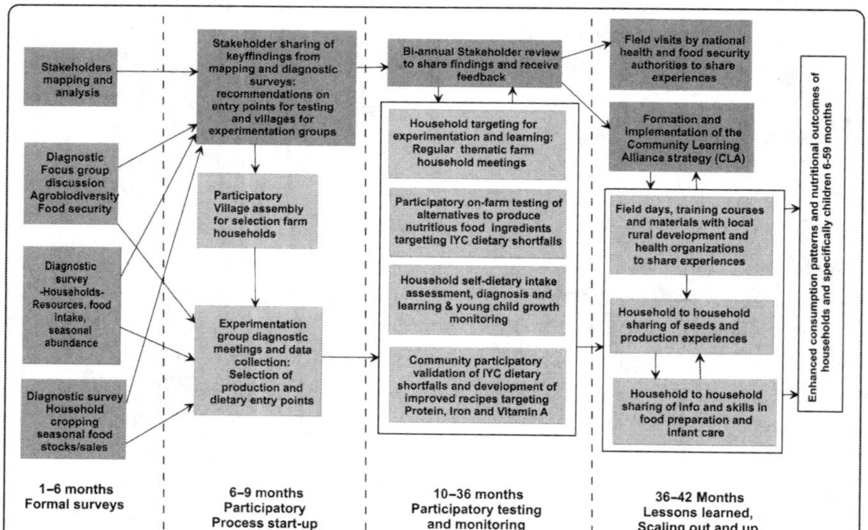

Figure 14.1 Project implementation framework based on two main participatory processes: (1) multi-level stakeholder approach for learning, feedback, sustainability, and scaling; and (2) household and community targeting for experimentation and learning.

agrobiodiversity-based alternatives grounded in local resource availability and emerging household priorities. The final three processes address specific technologies and approaches at household level for IYC feeding and food preparation and the production of diversity. In each case the process begins with formal and participatory diagnostics which serve to design and test on-farm production solutions. Each of the participatory processes in Figure 14.1 culminates in the identification of informal and formal scaling out strategies based on multi-stakeholder engagement.

This research-for-development project was implemented in Central Uganda (Kiboga district) and Northwestern Tanzania (Bukoba district) between August 2012 and August 2016. The target area has a bimodal rainfall distribution with a short and a long rainy season and a short and long dry season. Average annual total rainfall for the region is between 1000 and 1300mm/year. For the years of project implementation from 2013 to 2016 annual rainfall varied at the Kawanda station north of Kampala, a station with 1217mm/year annual rainfall. The target population group was children up to five years old from smallholder banana-dependent households. All households cultivate a perennial banana garden surrounding the household with between 200 and 600 mats. It is recommended for a banana mat to have three plants (a mat refers to the parent plant or fruit-bearing plant and its suckers) Mpiira *et al.* 2013).

Preliminary surveys indicate that some banana cultivars, currently less commonly grown and consumed in East Africa, could substantially contribute to vitamin A intake of children 6–59 months. Food consumption studies also indicate that bananas are often prepared or consumed along with small amounts of fresh beans, green leafy vegetables, ground nut paste or meat (flesh/offal). Such local food items are excellent dietary sources of micronutrients and proteins (Ekesa *et al.* 2011). Diagnostic studies have suggested that current cropping strategies, which already have high crop diversity, have a potential for greater intercropping of bananas with other short-term food crops, especially leguminous crops and feed sources for small ruminants and poultry which could enhance the supply and diversity of household foods (Mpiira *et al.* 2013, Ouma 2009). Despite the great potential in banana-based farming systems, Ruel (2002) shows that the diets of most children 6–59 months in banana-dependent regions of East Africa are based on starchy staples, with little or no protein, fruits, or vegetables. In their study of childhood dietary patterns in these banana regions, Kikafunda *et al.* (1996) found that children are weaned on boiled banana, despite the presence of documented on-farm crop diversity. Levels of malnutrition have remained above WHO thresholds, especially micronutrient deficiencies such as vitamin A and iron deficiency (International Food Policy Research Institute 2016). Our research and development therefore focused on combining diversified food crops and feed production for rearing small ruminants or poultry targeted specifically to the dietary shortfall of young children, the use of nutrient-dense food ingredients in traditional food dishes, and IYC health and nutrition monitoring. This approach integrated three spheres of rural development often otherwise addressed independently

The main research questions being answered were:

1. What proportion of children (6–59 months) from households in the target regions are meeting their vitamin A, iron, and protein needs through their current diets? What are the difficulties for those not meeting these needs, and how can households mobilize learning and resources to escape IYC malnutrition?
2. What production strategies can increase agrobiodiversity, with the aim of improving households' dietary diversity and nutrition status among children below five years?
3. What food preparation and dietary combination strategies can be employed to optimize the retention and bio-accessibility of essential nutrients for IYC?
4. To what extent does promotion of a multi-sectoral community learning alliance contribute to project sustainability and scaling emerging best practices?

In the following sections, we describe how each of the participatory processes addressed the research questions, as well as the links between these processes. We will highlight in each section the contribution to the understanding of IYC nutrition, the role of dietary diversity, and the role of men and women in household food decisions, food preparation, and crop and animal production.

Multi-sector stakeholder engagement for learning, feedback, and scaling

During project implementation, public and civil society organizations engaged at the national, district, and community levels to contribute to planning and monitoring project research activities, to identify lessons learned, and to develop and begin implementing expanded training and outreach based on research results (see Figure 14.1).

A thorough stakeholder analysis and mapping was an essential starting point. The objective was to engage stakeholders from different sectors with a potential interest in working toward a common goal. Through secondary data and key informant interviews, both development- and research-oriented actors were identified from agriculture, nutrition, health, social services, and education sectors operating within the target regions. Actors working at the different levels within the districts were considered. For example, the case for Uganda mapped actors working at district, sub-county, parish, and village levels. A tool was developed to analyze this information and generate a list of the most viable actors to engage with and inform during project planning, implementation, and scaling. Below, we discuss the activities and achievements of the stakeholder platforms as well as the mechanism of a learning alliance which was established to articulate different stakeholders at the community level for formulating a strategy to build capacity for and initiate scaling out of project research results.

A multi-sector stakeholder platform for learning and feedback: District level

More than 15 actors were identified and together formed the district stakeholders' forum which met several times a year over the course of the project. During the initial phase, the stakeholder forum members met to learn more about the project and to identify the links between their organizational agenda and agrobiodiversity, dietary diversity, and IYC nutrition. A short baseline questionnaire showed that very few of the public and civil society organizations in agriculture, health, and education understood and identified with agrobiodiversity and IYC dietary diversity. It was therefore important for the stakeholders to identify and understand the potential benefits following their involvement in the project implementation process.

The stakeholder platform also collaborated in the identification of entry points in terms of agrobiodiversity alternatives for IYC dietary diversity to be tested within the community. The research team first shared the findings of the 240-household diagnostic survey, and then invited stakeholders to discuss the gaps and establish the most suitable strategies to increase on-farm agrobiodiversity, households' dietary diversity, and nutrition status among under-fives. The suggestions from the stakeholders' forum on food and feed crops and animals were the starting point for the household experimentation and learning group research process. The stakeholder platform also provided input on the selection of villages for the formation of the two household learning and action groups in both Uganda and Tanzania and two control villages in each site. This approach coupled with the initial participatory diagnosis carried out with household groups reversed the common methodology where researchers or development agencies visit the community with the interventions already identified.

During the regular feedback and learning sessions which were organized two to three times a year, the research team summarized activities and advances since the previous meeting and invited stakeholder feedback. In year three, the CGIAR Humid tropics Consortium Research Programme (CRP) began work in Central Uganda, one of 15 action sites globally, and selected Kiboga district as one of two work sites. The project stakeholders' forum for IYC dietary diversity was expanded to become the Kiboga integrated systems intensification Research for Development (R4D) platform. Although the focus of the platform expanded both in topics and participants, there was a general consensus among the R4D stakeholders that the platform approach increased the exchange of information and decreased the resources needed to meet a specific objective, such as capacity building. This was especially because there was a reduction of redundant activities, and the different stakeholders were able to capitalize on combined strengths – such as access to resources, opportunities, skills, and knowledge – for jointly identifying and solving problems.

A community learning alliance (CLA) approach for sustainability and scaling

While the stakeholder platform provided a mechanism for exchange, learning, and planning, a more targeted strategy was needed to make the results from the participatory learning and action process with household groups on production and IYC dietary diversity more widely known. The challenge faced by the project was how to move beyond those directly involved in the implementation who ended up with the skill, knowledge, and seeds, despite limited project resources for such outreach activities.

Bringing together agriculture, health, education, religion, and social services, the community-level action was designed and tested using a community learning alliance (CLA) model. A community learning alliance was defined as a group of community-level stakeholders from different sectors and disciplines with a vested interest in working together toward a specific goal. CLAs were initiated in Kiboga, Uganda and in Bukoba, Tanzania. The objective was to ensure sustainability and scaling of best practices emerging from the research for development work. The alliance also aimed at improving skills and knowledge management; enhancing collaboration; accelerating institutional change; and deepening the level of impact. This model provided a mechanism that enabled the community-level stakeholders to continue with the CLA activities beyond the timeline of an initiating project. Overall, having the CLA approach working within existing frameworks leads to reduced costs and requires that the CLA is comprised of members with vested interests. Each CLA member has a chance to bring information, skills, and knowledge from their organization that are in line with the common goal. Although the initial strategy of such an approach was discussed in the stakeholder platform, implementation focused more locally on women's groups, farmers' groups, religious leaders, local teachers, community extension service providers, and youth groups.

The CLA was implemented through a series of workshops to enable the members to develop their specific CLA operating frame with the following elements:

1 Articulate clear objectives, roles, and responsibilities for the learning alliance.
2 Identify priority topics of interest based on stakeholders' needs and priorities.
3 Develop messages and approaches for each topic of interest based on project results and other local experience.
4 Build capacity in alliance on key topics, on training approaches, and on double-loop learning cycle.
5 Monitor progress, re-strategize, reformulate, and build new skills as needed.

These key messages were put together to develop a CLA package that all members were able to convey to community members in their different capacities. Bringing together these messages increases ownership of the CLA goal and increases the individual member and organization benefits for participating in the CLA.

After introducing the CLA and identifying objectives, smaller clusters of stakeholders were identified depending on expertise, discipline, and/or interest. The clusters included:

i Health workers and village health teams
ii Community leaders

iii Community development offices and community-based facilitators
iv School head teachers
v Religious and cultural leaders

Each CLA member and each small cluster then identified the main message/information/skill (key message) that their respective organization would like to be part of the CLA package. This was presented to the whole project team – other CLA members, the research team, and representatives from the R4D platform. In light of the stakeholder messages, the objectives and main activities of the CLA were reviewed. The different CLA clusters then each harmonized their messages into one package for the community (see Figure 14.2).

An example of a successful CLA strategy developed by the community workers' cluster in Kiboga, Uganda involved: parish chiefs; community-based facilitators; village health teams; and community development officers. These stakeholders participated in three kinds of activities:

1 Quarterly **community meetings** with local leaders;
2 Monthly **home visits** with local leaders and health workers; and
3 **Government programs** involving local leaders, beneficiaries, and health workers.

These activities improved sanitation and hygiene by 70%, strengthened food security, and mobilized and implemented government programs. Once a package

Figure 14.2 Presentation of the CLA strategy by the community workers' cluster in Lwamata Sub-County, Kiboga District, Uganda.

of messages was assembled and debated, CLA stakeholders then assumed their own specific roles in outreach.

Monitoring and evaluation of the CLA was carried out by the project field team. The CLA workshops were monitored and evaluated based on the achievement of the set objectives. The attendance and minutes of all the workshops were noted and shared among the CLA members and with the initiating organization. CLA members' knowledge was also assessed before the start of the CLA and after the 5th CLA workshop. A monitoring template, developed by the CLA members was also used to track their interaction with the community: how many people were in attendance; what key message from the CLA package was shared; and notes on success/lessons or challenges. A second template was used by each cluster coordinator. This template was used to monitor progress of the CLA versus the developed strategy.

Household experimentation and learning to address IYC dietary shortfalls in the community

Undernutrition in rural smallholder communities is the immediate result of inadequate dietary intake and morbidity, and has often been addressed with feeding programs for the most needy. However, in nutrition-sensitive rural development, which seeks to improve dietary diversity and thereby optimize IYC nutrition, multiple factors which operate at the household level are taken into account. The allocation of time and other resources, as well as decision-making power among the members, influence the amount and type of food prepared and fed to children and the household in general. For example, who makes household food production and purchase decisions? Who decides which family resources can go toward healthcare and food expenditures? Because childcare is often culturally designated as a woman's responsibility, many child-nutrition intervention programs have focused their efforts on mothers. Since mothers already have multiple roles in the household, these intervention programs can negatively affect their workload, without addressing gender inequalities that affect IYC nutrition and household food production, such as the control of income and production resources. Focusing on mothers also ignores the role in decision-making that men and older women (e.g. grandmothers) often play in early childcare (Kerr *et al.* 2016). Studies have shown that household gender inequality is related to higher rates of child undernutrition (Brinda *et al.* 2015). Women's empowerment, in terms of control over resources, decision-making, and more equitable workloads are all pathways by which nutrition can be improved through agricultural interventions (Kerr *et al.* 2016). This suggests that nutrition-sensitive rural development should integrate factors that contribute to the kind and type of food that a household will produce, access off-farm, and consume. The role of men in making decisions regarding what is produced and reared within the household farms, allocation of household income and, in some cases, their influence on intra-household food distribution needs to be understood and brought into the identification of alternatives for testing.

Household experimentation and learning for integrated production, food preparation, and child feeding strategies

A participatory learning and action process designed to integrate the complex of factors intervening in lasting improved IYC nutrition was facilitated by the project team in communities proposed by the stakeholder forum. The process in each community brought together an average of 25 households with a child between one and two years of age. Over a period of two full years, the experimentation groups met at least bi-monthly. These households were identified in a community assembly in which all participants completed a simple profile of their household members, production resources, and food sufficiency status. The experimentation group was delegated by the community assembly to work in the name of the community on the challenges of improved dietary diversity for IYC. The learning and action process ended with a community field day to discuss the results and exchange seed and other planting material. Each year the experimentation group also joined the community to discuss activities being carried out and emerging lessons. For the inaugural meeting of the group itself, the household heads, their spouses (if applicable), and the identified children (reference child) were present for the introduction of the project. This ensured that the responsible parties in each household and the team were aware of the benefits and expectations for all members of the households. To create a learning environment and boost morale in each meeting, the research team provided healthy, affordable snacks for the children. Three hours was the agreed meeting length among participants to accommodate the meeting in a daily schedule which includes food preparation, childcare, and tending crops and animals.

During the first phase of group meetings, with participatory facilitation by the field team, participants documented their household resources, the cropping cycle and crop areas, available food, and household eating patterns. The field team had already conducted focus groups and formal household interviews among approximately 430 households over a larger area covering both a control zone and the participating village (Figure 14.1). This data informed the facilitation team's design process of the sequence of questions in each meeting. Prior to each group meeting, the field assistant visited each household to accompany data registry on the topic of the meeting. This included the self-registry of food intake, land area and area with different crops, documentation of available food stocks of crops in the field and stored in the house, and animal-based food stocks. During the diagnostic phase this data was reported back in the group meetings and participants discussed the variability among households and grouped households with similar situations. The data by individual households were not singled out in the discussion, although questions and discussion provided criteria for each household to reflect on their own situation.

The diagnostic phase culminated in a series of meetings to discuss useful alternatives to address the shortfalls which had been identified in the diets of IYC (see sections 3.3 and 3.4 for more details). At the same time, in these meetings, the group continued to review food intake data and child health status (section 3.2).

These meetings therefore provided a forum for looking at household-level interactions among household land, labor and other resources for production, the livelihood demands by the household on resources, and IYC food intake and health status. The household experimentation groups verified and expanded on the results which emerged from the formal surveys carried out at the beginning of the project. Their self-reported data showed that the consumption patterns in terms of type of food and number of meals consumed by young children were similar to those of the general household members (older children and adults). This opened an option of shifting priority food resources to younger children. In addition, despite the ease in production of green leafy vegetables, their consumption was negligible. It was therefore important to also focus on cultural changes – learning to eat different food groups. Some foods were found to be more common in the diets of more well-resourced households which indicates the importance of access of poor households to key resources, even, for example, a few chickens and an appropriate feeding strategy. Finally, there are certain food groups which are scarce in certain seasons even among wealthier households and special technologies or germ plasm may be needed to address this issue. These diverse entry points to mobilize crop and animal diversity to address IYC dietary shortfalls were taken up in the participatory learning and action program developed with each household group.

The meetings continued over the course of two full years involving four cropping cycles. In each meeting, households discussed data on household food intake, IYC health status, use of experimental recipes which were proposed, crop plots, and small animal modules which each household had established. Figure 14.3 illustrates meetings showing the discussion of data, monitoring of crop plot status, and testing of experimental recipes with more nutrient-dense ingredients.

In addition, growth monitoring of the reference children took place at each household meeting. The parents or guardians tracked anthropometric measurements and other health and dietary indicators of the child and any issues identified were discussed. This increased parent interest in the growth of the child such that they were more observant of any circumstances affecting growth in between meetings. The growth monitoring also improved the uptake of practices around meal planning, child feeding, balanced diet, and health. It also enhanced the farmer's understanding of the linkages between consumption, nutrition outcome, and health status.

Figure 14.3 Community meetings involving the experimentation farmer households.

Given the need to have the reference child present for each meeting, the mothers or female household heads were the main participants. Over time, however, there was an increase in participation of the male heads. It was observed that the farmer experimentation groups in Bukoba, Tanzania had more households with both the husband and wife present in a meeting compared to those in Kiboga, Uganda. Nonetheless, households in both sites identified with the project such that if the mother was not available, commonly the husband brought the reference child to the meeting for monitoring and actively participated in the meeting.

Feedback from the male household heads revealed that the vast array of topics discussed and their application to the household as a whole were some of the factors that influenced their continuous attendance. For example, skills in crop production applied to their existing farms positively affected their yields; the food prepared in the home; and in some cases income.

Although this approach could be expensive compared to that targeting specific household members at only one point in time, more robust outcomes and impact are more likely to be realized because all key individuals within the household who influence decisions regarding production, consumption, and income are involved and informed. This approach also provides an environment for participatory learning and feedback; creation of partnerships between households, and between the research team and the households. It leads to not only useful agrobiodiversity-based components, production techniques, and recipes for wider use, but also an experienced community and member households ready to share experiences through informal pathways.

In the last months of the project, the participatory learning and action cycle was expanded to include exchange evaluation visits among the household groups. The two groups in Uganda carried out mutual evaluations and provided feedback to each other. Participants in the exchange astutely assessed the design and management of the test crop plots and chicken enclosures and analyzed the contribution of these alternatives to IYC nutrition. Drought and seed availability were two major challenges identified by participants to greater uptake and performance of on-farm production alternatives. In the exchange evaluations organized across national borders with mutual visits between Tanzania and Uganda, the Tanzanian evaluators to Uganda identified possible improvements in chicken enclosures – improved ventilation and the addition of feed troughs and laying boxes which they found missing. Their visit coincided with a drought and they questioned whether amaranth plots could have been irrigated. In their visit to Bukoba, Ugandan households appreciated the crop management advances achieved by households there – expanded sweet potato plantings, seed saving for amaranth and pumpkin, and increased numbers of chickens. However, they did suggest improved ventilation in the chicken enclosures to reduce dampness and associated parasites and diseases. While both evaluation visits found that children evidenced good nutrition, they also agreed that plots generating dietary diversity and chicken enclosures needed to be expanded for a more long-lasting harvest covering the numerous children in the household.

Household self-dietary intake recording, diagnosis, and learning (HS-DIRDL)

Accurately measuring dietary intake is crucial to understanding the role of diet in causing and preventing diseases as well as promoting growth and development. Although every individual eats food, the amount and kind of food consumed varies between subjects and people rarely perceive what they eat and how much they eat (Shim *et al.* 2014). Although dietary measurement is complex and multidimensional (Nogueira *et al.* 2016), additional efforts to increase accuracy and reduce costs are needed. Dietary recommendations aimed at encouraging people to manage dietary patterns to promote health and reduce disease risks are based in part on information about actual food consumption (National Cancer Institute 2017). The gold standard method for collecting dietary consumption data is the 24-hour quantitative recall. This method can be either enumerator-assisted or self-administered. The United States, Canada, and Australia have piloted a self-administered tool called the Automated Self-Administered 24-hour (ASA24®) dietary assessment tool (National Cancer Institute 2017). Use of the tool requires good IT infrastructure as well as high levels of literacy. While the use of this method works well in educated and affluent communities, it has not been tested in areas with low literacy levels and limited access to technology. To overcome these barriers, this project developed the Household-Self dietary intake recording, diagnosis and learning (HS-DIRDL) tool. First, a community food list was generated through focus group discussion which served to identify the many food options common in an area. Then a one-page/day recording sheet was prepared from the standard household level 24-hour recall. The focus was on the name of the meal consumed, the ingredients within the meal, and the source of the ingredients (farm, market, gift etc.). Due to illiteracy, differences in perception regarding weight, and the additional work and errors that would come with weighing food all the time, quantities of the consumed food was not included (see Appendix A for a sample filled-in tool).

Innovations of this method include involving households in self-reporting as compared to interviewer-assisted reporting. Interviewer-assisted reporting does not fully engage the household in self-reflection or problem identification. In enumerator-assisted methods, households often do not internalize the meaning of the data being collected and do not perform any self-analysis to connect the dietary patterns of the household with nutritional status of household members. Enumerator-assisted data collection can be viewed by the households as something needed by the researchers or the development agencies that has little direct application to them. With enumerator-assisted data collection, reporting bias is often common. Although the amount of bias present in the HS-DIRDL was not measured, our hypothesis is that reporting bias may be lessened when the respondent is requested to accurately self-report dietary intakes on their own time schedule and without the presence of an outside enumerator. The self-reporting method gives agency to households whereby they are empowered to take a more active role in reflecting on the household dietary intake, recording it, and reflecting on nutritional gaps in their eating patterns. Finally, the self-reported method is less invasive and more economical compared to enumerator-assisted methods.

The household experimentation groups involving 100 households used the HS-DIRDL tool. For this reason at least one person (whether older child or parent) in the household needed to be able to read and write either in English or the local language.

In the first meeting after the experimentation group was formed, the self-recording tool was introduced. Participants filled out a dietary recall for the last 24 hours for their households. This activity allowed the research team to carry out the first pre-test and familiarize the farmers with the tool. This exercise was accompanied by a discussion of the three main food groups commonly known to communities within East Africa as energy-giving foods, body-building foods, and protective foods. This was followed by generation of a community food list and placing of individual foods in their respective food groups. The three food groups were then broken down into smaller groups according to Food and Agriculture Organization (2011), with adjustments to emphasize on food groups targeted by the project such as vitamin A-rich foods and animal-source foods. The broader and specific food groups discussed with the experimentation farmer households included:

1 **Energy-giving foods** – grains and cereals; white roots, tubers and cooking bananas; fats and oils; sweets and sugars.
2 **Body-building foods** – legumes, pulses, and oil seeds; meats, poultry, fish, eggs, and insects; milk and milk products.
3 **Protective foods** – green leafy vegetables; deep orange, yellow, and red fruits and vegetables; deep orange, yellow, and red roots tubers, and bananas; and other fruits and vegetables.

During the pre-test, it was observed that the families did not consider breakfast and evening tea as a meal; it was assumed that only lunch and dinner qualified as meals. It was also noted that some ingredients such as sugar, salt, and cooking oil were easily forgotten and assumed to be unimportant. The importance of every meal and ingredient was therefore emphasized. Based on the observations and discussions during this meeting, the tool was modified to simplify data recording. The modified tool for seven days (seven pages) was delivered to the households through the farmer representatives for the 2nd pre-test.

The seven-day self-recording tool filled out by the farmers was assessed during the 3rd meeting. This evaluation served as the 2nd pre-test. Gaps in filling it out were identified and addressed. The discussions also provided a simple assessment of their diets to highlight the importance of the tool, food groups, dietary diversity, and implications for their household. Nine days prior to the 4th meeting, the field assistants and farmer representatives distributed data collection forms to experimental households. Households came to the 4th meeting with the filled out tools where further challenges in self-recording were addressed.

With continuous exposure to the tool and feedback from the research team, the quality of data provided by the households improved. Appendix A shows a filled-out tool for one day by one household in Uganda. Farm households were able to carry out the self-dietary recording, diagnosis, and learning every other month. The tools were given to the group leader one day before the expected start date of recording and distributed to households. The households filled out the tool every evening for seven days on the 3rd week of the month; the date and day of the week are clearly indicated. For validation purposes, after the field assistant collected the completed forms, (s)he also carried out a 24-hour recall capturing consumption of the 7th day. This information was used to validate the self-recorded information.

From this case study, we learned that this kind of approach only requires basic literacy levels when the time is taken in sequential meetings to explain, demonstrate, and practice completing the tool. There is the potential for improving the tool by having coded pictures in the community food catalogue so that the recording person only indicates codes of the consumed food items. The agency encouraged from household participation increased the likelihood of compliance. In this study a compliance of >80% was reported in both Uganda and Tanzania. Although the approach could lead to participation burden if not implemented carefully, with the right communication and accompanied by regular interaction with the research team it was observed to contribute to positive change in behavior related to consumption. At least four months of interaction with a community is needed to create a rapport with the community and ensure high compliance. A longer time is needed to capture seasonal differences in consumption and IYC as well. This would provide insight into ways quality IYC diets can be achieved and promoted throughout the year, especially for rural farming communities.

Community participatory validation of IYC dietary shortfalls and development of improved recipes targeting protein, iron, and vitamin A

The diagnostic survey carried out during the initiation process of this study included identification of the most common diets given to children 12–59 months and further follow up to establish the exact preparation procedure and a laboratory analysis to establish the nutrient content of these diets. Following this, six dishes were identified as the most common ones in both Tanzania and Uganda. These included: Ugali/posho (maize mush) with beans and stewed mukene (dried freshwater Lake Victoria sardine); boiled banana mixed with beans; boiled cassava mixed with beans; steamed mashed banana with groundnut sauce; steamed mashed banana, mukene, groundnut sauce, onions, tomatoes; boiled banana with groundnut sauce, tomatoes, and onions (locally known as Katogo). After a number of home visits and observation, the recipes were replicated in the laboratory, and analyses to establish the protein, vitamin A, and iron levels carried out. Findings showed that these dishes had between 0.31 and 2.28mg/100gfw of iron and between 2.48 and 43.68 RAEug/100gfw of vitamin A. Considering the recommended dietary allowances (RDAs) for children 1–3 years old as per the National Academy of Science (2001), these dishes are meeting only between 4%–32% of the iron RDAs and only between 0%–15% of the Vitamin A RDAs.

These above findings were shared with the district level R4D platform and the community members through the experimentation farmer groups. Collectively and using community participatory methods, the research team, key stakeholders, experimentation farmer group, and community members especially mothers with young children, explored the local food sources with potential to fill the nutrient gaps to bridge vitamin A, protein, and iron gaps. The focus for improved dishes was on sustainable utilization of the existing agrobiodiversity. These alternatives included; chicken to produce eggs to be included in diets of children, amaranth leaves, orange-fleshed sweet potato (OFSP), vitamin A-rich banana, papaya, and iron-rich kidney beans.

The community food lists generated through focus group discussions in the diagnostic survey were reviewed and the community members asked to think through ingredients or methodologies that could be utilized to fill the nutrient gaps. Feasibility, acceptability, and affordability were major factors considered. Some of the points emerging from the community were: inclusion of eggs in the porridge; addition of citrus fruit (orange/lemon) to the porridge; mixing several flours such as flour from OFSP or vitamin A banana with the maize porridge; mixing soya bean or groundnuts with the maize flour when preparing porridge; addition of green leafy vegetables to the banana-dominant dishes. During the discussions, scientists introduced additional points for consideration:

- i How to maintain the same time for food preparation.
- ii Potential to ferment certain flours used in making porridge to destroy anti-nutrients and promote bioavailability.
- iii Adding liquid foods (soup or milk) and mashing foods to make them more palatable for younger children.
- iv Criteria that any dish should contain in addition to a starchy food, substantial portions of vegetable food and a protein food.

Based on the points highlighted in the meetings on improved recipes, the research team, including graduate students, and stakeholders came up with five improved recipes for testing in the laboratory and with experimentation groups. The improvements were based on altering the cooking time and exclusion and inclusion of specific ingredients. Ingredients of focus included groundnuts, vitamin A-rich banana, OFSP, fish (silverfish), and iron-rich beans.

The feasibility of the improved dishes was then tested through the participatory preparation of the recipes (including experimentation farmer group members and additional mothers with young children) and organoleptic tests by mother–child pairs.

The modified dishes were markedly improved in iron content compared to their original common versions. Modified porridge dishes had improved with iron content of 5.84mg/200gfwt compared to untraceable amounts in traditional versions. The inclusion in the porridge of OFSP (Hotz *et al.* 2012, Low *et al.* 2007) and the vitamin A-rich banana "Biira" (Ekesa *et al.* 2015) increased the pro-vitamin A carotenoids' (pVACs') content compared to the common recipe which had no traceable amount of pVAC.

Participatory on-farm testing of alternatives to produce nutritious food ingredients targeting IYC dietary shortfalls

In response to the dietary shortfalls identified in the formal survey and by the household groups and verified in the stakeholder learning alliance, the field team elaborated alternative production options based on locally-applicable technologies for iron-rich food: green leafy vegetables and iron-enriched beans; orange-fleshed fruits and vegetables – orange-fleshed sweet potatoes, papaya, pumpkin, and introduced vitamin A-rich banana; and protein: chicken eggs based on on-farm production of introduced and local feeds. For each option, the research team set a target household production level, identified germ plasm and seed sources, and compiled management practices from research documents, expert interviews, and experiences in other projects. These initial calculations provided the field team with experience to design a series of participatory group meetings to adapt possible production approaches to local circumstances. After each meeting, the field team in each site discussed production options with each household in the context of their land, labor, and other production resources. This approach drew on the experience of a prior project in Central Uganda to test and adapt the use of zero-grazed goats and fodder shrubs to produce manure to increase banana production (Bioversity 2014). By the end of the initial design phase, each household had their proposed set of small experiments to increase food availability for IYC diets. The household groups reviewed the status of experiments during the crop cycle, calculated the adequacy of food stocks from wet to dry season, and identified important points at the end of the crop cycle for the next cropping season. The available seed for the next cropping cycle was also calculated.

The group monitoring indicated a significant change in the production patterns of the experimental farm households. On-farm agrobiodiversity increased in all the 100 households. More iron-rich beans, orange-fleshed sweet potatoes, amaranth vegetable, papaya fruit, and chicken eggs were being produced (see section 3.2).

During the diagnostic survey, the consumption of animal-source foods, fruits, and vegetables was negligible among the general household and especially among the young children (Kizza 2014, Godson 2014). By the official close of the project, significant changes were observed with regard to this. Children from the experimentation groups showed inclusion of more healthy snacks in addition to the three main meals. This was observed for between 65% and 95% of the households, with the highest proportions observed during harvest season and the lowest during planting/weeding seasons (Beil 2016). Regarding the general community, the end-line survey showed that although only 49.5% of the 404 children had consumed at least one animal-source food, there was a significant increase in the proportion of households and children consuming more diverse diets especially animal-source foods (eggs, milk, meat, fish), vitamin A-rich fruits, and vitamin A-rich green leafy vegetables (Bioversity International 2016).

The mean weekly dietary diversity score (DDS) was 8.6, and was significantly higher (DDS = 9.1, P = 0.008) among individuals from households that reported direct involvement in organization programs and extension activities on agriculture and nutrition than those not directly involved (DDS = 8.4). Based on 12 food groups, the household DDS is categorized as low DDS (≤3 food groups), medium DDS (4–5 food groups), and high DDS (≥6) (Swindale and Bilinsky 2006). The significant difference observed was attributed to the fact that, although less than 32% of the general population had experienced any direct contact with organizations or extension service providers with regard to production and nutrition at the beginning of the intervention, by project end, more than 65% of the households had directly accessed training or information on nutrition and good consumption patterns, and between 28 and 42% had actually been involved in community interventions and activities that promote agriculture and nutrition (Bioversity International 2016).

Conclusion

The multi-level participatory approach implemented in four rural banana-based communities in Tanzania and Uganda provided positive evidence for the potential to mobilize on-farm food and feed diversity to address the dietary shortfalls of IYC. Work combined monitoring of food intake and child nutritional status, farmer test plots to produce more iron- and vitamin A-rich and protein foods, and participatory development of more nutrient-rich dishes for children. This novel combination resulted in increased food production and improved dietary diversity among children in collaborating communities compared to control communities. Farm households began to observe the relationship between the changes observed in weights of their children and their food consumption. They realized that small children respond positively to improved diets which they themselves can provide. The engagement of stakeholders from the agriculture, education, and nutrition sectors at the community and district level contributed to the uptake of this practical approach to breaking the vicious cycle of malnutrition in ongoing health, education, and community development programs in both project work sites.

Specific participatory learning and action methods developed and adapted by the project team were key to these results. Self-monitoring of food intake using HS-DIRDL, monitoring of child growth parameters, and health and group discussions on the data created motivation to target IYC with diets meeting their special nutritional requirements. The lab testing of traditional foods and the participatory development of dishes based on local ingredients with improved nutrition generated alternatives within the reach of household food preparation routines. The identification of production alternatives and the design of household-specific experiments required households to analyze their available production opportunities for increased production of nutrient-rich foods. Although the project provided small amounts of seeds and five chickens to facilitate the establishment of the experiments, on-farm seed supplies were increasingly important with each successive season of experimentation. The household learning and action process which focused on IYC and the role of all household members in their wellbeing proved to be a useful space to integrate the resources and roles of both men and women and other providers of IYC, including older siblings and other women and men in the household. Of even greater significance for the uptake of these methods and the results they have the potential to generate was the multi-stakeholder approach demonstrating at the local level the multi-sectoral approach described by Garret *et al.* (2014). The process of stakeholder review and feedback not only served to integrate local lessons into the project approach and to create ownership for emerging results, but it also set the stage for the formation of a community learning alliance with the specific task to mainstream lessons into the ongoing work of a diversity of stakeholders with ongoing presence in rural communities with similar challenges of IYC dietary shortfalls.

The project identified a number of opportunities for a more effective multi-level participatory process:

1. More timely processing of data generated in baseline and monitoring surveys and self-monitoring both for feedback to stakeholders and farm households. The project team and the multiple participatory processes often generated data which was not analyzed for the full benefit to the challenges of addressing IYC dietary shortfalls. More attention to database formats, staff time, equipment for data analysis, and methods to visualize changing status of the diverse processes being tracked should all be addressed in project design and budgeting.
2. Additional studies and improved methods to document and understand how households calculate and track their food stocks and plan new crop cycles, especially the availability of planting material. Households face complex decisions under great uncertainty and these methods should help to clarify how production and food preparation alternatives can be proposed in an improved household decision-making framework which addresses both men and women members in households.
3. Additional studies and improved methods to identify and take into account poorest households which face particular challenges in terms of resources and vulnerabilities due to health, weather, and lack of income alternatives.

In closing, this multi-level participatory approach is best applied in projects to allow time to modify and contextualize the tools used. The participatory approaches require sufficient planning to ensure skills development and adequate attention to the learning cycle. This ensures that each activity is adequately carried out; sufficient information is collected and resources are well appropriately utilized. As our experience has shown, the additional time for implementation pays off in the quality of the results and the engagement of stakeholders who will move forward to broader impacts.

Acknowledgements

The authors of this chapter acknowledge the Austrian Development Agency (ADA) for providing financial support through Bioversity International that enabled identification and implementation of the approaches described in this chapter. The authors also express their gratitude to postgraduate students in Ugandan, Tanzanian, Kenyan, Austrian, and United States of America universities – Ms. Stephanie Beil, Ms. Edith Diigo, Ms. Phillipa Erlacher, Mr. Francis Kidake, Ms. Christine Kizza, Ms. Godson Namsiifu, Ms. Deena Cowans, and Ms. Linda Pamminger, and their university supervisors for the great role they played during data collection and analysis. The project staff in Uganda and Tanzania – Mr. Francis Kalyango, Mrs. Jojianas Kibura, Mr. Joseph Kimisha, and Mr. Innocent Ndyetabura who worked in this project under the framework of the National Agricultural Research Organization (NARO) of Uganda and the Agricultural Research and Developments Institute (ARDI), Maruku Tanzania are thanked for their great role during the implementation of these approaches. The many other people who formed part of the stakeholders' forum, community leaders, experimentation farmer households, and the community members from both Tanzania and Uganda are thanked for their willingness and commitment during the whole process.

References

Biggs, S., 1989. Resource-poor farmer participation in research: A synthesis of experiences from nine agricultural research systems. *On-Farm Client-Oriented Research Comparative Study Paper No. 3*. The Hague, the Netherlands: International Service for National Agricultural Research.

Bioversity International, 2014. How can we produce more manure on-farm for our bananas? [Online]. Available from http://banana-networks.org/barnesa/files/2015/03/11a-Extension-Newsletter-manure-English.pdf [Accessed 13 July 2017].

Bioversity International, 2016. Developing agrobiodiversity-based strategies for the alleviation of micronutrient and protein deficiencies among smallholder households in banana growing regions of East Africa. *End of project report.*

Beil, S., 2016. *Assessment of dietary patterns and health status – with focus on dependency on agricultural diversity and seasonality – among preschool children from smallholder households in Kiboga district, central Uganda*. Master of Science Thesis. University of Vienna, Austria.

Black, M.M. and Dewey, K.G., 2014. Promoting equity through integrated early child development and nutrition interventions. *Annals of the New York Academy of Sciences*, 1308, 1–10.

Brinda, E.M., Rajkumar, A.P., and Enemark, U., 2015. Association between gender inequality index and child mortality rates: A cross-national study of 138 countries. *BMC Public Health*, 15, 97.

Charlie, 2011. What are East African Highland Banana's (EAHBs) and why are they important for food security? [Online] *East African highland bananas, a staple crop of the poor in the Great Lakes region of Africa*. Available from http://eastafricanhighlandbananas.org/ [Accessed 13 July 2017].

Ekesa, B.N., Blomme, G., and Garming, H., 2011. Dietary diversity and nutritional status of pre-school children from Musa-dependent households in Gitega (Burundi) and Butembo (Democratic Republic of Congo). *African Journal of Food, Agriculture, Nutrition and Development*, 11 (4).

Ekesa, B., Nabuuma, D., Blomme, G., and Van Den Bergh, I., 2015. Provitamin A carotenoid content of unripe and ripe banana cultivars for potential adoption in eastern

Africa. *Journal of Food Composition and Analysis*, 43. doi:10.1016/j.jfca.2015.04.003. https://www.sciencedirect.com/science/journal/08891575/43

FAO, IFAD, and WFP, 2014. The state of food insecurity in the world 2014. Strengthening the enabling environment for food security and nutrition [Online]. Rome. Available from http://www.fao.org/publications/sofi/2014/en/ [Accessed 29 July 2017].

Food and Agriculture Organization, 1993. *Guidelines for participatory nutrition projects*. Rome.

Food and Agriculture Organization, 2016. *Farmer field school guidance document – Planning for quality programmes*. Plant Production and Protection Division, FAO: Rome.

Garrett, J., Kadiyala, S., and Kohli, N., 2014. Working multisectorally to improve nutrition: Current status in India and global lessons. *POSHAN Policy Note #1*. New Delhi, India: International Food Policy Research Institute.

Gewa, C.A. and Leslie, T.F., 2015. Distribution and determinants of young child feeding practices in the East African region: Demographic health survey data analysis from 2008–2011. *Journal of Health, Population and Nutrition*, 34, 6.

Global Panel on Agriculture and Food Systems for Nutrition. 2016. *Food systems and diets: Facing the challenges of the 21st century*. London, UK.

Godfray, J. *et al.*, 2010. Food security: The challenge of feeding 9 billion people. *Science*, 327 (5967), 812–818.

Godson, N., 2014. *Nutrient content of popular dishes consumed by children below five years of age in banana growing communities: A case study of Bukoba rural district, Tanzania*. Master of Science Thesis. Sokoine University of Agriculture, Tanzania.

Gonsalves, J.F. (ed.), 2005. *Participatory research and development for sustainable agriculture and natural resource management: A sourcebook* (Vol. 1–3). Ottawa, Canada: International Development Research Centre.

Hotz, C., Loechl, C., de Brauw, A., Eozenou, P., Gilligan, D., Moursi, M., Munhaua, B., van Jaarsveld, P., Carriquiry, A., and Meenakshi, J.V., 2012. A large-scale intervention to introduce orange sweet potato in rural Mozambique increases vitamin A intakes among children and women. *Br J Nutr*, 108 (1), 163–176. doi: 10.1017/S0007114511005174. Epub 2011 Oct 10.

International Food Policy Research Institute, 2016. *Global nutrition report 2016: From promise to impact; ending malnutrition by 2030*. Washington, DC.

Kerr, R.B., Chilanga, E., Nyantakyi-Frimpong, H., Luginaah, I., and Lupafya, E., 2016. Integrated agriculture programs to address malnutrition in northern Malawi. *BMC Public Health*, 16, 1197.

Kikafunda, J.K., Walker, A.F., Kajura, B.R., and Basalirwa, R., 1996. The nutritional status and weaning foods of infants and young children in Central Uganda. *The Proceedings of the Nutrition Society*, 56 (1A), 16A.

Kizza, 2014. *Dietary protein, iron and vitamin A intake, dietary diversity and nutrition status of preschool children (12–59 months). From farmer households; case study of Kiboga district, Uganda*. Master of Science Thesis. Department of Food Technology and Nutrition, Makerere University, Uganda.

Low, J.W., Arimond, M., Osman, N., Cunguara, B., Zano, F., and Tschirley, D., 2007. A food-based approach introducing orange-fleshed sweet potatoes increased vitamin A intake and serum retinol concentrations in young children in rural Mozambique. *Journal of Nutrition*, 137 (5), 1320–1327. https://doi.org/10.1093/jn/137.5.1320

Meynard, J.M., Dedieu, B., and Bos, A.B., 2012. Re-design and co-design of farming systems. An overview of methods and practices. In *Farming systems research into the 21st century: The new dynamic*. Dordrecht, the Netherlands: Springer, 405–429.

Mpiira, S. *et al.*, 2013. 19 The use of trees and shrubs to improve banana productivity and production in Central Uganda: An analysis of the current situation. *Banana systems in the humid highlands of sub-Saharan Africa*, 150.

National Cancer Institute, 2017. *Automated self-administered 24-hour (ASA24®). dietary assessment tool* [Online]. Available from https://epi.grants.cancer.gov/asa2/ [Accessed 13 July 2017].

Nogueira, Previdelli, Á., de Andrade, S.C., Fisberg, R.M., and Marchioni, D.M., 2016. Using two different approaches to assess dietary patterns: Hypothesis-driven and data-driven analysis. *Nutrients*, 8, 593.

Ouma, G., 2009. Intercropping and its application to banana production in East Africa. *Journal of Plant Breeding and Crop Science*, 1 (2), 13–15.

Pretty, J., Guiit, I., Scoones, I., and Thompson, J., 1995. *A trainer's guide for participatory learning and action*. IIED Participatory Methodology Series. International Institute for Environment and Development, London, UK.

Ruel, M.T., 2002. *Is diversity an indicator of food security or dietary quality? a review of measurement issues and research needs*. FCND Discussion Paper No. 140, Food Consumption and Nutrition Division, International Food Policy Research Institute, Washington, DC. Available from www.ifpri.org/sites/default/files/pubs/divs/fcnd/dp/papers/fcnbr140.pdf [Accessed 4 April 2018].

Ruel, M.T. and Alderman, H., 2013. Nutrition-sensitive interventions and programmes: How can they help to accelerate progress in improving maternal and child nutrition? *Lancet*, 382, 536–551.

Shim, J.S., Oh, K., and Kim, H.C., 2014. Dietary assessment methods in epidemiological studies. *Epidemiology and Health*, 36, e2014009.

Sibhatu, K.T., Krishna, V., and Qaim, M., 2015. Production diversity and dietary diversity in smallholder farm households. *Proceedings of the National Academy of Sciences USA*, 112, 10657–10662.

Signorellia, S., Hailea, B., and Kotub, B., 2016. *Exploring the link between agricultural production diversity and dietary quality: Evidence from Ghana* (Draft). Available from https://editorialexpress.com/cgi-bin/conference/download.cgi?db_name=CSAE2016&paper_id=1068 [Accessed 13 July 2017].

Swindale, A. and Bilinsky, P., 2006. *Household dietary diversity score (HDDS) for measurement of household food access: Indicator guide*. Washington, DC: Food and Nutrition Technical Assistance Project, Academy for Educational Development.

Uganda Bureau of Statistics (UBOS) and ICF International Inc., 2012. *Uganda demographic and health survey 2011*. Kampala, Uganda: UBOS and Calverton, Maryland: ICF International Inc.

Uganda Bureau of Statistics (UBOS) and ICF, 2017. *Uganda demographic and health survey 2016:* Key Indicators Report. Kampala, Uganda: UBOS, and Rockville, Maryland, USA: UBOS and ICF.

UNICEF, 2014. *The importance of multi-sectoral approaches to nutrition* [Online]. Available from https://www.unicef.org/eapro/Brief_Nutrition_Overview.pdf [Accessed 14 July 2017].

USAID, 2016. *Uganda nutrition profile* [Online]. Available from www.usaid.gov/../countries/uganda-nutrition-profile [Accessed 24 July 2017].

Appendix A

Date:
Name: **Village:**

Period	Type of Meal taken	Ingredients in the meal	Source of food
Early morning			
Mid-Morning			
Lunch			
Mid-Afternoon			
Dinner			
After Dinner			

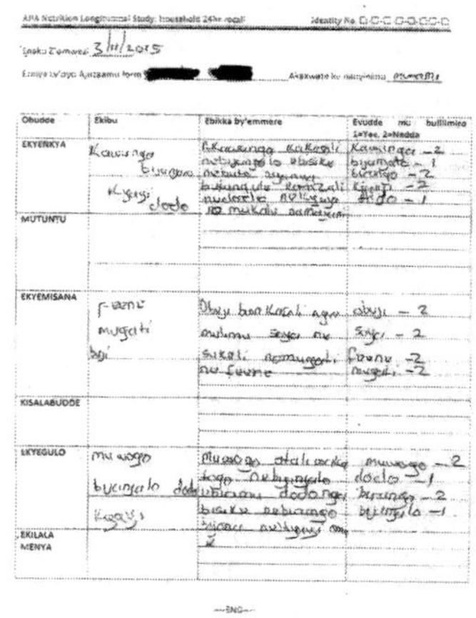

Figure 14.4 Seven-day self-recording dietary assessment tool.

Index